Introduction
to
Advanced Mathematics

Introduction
to
Advanced Mathematics

William Barnier
Norman Feldman

Sonoma State University

Prentice Hall, Englewood Cliffs, New Jersey 07632

Library of Congress Cataloging-in-Publication Data

Barnier, William
    Introduction to advanced mathematics / William J. Barnier, Norman
Feldman.
        p.   cm.
    Includes index.

    1. Mathematics.   I. Feldman, Norman.   II. Title.
QA39.2.B3686 1990
510–dc20   89-8600
                                                                            CIP

Editorial/production supervision: Karen Winget
Cover design: Megan Reid
Manufacturing buyer: Paula Massenaro

ⓒ1990 by Prentice-Hall,Inc.
A Division of Simon & Schuster
Englewood Cliffs, New Jersey 97632

Printed in the United States of America
10   9   8   7   6   5   4   3   2

ISBN 0-13-477084-6

PRENTICE-HALL INTERNATIONAL (UK) LIMITED, *London*
PRENTICE-HALL OF AUSTRALIA PTY. LIMITED, *Sydney*
PRENTICE-HALL CANADA INC., *Toronto*
PRENTICE-HALL HISPANOAMERICANA, S.A., *Mexico*
PRENTICE-HALL OF INDIA PRIVATE LIMITED, *New Delhi*
PRENTICE-HALL OF JAPAN, INC., *Tokyo*
SIMON & SCHUSTER ASIA PTE. LTD., *Singapore*
EDITORA PRENTICE-HALL DO BRASIL, LTDA., *Rio de Janeiro*

To my MOTHER and MOTHER-IN-LAW

W.B.

To BEVERLY, TAMAR, and ILANA

N.F.

Contents

PART I

Preface

This text is intended for a sophomore-level course similar to the course in "Logic and Proof" that we have taught at Sonoma State University since 1971. It is a transitional course supplying background for students going from calculus to the more abstract upper division mathematics courses. A prerequisite for the course is at least one semester of calculus. The text, which is an outgrowth of notes collected over the years teaching "Logic and Proof," is designed for use in either a three or four unit semester course or a four unit quarter course. It can also be used as a supplement to junior-level courses such as abstract algebra or real analysis.

Introduction to Advanced Mathematics covers material indispensable to any mathematician: it could be subtitled "What Every Mathematician Needs to Know." The text includes the material necessary for students to succeed in upper-division mathematics courses, and more importantly, the analytical tools necessary for thinking like a mathematician.

The book is naturally divided into Part I (Chapters 1–5) and Part II (Chapters 6–9). The first five chapters are the core of the book. The topics included form a natural progression: elementary logic, methods of proof, set theory, relations, and functions. Each chapter contains a full exposition of topics with many examples. In addition to the examples, practice problems are included to reinforce the concepts as they are introduced. Solutions to the practice problems appear at the end of each chapter. Solutions and hints for the odd numbered exercises appear at the back of the book.

Chapters 6–9 are a rich source of examples, theorems and projects. They provide an opportunity for the student to apply what is learned in the core chapters. Many theorems have no proof or only a hint or outline for the proof. Likewise, the examples may have no solutions or just a hint for the solution. The intent is that the material be used as a basis for students to construct their own proofs or solutions and perhaps present them to the class.

The following figure, in which the arrow means "is a prerequisite for," shows the relationship among chapters:

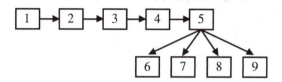

CHAPTER PREREQUISITES

We recommend that Chapters 1–5 be covered in order. Sections 1.5 and 2.5 may be omitted without loss of continuity. Section 5.3 is useful in Chapters 7 and 8 but may be omitted otherwise. Chapters 6–9 are independent; if time permits only partial coverage of any of these chapters, we recommend Sections 6.1–6.2, 7.1, 8.1–8.3, or 9.1–9.2.

A note about our pedagogical approach

We have attempted to include clearly written examples, practice problems that involve the student, and many exercises with solutions. In the early part of the book, where logic is developed and the techniques of proof are presented, the mathematical setting is clearly laid out so that students know exactly what they may assume for the construction of proofs. The development of the material here is detailed, with many examples and suggestions on how to develop proofs. From our experience teaching the course, we have tried to anticipate many of the questions students might have. It is our belief that an initial development of the subject done slowly and carefully yields great rewards. Students are then able to work more independently and with much greater understanding of the material.

To encourage further independence, there are a number of supplementary exercises that extend or are related to some of the concepts discussed in the text, but which are not necessary for the continuity of the subject matter.

To the student

> Nothing can permanently please
> which does not contain in itself the
> reason why it is so and not
> otherwise.
>
> Samuel Taylor Coleridge
> (1772-1834)

There are two fundamentally important aspects to mathematics. One is the discovery of plausible statements and the other is proving (or disproving) them. Without the first, the second is not possible. However, without the second, how are we to know that a plausible statement truly follows from a given set of hypotheses? The principal aim of this book is to develop your proficiency in reading and creating proofs.

Mathematics is not a spectator sport. Mastering the topics covered in this text is important; but more important is enhancing your ability to think. Reading mathematics is prerequisite to thinking mathematically. For each assigned section you should:

1. Read the section from beginning to end,
2. Carefully study each example, while working out the details with paper and pencil,
3. Work each practice problem before looking up the solution,
4. Finally, work the assigned exercises.

Acknowledgments

We gratefully acknowledge the assistance of Professor Jean Chan, Ilana Feldman, and Professor James Jantosciak. In addition, the students in our Fall 1988 and Spring 1989 classes offered many helpful suggestions. We also thank the reviewers for their helpful suggestions and the editors at Prentice-Hall for their support in this project.

1

Introduction to Logic

The concepts in this chapter provide a foundation for all the techniques of proof we will use in this book. First, we informally discuss how to read and understand proofs. Then we introduce propositional logic and formal proof techniques for propositional logic called rules of inference. Finally, we use the rules of inference to derive many laws of logic. So the chapter is an introduction to both logic and methods of logic. These methods of logic are used throughout the rest of the book to prove many mathematical facts.

1.1 UNDERSTANDING PROOFS

Mathematicians and scientists need the ability to read and understand valid arguments (or proofs) and to recognize invalid arguments. A basic knowledge of logic is indispensable for analyzing and constructing proofs. In this section, we introduce elementary logic and discuss proofs, emphasizing how to read and understand proofs and how they are created. The section is intended as an overview; a more detailed discussion of logic and its application to proofs will follow in subsequent sections.

In all mathematical proofs, there is a collection of statements called **hypotheses** and a statement, called the **conclusion**, which must be proved to follow logically from the hypotheses. There are many forms a proof can take. However, every proof must be a finite sequence of statements that are either hypotheses, previously proved statements, or statements that follow logically from previous statements in the proof. The final statement in a proof should be the statement to be proved.

1

Two major features of a proof are the idea behind the proof—the creative part, what really makes it work—and the written part. The written part must be done so that other mathematicians can read and understand the proof. To become a successful mathematician, one must learn to communicate ideas, both verbally and in written form, and also to understand proofs created by others.

To understand a proof, it is necessary to know what is to be proved. That is, we must know what the proof is "all about." For example, is it a proof about sets or about functions in calculus? A proof about sets will require facts and definitions of set theory. A proof involving functions in calculus will require definitions such as that of a differentiable function and facts such as the mean value theorem for derivatives.

When reading any proof, we must always ask:

What is the goal of the proof?

What are the hypotheses?

What definitions are necessary?

What previously proved facts or laws of logic are used in the proof?

To illustrate these ideas let us consider several examples of proofs. We need the following definitions: an **even integer** is an integer that can be expressed in the form $2k$, where k is an integer; an **odd integer** is an integer that can be expressed in the form $2k + 1$, where, again, k is an integer. For example, if p and q are integers, then $2(p + q)$ is even, whereas $2(p^2 - q) + 1$ is odd. We assume that the following is known:

An integer is either even or odd, but not both. (I)

Example 1

Let n be an odd integer. Prove that n^2 is an odd integer.

Proof	Let n be an odd integer.	This is a hypothesis.
	Hence, $n = 2k + 1$ for some integer k.	Here, the definition of odd integer is used.
	Hence, $$n^2 = (2k + 1)^2$$ $$= 4k^2 + 4k + 1$$ $$= 2(2k^2 + 2k) + 1$$	These steps are assumed to be previously proven statements of algebra.

Since $2k^2 + 2k$ is an integer, we conclude that n^2 is odd.

Practice Problem 1. Let n be even. Prove that n^2 is even.

Example 2

Prove: If n is an integer and n^2 is even, then n is even.

Proof To allow us to use the result proved in Example 1, we prove a statement that has the same meaning as the given statement, namely,

$$\text{If } n \text{ is not even, then } n^2 \text{ is not even.} \tag{II}$$

Assume that n is not even.	This is the hypothesis of (II).
Therefore, n is odd.	This is a consequence of (I).
Therefore, n^2 is odd.	By the result proved in Example 1.
Hence, n^2 is not even.	A consequence of (I) again.
Therefore, we have proved (II).	
Therefore, if n^2 is even, then n is even.	This has the same meaning as (II).

This proof illustrates logical derivation, an important idea that is explored in detail in Section 1.4. Logical derivation is the structure upon which mathematical proofs are built.

Practice Problem 2. Prove: If n is an integer and n^2 is odd, then n is odd.

In the next example, the proof is more difficult to obtain. To assist us, we introduce the idea of a "needs assessment."

Example 3

Let x and y be nonnegative numbers. Prove that $\sqrt{xy} \leq (x+y)/2$. Let us see how a needs assessment will help us to construct a proof.

We need to prove that $\sqrt{xy} \leq (x+y)/2$.

So we need to prove that $2\sqrt{xy} \leq (x+y)$. But all quantities are nonnegative. Hence, an equivalent statement is $[2\sqrt{xy}]^2 \leq (x+y)^2$.

So we need to prove that $4xy \leq x^2 + 2xy + y^2$. Subtracting $4xy$ from both sides, we obtain the equivalent statement, $0 \leq x^2 - 2xy + y^2 = (x-y)^2$. This statement is true, since the square of any real number is nonnegative.

Thus, by working backwards from $\sqrt{xy} \leq (x+y)/2$, we arrive at the (true) statement $0 \leq (x-y)^2$.

After this scratch work, we can construct the following proof. The reasoning for the steps is given in a less formal manner than in the previous proofs.

Proof Since the square of every real number is nonnegative, we have $0 \leq (x-y)^2$. Hence, $0 \leq x^2 - 2xy + y^2$. Adding $4xy$ to both sides yields $4xy \leq x^2 + 2xy + y^2 = (x+y)^2$, and taking the square root of both sides gives $2\sqrt{xy} \leq (x+y)$. Finally, dividing both sides by 2 yields $\sqrt{xy} \leq (x+y)/2$.

By comparison with the finished proof, the scratch work in Example 3 is far less polished, but contains information that makes the proof easier to create and understand. The proof is correct and logically complete without any reference to the scratch work, but it lacks the information that motivates each step. For example, the first step of the proof seems to come "out of the blue." However, it becomes clear why that first step is taken after the proof is completely read and the total picture absorbed.

The main goal of the proof writer is to convince others that the statement being proved follows logically from certain assumptions. A secondary goal is to write an elegant, concise proof. In some ways, these goals compete with one another because what is clear, concise, and elegant to some readers will be a terse muddle to others. The less experienced a person is at reading proofs, the more important it is for that person to do scratch work in order to absorb the ideas behind the proof.

The following example presents a proof in its final form first and then outlines an analysis of it.

Example 4

Let a, b, and c be the lengths of the sides and the hypotenuse, respectively, of a right triangle (see Figure 1–1). Prove that if the area of the triangle is $c^2/4$, then the triangle is isosceles.

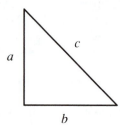

Figure 1-1

Proof Since the area is $c^2/4$, we have $c^2 = 2ab$. By the Pythagorean theorem, $a^2 + b^2 = 2ab$. Hence, $(a - b)^2 = 0$. Therefore, the triangle is isosceles.

Discussion This proof is very terse—almost "bare bones." Although proofs in the subsequent chapters will have more "flesh" on them, the preceding proof is logically correct, and with a little close reading, we can completely understand it. Let us consider each of the questions that should be asked when reading a proof.

What is the goal of the proof? Here, the goal is to prove that a triangle is isosceles. A short needs assessment using Figure 1–1 tells us that we need to prove that $a = b$.

What is the hypothesis? We are given that the area of a right triangle is equal to $c^2/4$, where c is the length of the hypotenuse of the triangle.

What definitions are necessary? The definition of *isosceles* was used in the needs assessment.

What previously proved facts or laws of logic are used in the proof? The area of a right triangle with legs of length a and b is $ab/2$. The Pythagorean theorem says that, for a right triangle, $a^2 + b^2 = c^2$.

Now let us fill in some of the missing details for this proof. We have the following annotated proof:

Since the area is $c^2/4$, we have $c^2 = 2ab$: We can assume that the area is $c^2/4$. But the area of such a triangle is $ab/2$. Hence, $ab/2 = c^2/4$. Multiplying both sides by 4 yields $c^2 = 2ab$.

By the Pythagorean theorem, $a^2 + b^2 = 2ab$: The Pythagorean theorem says that $a^2 + b^2 = c^2$ for a right triangle with hypotenuse c. Using $c^2 = 2ab$ from the preceding step then gives us $a^2 + b^2 = 2ab$.

Hence, $(a - b)^2 = 0$: Subtracting $2ab$ from both sides of $a^2 + b^2 = 2ab$ yields $a^2 - 2ab + b^2 = 0$. So $(a - b)^2 = a^2 - 2ab + b^2 = 0$.

Therefore, the triangle is isosceles. Since $(a - b)^2 = 0$, it follows that $a - b = 0$, and hence, $a = b$.

The annotated version of the original proof in Example 4 helps make the proof easier to understand. However, it still may not be clear why or how we might think of the steps in the first place. What went into the creation of this proof? For example, why was the Pythagorean theorem considered important for the proof? Perhaps because c^2 appears in the hypothesis. Perhaps because the triangle is a right triangle with legs of length a and b and hypotenuse of length c. Of course, we may have followed a couple of bad leads before finding the proof.

If we were creating the proof in Example 4, we might just as well have done a needs assessment that brought us to the goal of proving that the angles adjacent to the hypotenuse are equal. However, we would soon see that it is difficult to use the given information to arrive at any conclusion regarding angles. If we were not too discouraged, we would go back and revise our needs assessment so as to use the facts in the hypothesis.

The creation of a proof sometimes requires that we work backwards from the original goal, by means of a needs assessment, to a new goal that is closer to the given facts. Also, we may have to rewrite (but not alter) the given facts in the hypothesis so that they are closer to the new goal. This parallel process is sometimes called the **backwards/forwards method**.

EXERCISE SET 1.1

In each of the following exercises, a proof will be given or you will be asked to supply a proof. If a proof is given, outline the proof and fill in any details (for example, reasons for each step) that will make the proof easier to understand. Hints are given for the exercises in which you must supply the proof. Answer each of the following questions in every exercise:

What is the goal of the proof?

What is the hypothesis?

What definitions are necessary?

What previously proven facts or laws of logic are used in the proof? (You may need to look up relevant definitions or theorems elsewhere in this book or in a calculus book.)

1. Let x and y be nonnegative numbers. Prove that if $\sqrt{xy} = (x + y)/2$, then $x = y$.
Proof: Assume $\sqrt{xy} = (x + y)/2$. Hence, $2\sqrt{xy} = x + y$. So $(2\sqrt{xy})^2 = (x + y)^2$. This yields $4xy = x^2 + 2xy + y^2$. After subtracting $4xy$ from both sides, we obtain $0 = x^2 - 2xy + y^2 = (x - y)^2$. Therefore, $x = y$.

2. Let x and y be nonnegative numbers. Prove that if $x = y$, then $\sqrt{xy} = (x + y)/2$.

3. Let a and b be the lengths of the sides and c the length of the hypotenuse of a right triangle (see Figure 1–1). Prove that if the triangle is isosceles, then the area of the triangle is $c^2/4$. (*Hint:* The steps of this proof will be the reverse of the steps in the proof of Example 4.)

4. Let a, b, and c be the lengths of the sides of an isosceles triangle. Assume $a = b$ and the area of the triangle is $c^2/4$. Prove that the triangle is a right triangle with hypotenuse of length c. (Recall that the Pythagorean theorem states that in a right triangle with legs of length a and b and hypotenuse of length c, $a^2 + b^2 = c^2$. It is also true that a triangle with sides of length a, b, and c such that $a^2 + b^2 = c^2$ is a right triangle with hypotenuse of length c. This is the converse of the Pythagorean theorem.)

   Proof: Draw a triangle with sides of length a and b and base of length c, with $a = b$. Since the area is $c^2/4$, we have $c^2/4 = (c/4)\sqrt{4a^2 - c^2}$. Hence, $c = \sqrt{4a^2 - c^2}$ and $c^2 = 2a^2$. Since $a = b$, we have $c^2 = a^2 + b^2$. Therefore, by the converse of the Pythagorean theorem, the triangle is a right triangle.

5. Let x and y be integers. Prove that if x and y are odd, then $x + y$ is even.

   Proof: Assume x and y are odd integers. Then $x = 2j + 1$ and $y = 2k + 1$ for some integers j and k. Hence, $x + y = 2j + 2k + 2 = 2(j + k + 1)$. Therefore, $x + y$ is an even integer.

6. Let x and y be integers. Prove that if $x + y$ is odd, then x is even or y is even. (*Hint:* Use Exercise 5 to prove this result in the same way that Example 1 was used to prove Example 2.)

7. Let x be a positive real number. Prove that $x + 1/x \geq 2$. *Hint:* Do a "needs assessment" and work backwards. Start by multiplying both sides of the inequality by x.

8. Let f be any real-valued function. Prove that if f is a strictly increasing function, then f is a one-to-one function.

   Proof: Assume $x \neq z$. Then either $x < z$ or $z < x$. Hence, $f(x) < f(z)$ or $f(z) < f(x)$, since f is a strictly increasing function. In either case, $f(x) \neq f(z)$. Therefore, f is a one-to-one function.

   Exercises 9 and 10 require a knowledge of first semester calculus.

9. Prove that $\sin x \leq x$ for all nonnegative real numbers x.

   Proof: Consider the function $g(x) = x - \sin x$. The derivative $g'(x) = 1 - \cos x$, and so $g'(x) \geq 0$. Hence, g is a nondecreasing function. But $g(0) = 0$. Therefore, $g(x) \geq 0$ for all $x \geq 0$, and the result follows.

10. Prove that $\cos x = x$ for some nonnegative real number x.

    Proof: Consider the function $f(x) = x - \cos x$. It follows that $f(0) = -1 < 0$ and $f(\pi/2) = \pi/2 > 0$. Therefore, by the intermediate value theorem, there is a number x, with $0 < x < \pi/2$, such that $f(x) = 0$. The result follows.

1.2 INTRODUCTION TO PROPOSITIONAL LOGIC

Symbolic logic can be described as the analytical study of the art of reasoning. The two most important branches of symbolic logic are propositional logic and predicate logic. Although interesting in its own right, an extensive study of logic is beyond our immediate purpose. Rather, for this text, logic is important because it forms the basis for proof techniques and, therefore, has special utility for mathematics.

Propositional logic, also called the statement calculus or propositional calculus, is the study of a certain kind of statement. In propositional logic the actual content of such statements is unimportant; of primary interest is their truth or falsity.

A **statement** may either be true or false, but not both. The **truth value** of a statement is "true" if the statement is true and "false" if the statement is false.

We wish to form complex statements from simple statements and to determine the truth value of the complex statements from the truth values of the simple ones. For example, "2 is an even integer and 3 is an odd integer" is true since "2 is an even integer" and "3 is an odd integer" are both true. In this case, a complex statement is formed from two simple statements by placing the connective "and" between them. The truth value of the complex statement is determined by the truth values of the two simple statements.

To facilitate our discussion of propositional logic, we will use P, Q, R, and other letters to symbolize statements. So, for example, P may have one of two truth values (T for true or F for false). Complex statements can be constructed from simple ones (or other complex ones) by means of **statement connectives.** For example, if P and Q are statements, then P AND Q is also a statement, as is P OR Q. Besides AND and OR, some other statement connectives are IMPLIES, IF AND ONLY IF, and IT IS NOT THE CASE THAT, usually written simply as NOT. NOT is applied to only one statement to obtain a new one. For example, let P be "it is raining." Then "it is not raining" is also a statement. Because of the complicated way statements are negated in English, we choose a uniform but more cumbersome form: "It is not the case that it is raining." In this way the connective IT IS NOT THE CASE THAT always comes before the statement.

Statements without statement connectives are called **prime statements** and are denoted by lowercase letters such as p, q, and r. Statements with connectives are called **composite statements.** Symbolically, we abbreviate connectives in the following way:

Statement connective	Abbreviation
AND	$\land$
OR	$\lor$
IMPLIES	$\Rightarrow$
IF AND ONLY IF	$\Leftrightarrow$
IT IS NOT THE CASE THAT	$\neg$

The first four of these connectives are called **binary connectives** because they combine two statements to make one resulting statement. The last connective is called a **unary connective.**

Syntactics

Syntactics is a set of rules used to determine whether a sequence of symbols is a statement. The syntactical rules for constructing statements are as follows:

S1. All prime statements $p, q, r, \ldots$ are statements.

S2. If P and Q are statements, then $(P \wedge Q)$ is a statement.

S3. If P and Q are statements, then $(P \vee Q)$ is a statement.

S4. If P and Q are statements, then $(P \Rightarrow Q)$ is a statement.

S5. If P and Q are statements, then $(P \Leftrightarrow Q)$ is a statement.

S6. If P is a statement, then $\neg P$ is a statement.

No other sequence of symbols is a statement.

The main connective of a statement is the last one used to construct the statement.

Example 1

Show that $((p \wedge \neg q) \vee r)$ is a statement. What is the main connective?

Solution

1. p, q, and r are statements by S1.
2. $\neg q$ is a statement by S6 and Step 1.
3. $(p \wedge \neg q)$ is a statement by S2 and Steps 1 and 2.
4. $((p \wedge \neg q) \vee r)$ is a statement by S3 and Steps 3 and 1.

The main connective is $\vee$ since it is the one introduced in Step 4.

In what follows, we often omit outside parentheses, since doing so will not cause any confusion in reading and interpreting a statement correctly. Also, when a statement is complicated, pairs of brackets, [and], may be used in place of some pairs of parentheses.

Example 2

a. Show that $\neg(p \wedge q) \Rightarrow (\neg p \wedge \neg q)$ is a statement. What is the main connective?

b. If P and Q are statements, show that $\neg(P \wedge Q) \Rightarrow (\neg P \wedge \neg Q)$ is a statement. What is the main connective?

Solution

a. 1. p and q are statements. (S1)
   2. $(p \wedge q)$ is a statement. (1, S2)
   3. $\neg(p \wedge q)$ is a statement. (2, S6)
   4. $\neg p$ is a statement. (1, S6)
   5. $\neg q$ is a statement. (1, S6)
   6. $(\neg p \wedge \neg q)$ is a statement. (4, 5, S2)
   7. $\neg(p \wedge q) \Rightarrow (\neg p \wedge \neg q)$ is a statement. (3, 6, S4)

The main connective is $\Rightarrow$.

b. 1. P and Q are statements. (Given)

The rest of this proof is exactly the same as that of part a, except that uppercase letters replace lowercase letters.

Practice Problem 1. Show that $\neg p \Rightarrow (q \wedge r)$ is a statement. What is the main connective?

The sequences of symbols $(p\wedge \Rightarrow q)$, $(\neg \wedge p)$, and $\neg p \vee q$ are not statements since there is no way to construct them from Rules S1–S6.

Remarks on Punctuation

The meaning of a phrase in English can differ wildly depending on punctuation. Consider, for example, the following two phrases:

What do you think? You got an A on the final!

What? Do you think you got an A on the final?

For statements expressed symbolically, punctuation is accomplished by using parentheses. To keep statements from looking cluttered, we will use certain conventions for leaving out some parentheses. The most important of these are illustrated in the next example.

Example 3

The NOT ($\neg$) connective is applied first to the next letter or to the next parenthesized expression. The AND ($\wedge$) and OR ($\vee$) connectives are applied next. The implication symbol ($\Rightarrow$) is applied only after the NOT ($\neg$), AND ($\wedge$), and OR ($\vee$) connectives have been applied. Finally, the biconditional ($\Leftrightarrow$) is applied.

  a. $\neg p \wedge q$ does not stand for $\neg (p \wedge q)$.

  b. $\neg p \Rightarrow q$ does not stand for $\neg (p \Rightarrow q)$.

  d. $p \wedge q \Rightarrow r$ stands for $(p \wedge q) \Rightarrow r$, not for $p \wedge (q \Rightarrow r)$.

  e. $\neg p \vee q \Rightarrow r \wedge q$ stands for $(\neg p \vee q) \Rightarrow (r \wedge q)$, and not for $\neg p \vee (q \Rightarrow r \wedge q)$ or any other way of placing the parentheses.

  f. $p \Rightarrow q \Leftrightarrow p \vee r$ stands for $(p \Rightarrow q) \Leftrightarrow (p \vee r)$, and not for $p \Rightarrow (q \Leftrightarrow p \vee r)$ or any other way of placing the parentheses.

  g. $p \vee r \wedge t$ is ambiguous, since it is not clear whether to apply $\vee$ or $\wedge$ first.

  h. $p \Rightarrow q \Rightarrow r$ is ambiguous, since either occurrence of $\Rightarrow$ can be applied first.

Do not omit parentheses when there is any chance of confusion, even though the above conventions permit it. The hierarchical order of the connectives is as follows: NOT and then OR and AND have the same order of precedence, followed by IMPLIES and then IF AND ONLY IF. In brief, we have:

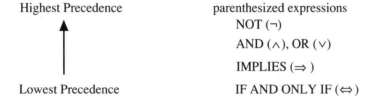

Highest Precedence	parenthesized expressions
	NOT ($\neg$)
	AND ($\wedge$), OR ($\vee$)
	IMPLIES ($\Rightarrow$)
Lowest Precedence	IF AND ONLY IF ($\Leftrightarrow$)

Example 4

Each of the following is unambiguous as written. Some parentheses have been omitted. Put in all missing parentheses:

a. $p \land q \Rightarrow \neg q \lor r$

b. $r \Rightarrow t \Leftrightarrow r$

c. $\neg (p \Rightarrow q \land r) \Rightarrow (r \land p \Rightarrow q)$

Solutions

a. $(p \land q) \Rightarrow (\neg q \lor r)$

b. $(r \Rightarrow t) \Leftrightarrow r$

c. $\neg (p \Rightarrow (q \land r)) \Rightarrow ((r \land p) \Rightarrow q)$

Practice Problem 2. Each of the following is unambiguous as written. Some parentheses have been omitted. Put in all missing parentheses.

a. $p \Rightarrow \neg q \lor r$

b. $s \land \neg q \Rightarrow t \lor \neg r$

c. $p \Rightarrow q \Leftrightarrow r \Rightarrow s$

d. $(p \Rightarrow q) \lor (s \land q) \Rightarrow r$

Example 5

Each of the following is ambiguous. Parenthesize in at least two ways.

a. $(p \land q) \Rightarrow s \Rightarrow r$

b. $p \land q \lor r \Rightarrow r \land p$

Solutions

a. $(p \land q) \Rightarrow (s \Rightarrow r)$, $((p \land q) \Rightarrow s) \Rightarrow r$

b. $(p \land q) \lor r \Rightarrow r \land p$, $p \land (q \lor r) \Rightarrow r \land p$

Practice Problem 3. Each of the following is ambiguous. Parenthesize in at least two ways.

a. $p \land q \Rightarrow r \lor q \land \neg r$

b. $p \Rightarrow q \lor r \Rightarrow q$

Semantics

Semantics is a set of rules used to determine the meanings of statements. We are interested here in studying only the truth values of statements. In particular, how can we determine the truth value of a composite statement if we are given the truth values of all its constituent prime statements? To answer this question, we need the truth tables for the statement connectives AND, OR, and NOT.

	AND			OR			NOT	
P Q	$P \wedge Q$		P Q	$P \vee Q$		P	$\neg P$	
T T	T		T T	T		T	F	
T F	F		T F	T		F	T	
F T	F		F T	T				
F F	F		F F	F				

In the first two truth tables, the four rows correspond to the four ways of assigning truth values to the two statements P and Q. In the truth table for NOT, there are only two ways of assigning truth values to the statement P, and hence, there are two rows in the truth table.

From now on, we will also say "P is true" in place of "P has truth value true (T)," and similarly for "P is false (F)." Note that an AND statement, $P \wedge Q$, is true when both the statements P and Q are true, and false otherwise. An OR statement, $P \vee Q$, is false when both the statements P and Q are false, and true otherwise. An OR statement is sometimes called an **inclusive-or** statement, since it is true when both constituent statements are true. A NOT statement, $\neg P$, is true when the statement P is false, and false when P is true.

The truth table for $P \Rightarrow Q$, or P *implies* Q, can be understood by the following example. Suppose you buy a washing machine, and it has a guarantee. Essentially, the guarantee says: "If the machine breaks, then a repairman will come to fix it." (or "The machine breaks implies that a repairman will come to fix it.") Let p be "the machine breaks" and let q be "the repairman will come to fix it." Then to say that the guarantee is not valid is to say that the statement $p \Rightarrow q$ is false. The only way that the guarantee can be invalid is when the machine breaks (p is true) and no repairman comes to fix it (q is false).

So $P \Rightarrow Q$ is false when, and only when, P is true and Q is false. This gives rise to the following truth table, which is the definition of IMPLIES:

P Q	$P \Rightarrow Q$
T T	T
T F	F
F T	T
F F	T

The washing machine example illustrates row 2 of the table. The other rows may seem strange, but since with their truth values, $P \Rightarrow Q$ cannot be false (the only way it can is given in row 2), they must be true. Let us examine, for example, row 3. It says, in the terminology of our example, "The machine is not broken, but the repairman comes to fix it." This in no way invalidates the guarantee, and similarly, neither do rows 1 and 4.

In the statement $P \Rightarrow Q$, P is called the **hypothesis** while Q is called the **conclusion.** Note that rows 3 and 4 of the truth table define an implication as being true (having truth value T) whenever the hypothesis is false (has truth value F), no matter what truth value the conclusion has. Because of this, these rows are called the **vacuously true cases.** To illustrate this idea with another example, consider the statement "If I play with my new tennis racket, then I'll win the match." As long as the person making this statement plays with an old racket, that person cannot be called a liar, no matter how the match turns out.

A statement that is true because it satisfies a vacuously true case is said to be a **vacuously true statement.**

The definition of an IF AND ONLY IF statement, is given by the following truth table:

P	Q	$P \Leftrightarrow Q$
T	T	T
T	F	F
F	T	F
F	F	T

Note that $P \Leftrightarrow Q$ is true whenever the truth values of P and Q are the same.

In order to get a better sense of the meaning of each of the sentence connectives, especially implication, consider the following example.

Example 6

Let

p stand for "Today is Thursday."

q stand for "Tomorrow is Friday."

r stand for "It is raining today."

s stand for "Today is Wednesday."

t stand for "Tomorrow is Thursday."

Assume today is a rainy Thursday. When we let p stand for "Today is Thursday," we are assigning a value of T to p, since we assume that today is Thursday. To let s stand for "Today is Wednesday" really means assigning the truth value F to s. Similarly, for the others.

Statement	Truth Value
$p \wedge r$	True
$q \vee r$	True
$r \wedge s$	False
$\neg t \wedge r$	True
$p \Rightarrow q$	True
$s \vee t$	False
$r \Rightarrow t$	False
$s \Rightarrow r$	(Vacuously) True
$s \Rightarrow t$	(Vacuously) True
$p \Leftrightarrow r$	True
$s \Leftrightarrow r$	False

Practice Problem 4. Using the statements in Example 6, supply the truth value for each of the following statements:

a. $s \vee \neg t$

b. $r \wedge \neg s$

c. $p \Rightarrow r$

d. $r \Rightarrow s$

e. $s \Rightarrow \neg q$

f. $p \Leftrightarrow q$

g. $\neg r \Leftrightarrow \neg p$

The truth value of any composite statement is determined by the truth table for the statement. Following is an example of how to construct the truth table for a composite statement.

Example 7

Construct the truth table for $\neg(P \wedge Q) \Rightarrow \neg P \wedge \neg Q$

Solution

P	Q	$\neg$	$(P \wedge Q)$	$\Rightarrow$	$\neg P$	$\wedge$	$\neg Q$
T	T	F	T	T	F	F	F
T	F	T	F	F	F	F	T
F	T	T	F	F	T	F	F
F	F	T	F	T	T	T	T
(1)	(2)	(4)	(3)	(8)	(5)	(7)	(6)

First, notice that all letters occurring in the statement are listed in alphabetical order at the top of the two leftmost columns. Alphabetical order is not essential, but it is easier to compare truth tables if they are all done this way. The listing of the truth values under the letters is also done in a standard way: In column (1), there is a block of two T's followed by a block of two F's. In column (2), T's and F's are alternated in turn.

Column (3) is the first column calculated. The first entry, T, in (3) is obtained by observing that if P and Q are both true (columns (1) and (2)) then $P \wedge Q$ is true. The second-row entry is F because if P is true and Q is false, then $P \wedge Q$ is false. The rest of (3) is obtained in a similar manner.

Next, look at (4). The first entry is F, since $P \wedge Q$ is true (recorded in (3)), and (4) records the negation of the truth values in (3). Similarly, for the rest of (4). Next, (5) is obtained using the negations of the truth values in column (1). The values in (6) are then obtained by negating the truth values in (2). Column (7) is obtained from (5) and (6) using the truth table for $\wedge$. Finally, (8) is obtained from (4) and (7) using the truth table for $\Rightarrow$. A truth table formed in this manner is called a **standard table.**

Observe that the order in which the columns are done is determined by the way the statement is constructed. In Example 2, we proved that the expression $\neg(P \wedge Q) \Rightarrow \neg P \wedge \neg Q$ is a statement. In the truth table in Example 7, columns (1) and (2) are set down first, corresponding to Step 1 of Example 2. Next, (3) is done, which corresponds to Step 2 of Example 2. Then (4), (5), (6), (7), and (8) are done corresponding to Steps 3, 4, 5, 6, and 7, respectively. The main column of the truth table is (8), the last one done. It tells us, for example, that when P is true and Q is false, then $\neg(P \wedge Q) \Rightarrow \neg P \wedge \neg Q$ is false.

Example 8

Construct the truth table for $(P \wedge R \Rightarrow \neg Q) \Rightarrow (P \wedge Q \Rightarrow \neg R)$.

Solution

P	Q	R	$(P \wedge R$		$\Rightarrow \neg Q)$	$\Rightarrow$	$(P \wedge Q$		$\Rightarrow \neg R)$
T	T	T	T	F F	T	T	F F		
T	T	F	F	T F	T	T	T T		
T	F	T	T	T T	T	F	T F		
T	F	F	F	T T	T	F	T T		
F	T	T	F	T F	T	F	T F		
F	T	F	F	T F	T	F	T T		
F	F	T	F	T T	T	F	T F		
F	F	F	F	T T	T	F	T T		
(1)	(2)	(3)	(4)	(6)(5)	(10)	(7)	(9)(8)		

When there are three letters in a statement, the truth table is formed in the same way as when there are two letters, except that there are more combinations for assigning truth values to the letters. There are eight ways to assign truth values to three letters, and hence, there are eight rows in the truth table. In the first column, under P, there is a block of four T's and then a block of four F's. In the next column, under Q, there is a block of two T's and then a block of two F's, followed by two more T's and two F's. In the last column, under R, starting with T, T's and F's alternate down the column.

Practice Problem 5. Construct the truth table for $\neg(P \wedge Q) \vee R \Rightarrow \neg P \vee R$.

EXERCISE SET 1.2

Use the syntax rules S1–S6 to verify that the sentences in Exercises 1–6 are statements. Specify the main connective in each case.

1. $(p \Rightarrow \neg q) \wedge r \Rightarrow q$
2. $p \Rightarrow \neg q \wedge r$
3. $(p \Rightarrow q) \Rightarrow [(p \wedge r) \Rightarrow (q \wedge r)]$
4. $(p \Rightarrow q) \Rightarrow [(q \Rightarrow r) \Rightarrow (p \Rightarrow r)]$
5. $[(p \Rightarrow q) \wedge (r \Rightarrow s)] \Rightarrow [(p \vee r) \Rightarrow (q \vee s)]$
6. $[(p \Rightarrow q) \wedge (r \Rightarrow s)] \Rightarrow [(\neg q \vee \neg s) \Rightarrow (\neg p \vee \neg r)]$

Construct truth tables for the statements in Exercises 7–18.

7.   (a) $P \wedge Q \Rightarrow P$
     (b) $P \Rightarrow P \vee Q$
8.   (a) $P \wedge (P \Rightarrow Q) \Rightarrow Q$
     (b) $Q \wedge (P \Rightarrow Q) \Rightarrow P$
9.   (a) $P \wedge (\neg P \vee Q) \Rightarrow Q$
     (b) $\neg P \wedge (P \Rightarrow Q) \Rightarrow Q$
10. $(P \Rightarrow Q) \wedge (Q \Rightarrow R) \Rightarrow (P \Rightarrow R)$
11. $(P \Rightarrow Q) \Rightarrow [(P \vee R) \Rightarrow (Q \vee R)]$
12. $(P \Rightarrow Q) \Rightarrow [(P \wedge R) \Rightarrow (Q \wedge R)]$
13. $(P \vee Q \Rightarrow R) \Rightarrow (P \Rightarrow R) \wedge (Q \Rightarrow R)$
14. $(P \wedge Q \Rightarrow R) \Rightarrow (P \Rightarrow R) \wedge (Q \Rightarrow R)$
15. $(P \wedge Q \Rightarrow R) \Rightarrow (P \Rightarrow R) \vee (Q \Rightarrow R)$
16. $[(P \vee Q) \wedge R] \Rightarrow (\neg P \vee Q)$
17. $\neg(P \wedge \neg Q) \vee (R \Rightarrow Q)$
18. $[\neg P \vee (P \Rightarrow \neg R)] \wedge R$

The connective $\underline{\vee}$ is called the **exclusive–or connective** and is defined by the following truth table:

P	Q	$P \underline{\vee} Q$
T	T	F
T	F	T
F	T	T
F	F	F

19. Construct the truth tables for
    (a) $P \underline{\vee} P$
    (b) $P \underline{\vee} \neg P$

20. Construct the truth tables for
  (a) $P \vee (Q \vee R)$
  (b) $(P \vee Q) \vee R$

21. Construct the truth tables for
  (a) $P \vee (Q \wedge R)$
  (b) $(P \vee Q) \wedge (P \vee R)$

22. Construct the truth tables for
  (a) $P \wedge (Q \vee R)$
  (b) $(P \wedge Q) \vee (P \wedge R)$

23. Write a statement composed of P, Q and the connectives $\neg$, $\vee$, and $\wedge$ that is true when exactly one of P and Q is true and false otherwise.

24. Write a statement composed of P, Q, R and the connectives $\neg$, $\vee$, and $\wedge$ that is true when exactly one of P, Q, and R is true and false otherwise.

25. Let p stand for "This statement is false." What can be said about the truth value of p? (See Example 6.) (*Hint:* Did we really assign a truth value to p?)

1.3 LOGICAL EQUIVALENCE AND TAUTOLOGIES

By examining the four combinations of truth values for P and Q, we can show that $P \Rightarrow Q$ and $\neg P \vee Q$ have the same truth tables:

P	Q	$P \Rightarrow Q$	P	Q	$\neg P$	$\vee Q$
T	T	T	T	T	F	T
T	F	F	T	F	F	F
F	T	T	F	T	T	T
F	F	T	F	F	T	T

We say that $P \Rightarrow Q$ is logically equivalent to $\neg P \vee Q$ and write $(P \Rightarrow Q) \equiv (\neg P \vee Q)$. We call $\neg P \vee Q$ the **or-form** of $P \Rightarrow Q$.

Definition.   Two statements P and Q are **logically equivalent,** written $P \equiv Q$, if P and Q have the same truth values whenever all constituent prime statements in one have the same values as the corresponding prime statements in the other.

Two statements composed of the same prime statements are logically equivalent whenever the main columns of their standard truth tables are identical.

In determining whether to omit parentheses, $\equiv$ has the lowest precedence. For example, $P \Rightarrow Q \equiv \neg P \vee Q$, when parenthesized, is $(P \Rightarrow Q) \equiv (\neg P \vee Q)$.

Several logical equivalences, including the or-form of an implication, are given in the following list. They are all important, and we will use them later.

   a. $\neg\neg P \equiv P$ [Rule of **double negation**]

   b. $P \Rightarrow Q \equiv \neg P \vee Q$ [**Or-form** of an implication]

   c. $P \Rightarrow Q \equiv \neg Q \Rightarrow \neg P$ [**Contrapositive** of an implication]

   d. $\neg(P \vee Q) \equiv \neg P \wedge \neg Q$ $\Big\}$

   e. $\neg(P \wedge Q) \equiv \neg P \vee \neg Q$ [**De Morgan's laws**]

   f. $(P \wedge R \Rightarrow Q) \equiv (R \Rightarrow (P \Rightarrow Q))$ [Rule for direct proof]

   g. $(P \wedge \neg Q \Rightarrow \mathcal{O}) \equiv (P \Rightarrow Q)$ [Rule for proof by contradiction]

The or-form and the contrapositive are useful alternative ways of writing an implication. Roughly speaking, De Morgan's laws can be paraphrased as negating an OR statement makes it an AND statement, while negating an AND statement makes it an OR statement. Both direct proof and proof by contradiction are discussed in Section 1.4. The symbol $\mathcal{O}$ used in the rule for proof by contradiction stands for any statement that is always false and is discussed shortly.

Note that uppercase letters are used in the Laws a–g. This means the letters stand for any statement, prime or composite. So, for example, if P is $\neg q \vee r$ and Q is $s \wedge t$, then law b becomes $(\neg q \vee r) \Rightarrow (s \wedge t) \equiv \neg(\neg q \vee r) \vee (s \wedge t)$.

Practice Problem 1. Compute the truth tables for $\neg(P \vee Q)$ and $\neg P \wedge \neg Q$ to verify De Morgan's law d.

Tautologies

The truth tables for $P \vee \neg P$ and $Q \Rightarrow P \vee Q$ are as follows:

P	$P \vee \neg P$
T	T
F	T

P	Q	$Q \Rightarrow P \vee Q$
T	T	T
T	F	T
F	T	T
F	F	T

The statements $P \vee \neg P$ and $Q \Rightarrow P \vee Q$ are two examples of tautologies.

Definition. A **tautology** is a statement that is true no matter what truth values its constituent prime statements have.

Every row in the main column of a tautology is T. The symbol $\mathcal{I}$ will denote a statement that always has truth value T. Hence, $P \vee \neg P \equiv \mathcal{I}$ and $(Q \Rightarrow P \vee Q) \equiv \mathcal{I}$.

Contradictions

The truth tables for $P \wedge \neg P$ and $(\neg P \vee Q) \wedge (P \wedge \neg Q)$ are as follows:

P	$P \wedge \neg P$		P	Q	$(\neg P \vee Q) \wedge (P \wedge \neg Q)$
T	F		T	T	F
F	F		T	F	F
			F	T	F
			F	F	F

The statements $P \wedge \neg P$ and $(\neg P \vee Q) \wedge (P \wedge \neg Q)$ are examples of contradictions.

Definition. A **contradiction** is a statement that is false no matter what truth values its constituent prime statements have.

Every row in the main column of a contradiction is F. The symbol $\mathcal{O}$ will be used to denote a statement that always has truth value F. Hence, $P \wedge \neg P \equiv \mathcal{O}$ and $(\neg P \vee Q) \wedge (P \wedge \neg Q) \equiv \mathcal{O}$.

We can summarize the three concepts just introduced:

Two statements P and Q are logically equivalent (written $P \equiv Q$) if they have the same truth values whenever all prime statements in one have the same values in the other.

A statement P is a tautology whenever the truth table for P has a T in every row of its main column (that is, $P \equiv \mathcal{I}$).

A statement Q is a contradiction whenever the truth table for Q has an F in every row of its main column (that is, $Q \equiv \mathcal{O}$).

Following are some important logical equivalences called the **Boolean laws** of logic.

Boolean Laws of Logic

1a $P \vee Q \equiv Q \vee P$	1b $P \wedge Q \equiv Q \wedge P$	[Commutative law]
2a $(P \vee Q) \vee R \equiv P \vee (Q \vee R)$	2b $(P \wedge Q) \wedge R \equiv P \wedge (Q \wedge R)$	[Associative law]
3a $P \vee (Q \wedge R) \equiv (P \vee Q) \wedge (P \vee R)$	3b $P \wedge (Q \vee R) \equiv (P \wedge Q) \vee (P \wedge R)$	[Distributive law]
4a $P \vee \mathcal{O} \equiv P$	4b $P \wedge \mathcal{I} \equiv P$	
5a $P \vee \neg P \equiv \mathcal{I}$	5b $P \wedge \neg P \equiv \mathcal{O}$	

Each law can be verified by a calculation of the truth tables for the statements to the left and right of $\equiv$ in each logical equivalence.

Example 1

The following table confirms that Boolean law 3b is a logical equivalence, since the main columns under $P \wedge (Q \vee R)$ and $(P \wedge Q) \vee (P \wedge R)$ have the same truth values in every row:

$P \; Q \; R$	$P \wedge (Q \vee R)$	$(P \wedge Q) \; \vee \; (P \wedge R)$
T T T	T	T
T T F	T	T
T F T	T	T
T F F	F	F
F T T	F	F
F T F	F	F
F F T	F	F
F F F	F	F

By the commutative law of the Boolean laws of logic, $P \vee Q$ is logically equivalent to $Q \vee P$, and $P \wedge Q$ is logically equivalent to $Q \wedge P$. So the order of the constituent statements is irrelevant to the truth value of the statement containing them in either an OR statement or an AND statement. Also, by the Associative law 2a, $(P \vee Q) \vee R$ is logically equivalent to $P \vee (Q \vee R)$. Hence, we frequently omit the parentheses and write $P \vee Q \vee R$ for either statement. Similarly, we write $P \wedge Q \wedge R$ without parentheses.

The **biconditional** $P \Leftrightarrow Q$ is logically equivalent to $(P \Rightarrow Q) \wedge (Q \Rightarrow P)$ (see Practice Problem 2). Hence, the biconditional $P \Leftrightarrow Q$ is true whenever the implications $P \Rightarrow Q$ and $Q \Rightarrow P$ are both true. The implication $Q \Rightarrow P$ is the **converse** of $P \Rightarrow Q$. So the biconditional $P \Leftrightarrow Q$ is true whenever the implication $P \Rightarrow Q$ and its converse are both true.

Practice Problem 2.   Verify that $(P \Rightarrow Q) \wedge (Q \Rightarrow P) \equiv P \Leftrightarrow Q$.

In the implication $P \Rightarrow Q$, we say that P is a **sufficient condition** for Q or that Q is a **necessary condition** for P. To see why the terms "sufficient" and "necessary" are used in this context, suppose that a given implication is true. Then the conclusion is necessarily true when the hypothesis is true, and the hypothesis being true is sufficient to make the conclusion true. For the biconditional $P \Leftrightarrow Q$, we say that Q is a **necessary and sufficient condition** for P (or that P is a necessary and sufficient condition for Q).

Counterexamples

By definition, a statement is not a tautology whenever at least one row of the main column of its truth table is F. Accordingly, to show that a statement is not a tautology, a row in the main column of the truth table for that statement must be found with a truth value of F. The combination of truth values assigned to the prime statements in any row that produces an F in the main column of the table is called a **counterexample**.

Example 2

Show that $q \wedge (p \Rightarrow q) \Rightarrow p$ is not a tautology.

p q	$q \wedge (p \Rightarrow q) \Rightarrow p$	
T T	T	
T F	T	
F T	F	$\longleftarrow$ Counterexample
F F	T	

The counterexample is p false (pF) and q true (qT).

Claiming that the statement $q \wedge (p \Rightarrow q) \Rightarrow p$ is a tautology is a well-known fallacy of logic. It is called the **fallacy of asserting the conclusion**.

In Example 2, we used lowercase letters for statements. Let us examine what would happen if we had used uppercase letters. Let Q be $r \wedge \neg r$, and let P be $r \vee \neg r$. Then, in order to find a counterexample and prove that $Q \wedge (P \Rightarrow Q) \Rightarrow P$ is not a tautology, we require Q to be true and P to be false. However, we see that Q is always false and P is always true. Hence, there is no way to make Q true and P false, and in that particular case the statement that results from substituting $r \wedge \neg r$ for Q and $r \vee \neg r$ for P is a tautology (though $Q \wedge (P \Rightarrow Q) \Rightarrow P$ is not, because other substitutions do yield a counterexample).

Practice Problem 3. Show that $(p \vee q) \Rightarrow p$ is not a tautology.

We can also use the idea of a counterexample to show that two statements are not logically equivalent. For example, is order important for an implication? In other words, is $q \Rightarrow p$ logically equivalent to its converse $p \Rightarrow q$?

Example 3

Show that $q \Rightarrow p$ is not logically equivalent to $p \Rightarrow q$.

p q	$p \Rightarrow q$	$q \Rightarrow p$	
T T	T	T	
T F	F	T	$\longleftarrow$ Counterexample
F T	T	F	$\longleftarrow$ Counterexample
F F	T	T	

The main columns of the truth tables are not the same, which means that $p \Rightarrow q$ is not logically equivalent to $q \Rightarrow p$. One counterexample is pT and qF, another is pF and qT.

Example 3 shows that the order of the constituent statements of an implication is crucial. That is, the converse of an implication is not logically equivalent to the implication. Example 4 illustrates this further.

Example 4

a. "If it is snowing, then there are clouds above" is true. However, its converse, "If there are clouds above, then it is snowing," is false.

b. The implication "If a function is differentiable, then it is continuous" is true. However, its converse, "If a function is continuous, then it is differentiable," is false.

EXERCISE SET 1.3

In Exercises 1–6 verify each of the given laws of logic by constructing the appropriate truth tables.

1.   (a) $\neg(P \vee Q) \equiv \neg P \wedge \neg Q$
      (b) $\neg(P \wedge Q) \equiv \neg P \vee \neg Q$

2.   (a) $\neg(\neg P) \equiv P$
      (b) $P \vee Q \Rightarrow Q \equiv \neg P \vee Q$

3.   (a) $P \vee (Q \wedge R) \equiv (P \vee Q) \wedge (P \vee R)$
      (b) $P \vee \neg P \equiv \mathcal{I}$
      (c) $P \wedge \neg P \equiv \mathcal{O}$

4.   (a) $(P \wedge R \Rightarrow Q) \equiv (P \Rightarrow (R \Rightarrow Q))$
      (b) $(P \wedge \neg Q \Rightarrow \mathcal{O}) \equiv (P \Rightarrow Q)$

5.   (a) $(P \wedge Q \Rightarrow P) \equiv \mathcal{I}$
      (b) $(P \vee Q \Rightarrow R) \equiv (P \Rightarrow R) \wedge (Q \Rightarrow R)$

6.   (a) $(P \Rightarrow P \vee Q) \equiv \mathcal{I}$
      (b) $(P \Rightarrow Q) \wedge (Q \Rightarrow R) \Rightarrow (P \Rightarrow R)$

7. Neither of the following is a tautology. Find a counterexample in each case.
    (a) $(p \vee \neg q) \vee (q \Rightarrow p)$
    (b) $(p \wedge q) \vee r \Rightarrow (p \vee q) \wedge r$.

8. Neither of the following is a tautology. Find a counterexample in each case.
    (a) $[(p \wedge q \Rightarrow r)] \Leftrightarrow (p \Rightarrow r) \wedge (q \Rightarrow r)$
    (b) $[(p \wedge q) \vee r] \Rightarrow p \wedge (q \vee r)$.

9. For each of the following statements, use truth tables to tell whether it is logically equivalent to the negation of $p \Rightarrow q$, or whether it is not. In each case, verify your answer.
    (a) $p \Rightarrow \neg q$
    (b) $\neg p \Rightarrow \neg q$
    (c) $\neg p \Rightarrow q$.

10. For each of the following statements, use truth tables to tell whether it is logically equivalent to the negation of $p \Rightarrow q$, or whether it is not. In each case, verify your answer.
    (a) $\neg q \Rightarrow p$
    (b) $\neg p \wedge q$
    (c) $p \vee \neg q$.

11. For each of the following statements, use truth tables to tell whether it is logically equivalent to the converse of $p \Rightarrow q$, or whether it is not. In each case, verify your answer.
    (a) $p \Rightarrow \neg q$
    (b) $\neg p \Rightarrow \neg q$

(c) $\neg p \Rightarrow q$.

12. For each of the following statements, use truth tables to tell whether it is logically equivalent to the converse of $p \Rightarrow q$, or whether it is not. In each case, verify your answer.

(a) $\neg q \Rightarrow p$

(b) $\neg p \wedge q$

(c) $p \vee \neg q$.

13. Reformulate each of the following so that the resulting logically equivalent statement has only NOT and OR as connectives.

(a) $P \Rightarrow Q \vee R$

(b) $P \wedge Q \Rightarrow R$.

14. Reformulate each of the following so that the resulting logically equivalent statement has only NOT and OR as connectives.

(a) The converse of $P \Rightarrow Q \vee R$

(b) The contrapositive of $P \wedge Q \Rightarrow R$.

15. Reformulate each of the following so that the resulting logically equivalent statement has only NOT and OR as connectives.

(a) The negation of $P \Rightarrow Q \vee R$

(b) The negation of $P \wedge Q \Rightarrow R$.

16. Show that $P \equiv Q$ if and only if $P \Leftrightarrow Q$ is a tautology.

17. What is wrong with the following argument? To verify that P is a tautology, it is sufficient to find a counterexample to $\neg P$.

1.4 RULES OF INFERENCE

In this section, we introduce proof techniques and apply them to proving statements from a collection of premises. We restrict ourselves to statements in propositional logic so that we can better concentrate on the form that a correct logical argument must take. In Chapter 2, we use the techniques introduced in this section and others to examine methods of proving other kinds of mathematical statements.

A **proof** is a step-by-step demonstration that a statement can be derived from a collection of premises. A **premise** is a statement that is assumed in the context of a proof. Each step of a proof is either a premise (Pr) or can be shown to be a consequence of previous steps using certain rules of inference. In what follows, we introduce some rules of inference and illustrate how each of them is used to construct proofs.

RULE OF INFERENCE   **Modus Ponens (MP).**   From P and $P \Rightarrow Q$, infer Q.

Modus ponens is based on the tautology $P \wedge (P \Rightarrow Q) \Rightarrow Q$.

Example 1

Premises: $R \wedge T$, $(R \wedge T) \Rightarrow S$. Prove S.

Solution

1. $R \wedge T$ (Pr)

2. $(R \wedge T) \Rightarrow S$ (Pr)

3. S (1, 2, MP)

Notice that the lines of the foregoing proof are numbered for easy reference. At the right of each step is an explanation for the step. The first two steps are premises. Step 3 follows from Steps 1 and 2 using modus ponens, where P is $R \wedge T$ and Q is S.

Practice Problem 1. *Premises:* $T \Rightarrow R$, $R \Rightarrow S$, T. Prove S.

Example 2

Premises: $R \vee T \Rightarrow (\neg R \Rightarrow T)$, $R \vee T$, $\neg R$. Prove T.

Solution

1. $R \vee T$ (Pr)
2. $R \vee T \Rightarrow (\neg R \Rightarrow T)$ (Pr)
3. $\neg R \Rightarrow T$ (1, 2, MP)
4. $\neg R$ (Pr)
5. T (3, 4, MP)

       Step 3 follows from Steps 1 and 2 by modus ponens, where P is $R \vee T$ and Q is $\neg R \Rightarrow T$.

Practice Problem 2. *Premises:* $\neg(R \wedge T) \Rightarrow \neg R \vee \neg T$, $\neg(R \wedge T)$, $\neg R$, $\neg R \vee \neg T \Rightarrow (\neg R \Rightarrow T)$. Prove T.

Notice that in a proof, you do not have to use all of the premises.

       RULE OF INFERENCE   **Adjoining Premises (AP).**   Any statement can be adjoined to the premises in a proof if it can be proved from the premises.

       Adjoining premises is based on the tautology $(P \Rightarrow Q) \wedge (P \wedge Q \Rightarrow R) \Rightarrow (P \Rightarrow R)$.

       For example, if Q has been previously proved from a set X of premises, then Q may be used as a step in a proof of R from a set of premises which includes the set X.

Example 3

Premises: $P \vee Q \Rightarrow (\neg P \Rightarrow Q)$, $P \vee Q$, $\neg P$, $Q \Rightarrow R$. Prove R.

Solution   Note that the premises of Example 2 are the first three premises (with different letters) of this example. Thus, we may use (AP) to obtain Q as a step in this proof.

1. Q (Example 2, AP)
2. $Q \Rightarrow R$ (Pr)
3. R (1, 2, MP)

Practice Problem 3. *Premises:* $P \vee Q \Rightarrow (\neg P \Rightarrow Q)$, $P \vee Q$, $\neg P$, $Q \Rightarrow R$, $R \Rightarrow S \wedge T$. Prove $S \wedge T$. (*Hint:* Look at the premises in Example 3.)

       We say that P is **provable** from a set of premises if there is a proof of P from the set of premises.

The following is, arguably, the rule of inference most often used.

RULE OF INFERENCE   **Direct Proof of an Implication (DPI).**   To prove an implication $P \Rightarrow Q$ from a set of premises, it is sufficient to assert the hypothesis (Hyp) P as an additional premise and show that the conclusion Q is provable from the augmented set of premises.

DPI is based on the tautology $(R \wedge P \Rightarrow Q) \Leftrightarrow (R \Rightarrow (P \Rightarrow Q))$.

Example 4

Premises: $S \vee P$, $P \Rightarrow \neg E$, $S \vee P \Rightarrow (\neg S \Rightarrow P)$. Prove $\neg S \Rightarrow \neg E$.

Solution

1. $\neg S$	(Hyp)
2. $S \vee P$	(Pr)
3. $S \vee P \Rightarrow (\neg S \Rightarrow P)$	(Pr)
4. $\neg S \Rightarrow P$	(2, 3, MP)
5. P	(1, 4, MP)
6. $P \Rightarrow \neg E$	(Pr)
7. $\neg E$	(5, 6, MP)
8. $\neg S \Rightarrow \neg E$	(1, 7, DPI)

In Step 1, $\neg S$, which is the hypothesis of $\neg S \Rightarrow \neg E$, is used as an additional premise. In Step 8, we prove $\neg S \Rightarrow \neg E$ by DPI, since we have proved $\neg E$ in Step 7 using $\neg S$ as an additional premise.

Practice Problem 4.   *Premises:* P, $Q \Rightarrow (P \Rightarrow \neg R)$. Prove $Q \Rightarrow \neg R$.

RULE OF INFERENCE   **Adjunction (Adj).**   If P and Q are provable from the same set of premises, then $P \wedge Q$ is provable from that set of premises.

Adjunction is based on the tautology $(R \Rightarrow P) \wedge (R \Rightarrow Q) \Leftrightarrow (R \Rightarrow P \wedge Q)$.

Example 5

Premises: $P \Rightarrow Q, Q \wedge (P \vee R) \Rightarrow S \wedge T$, P, $P \Rightarrow P \vee R$. Prove $S \wedge T$.

Solution

1. P	(Pr)
2. $P \Rightarrow Q$	(Pr)
3. Q	(1, 2, MP)
4. $P \Rightarrow P \vee R$	(Pr)
5. $P \vee R$	(1, 4, MP)
6. $Q \wedge (P \vee R)$	(3, 5, Adj)
7. $Q \wedge (P \vee R) \Rightarrow S \wedge T$	(Pr)
8. $S \wedge T$	(6, 7, MP)

Practice Problem 5.   *Premises:* $Q \Rightarrow P$, Q, $P \wedge Q \Rightarrow R \vee T$. Prove $R \vee T$.

RULE OF INFERENCE   Substitution (Sub).   Assume P_2 is obtained from P_1 by substituting R for *any* occurrence of S in P_1. Then we can derive P_2 from $S \Leftrightarrow R$ and P_1.

Example 6

Premises: $R \vee Q \Leftrightarrow P \vee Q$, $R \vee Q \Rightarrow S$, $P \Rightarrow P \vee Q$, P. Prove S.

Solution

   1. $R \vee Q \Rightarrow S$ (Pr)
   2. $R \vee Q \Leftrightarrow P \vee Q$ (Pr)
   3. $P \vee Q \Rightarrow S$ (1, 2, Sub)
   4. P (Pr)
   5. $P \Rightarrow P \vee Q$ (Pr)
   6. $P \vee Q$ (4, 5, MP)
   7. S (3, 6, MP)

In Step 3, Sub was used. In this case, S is $P \vee Q$ and R is $R \vee Q$.

Practice Problem 6.   *Premises:* $Q \vee P \Leftrightarrow (\neg Q \Rightarrow P)$, $\neg Q$, $R \Rightarrow (Q \vee P)$, R. Prove P.

Notice that in Sub we use $S \Leftrightarrow R$ and P_1 to prove P_2 by substituting R for any occurrence of S in P_1. That is, we substitute the expression on the right of $\Leftrightarrow$ for the expression on the left of $\Leftrightarrow$. We may also derive P_2 from P_1 by substituting S for any occurrence of R in P_1; that is, we substitute the expression on the left of $\Leftrightarrow$ for the expression on the right of $\Leftrightarrow$. From now on, we will use Sub in either form.

RULE OF INFERENCE   Contradiction (Contra).   To prove Q from a set of premises, it is sufficient to use $\neg Q$ (Neg of concl) as an additional premise and to prove a statement of the form $R \wedge \neg R$.

    Contra is based on the tautology $(P \wedge \neg Q \Rightarrow R \wedge \neg R) \Leftrightarrow (P \Rightarrow Q)$. Notice that $R \wedge \neg R \equiv \mathcal{O}$.

Example 7

Premises: $\neg K \Rightarrow M$, $K \Rightarrow J$, $M \Rightarrow \neg N$, N. Prove K.

Solution

   1. $\neg K$ (Neg of concl)
   2. $\neg K \Rightarrow M$ (Pr)
   3. M (1, 2, MP)
   4. $M \Rightarrow \neg N$ (Pr)
   5. $\neg N$ (3, 4, MP)
   6. N (Pr)
   7. $N \wedge \neg N$ (5, 6, Adj)
   8. K (1, 7, Contra)

In Step 1, we use the negation of the statement we want to prove as an additional premise. In Step 8, we prove the desired conclusion, since a contradiction, namely $N \wedge \neg N$, has been proved in Step 7.

Practice Problem 7. *Premises:* $\neg(P \vee Q) \Leftrightarrow \neg P \wedge \neg Q$, $\neg P \Rightarrow R$, $\neg Q \Rightarrow \neg R$, $\neg P \wedge \neg Q \Rightarrow \neg P$, $\neg P \wedge \neg Q \Rightarrow \neg Q$. *Prove* $P \vee Q$.

The following are several additional premises that may be used in proofs:

I $(P \Rightarrow S) \Leftrightarrow (\neg S \Rightarrow \neg P)$

II $\neg Q \wedge (S \vee Q) \Rightarrow S$

III $\neg \neg R \Leftrightarrow R$

IV $P \Rightarrow P \vee R$

Example 8

Premises: $Q \vee P$, $S \Rightarrow \neg P$. *Prove* $S \Rightarrow Q$.

Solution

1. S	(Hyp)	
2. $S \Rightarrow \neg P$	(Pr)	
3. $\neg P$	(1, 2, MP)	
4. $Q \vee P$	(Pr)	
5. $\neg P \wedge (Q \vee P)$	(3, 4, Adj)	
6. $\neg P \wedge (Q \vee P) \Rightarrow Q$	(Pr II)	
7. Q	(5, 6, MP)	

In Step 6, $\neg P \wedge (Q \vee P) \Rightarrow Q$ was used as a premise. This is number II in the list of additional premises, with a simultaneous substitution of P for Q and Q for S.

If we can prove P from a set of premises, then we say that P **logically follows**, or simply **follows**, from the set of premises. An **argument** is a sequence of steps proposed as a proof. If an argument is, in fact, a proof, then it is called a **valid argument**; if it is not a proof, then it is called an **invalid argument**.

Counterexamples and Invalid Arguments

It can be shown that $P_1 \wedge P_2 \Rightarrow Q$ is a tautology if and only if Q logically follows from the premises P_1 and P_2. Therefore, any argument with premises P_1 and P_2 and conclusion Q is **invalid** whenever $P_1 \wedge P_2 \Rightarrow Q$ is not a tautology. So a counterexample for an invalid argument with premises P_1 and P_2 and conclusion Q is a counterexample for $P_1 \wedge P_2 \Rightarrow Q$. Since P_1, P_2, and Q are usually composite statements, any combination of truth values of the constituent prime statements of P_1, P_2, and Q that makes P_1 true, P_2 true, and Q false is a counterexample to $P_1 \wedge P_2 \Rightarrow Q$. By extension, the discussion applies to an argument with any number of premises.

Recall the discussion in Section 1.3 about lowercase letters being used when we discuss counterexamples.

Example 9

Show that any argument with premises n and $m \Rightarrow n$, and conclusion m is invalid.

Solution   We need to show that $n \wedge (m \Rightarrow n) \Rightarrow m$ is not a tautology. The counterexample mF and nT was in fact already given in Example 2 of Section 1.3.

The point of a counterexample is to make the premises true and the conclusion false.

Example 10

Show that any argument with premises $s \vee b$, $b \Rightarrow \neg e$, and b and conclusion $\neg s$ is invalid.

Solution   We need to show that $(s \vee b) \wedge (b \Rightarrow \neg e) \wedge b \Rightarrow \neg s$ is not a tautology. Using a truth table to find a counterexample is straightforward and effective. In this case, we need an eight-row truth table. For some arguments, however, we would need a 16- or even 32-row truth table. Since the truth table method can be tedious, we will find a counterexample by means of a systematic method involving a diagram. In Figure 1–2 all the truth values shown are based on the fact that the implication representing the argument is false.

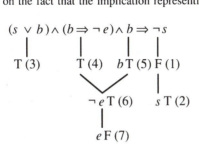

Figure 1–2

The goal is to find a combination of truth values for b, e, and s that makes the three premises true and the conclusion $\neg s$ false. Each step follows in order from that goal.

(1) $\neg s$ must be false, since the conclusion must be false.

(2) By (1), s must be true.

(3) $s \vee b$ must be true, since it is a hypothesis.

(4) $b \Rightarrow \neg e$ must be true, since it is a hypothesis.

(5) b must also be true, since it is a hypothesis.

(6) $\neg e$ must be true by (4) and (5).

(7) By (6), e must be false.

We must now check to make sure that bT, eF, and sT does not lead to any contradictions. We do so by constructing one row of the truth table with bT, eF, and sT.

b	e	s	$(s \vee b) \wedge (b \Rightarrow \neg e) \wedge b \Rightarrow \neg s$
T	F	T	T T T T T F F

None of the truth values in this row contradict any of the truth values in the diagram, and under the main connective we have an F. So the desired counterexample is bT, eF, and sT.

Practice Problem 8. Show that any argument with premises $p \wedge q$, $p \Rightarrow \neg r \vee s$, and $q \Rightarrow s$ and conclusion r is invalid.

EXERCISE SET 1.4

In Exercises 1–9, prove each of the given statements from the given premises. You may use any of the following statements, with all occurrences of a letter replaced by a particular statement, as premises (see Example 8):

I $(P \Rightarrow S) \Leftrightarrow (\neg S \Rightarrow \neg P)$

II $\neg Q \wedge (S \vee Q) \Rightarrow S$

III $\neg\neg R \Leftrightarrow R$

IV $P \Rightarrow P \vee R$

In Exercises 1–7, use direct proofs.
1. *Premises:* $P \vee Q \Rightarrow S \wedge T$, P
   Prove: $S \wedge T$
2. *Premises:* $\neg S$, $P \Rightarrow S$, $\neg P \vee Q \Rightarrow W$
   Prove: W
3. *Premises:* $P \Rightarrow \neg Q$, $S \vee Q$
   Prove: $P \Rightarrow S$
4. *Premises:* $P \Rightarrow M$, $M \Rightarrow T$, M
   Prove: $T \vee P$
5. *Premises:* $P \Rightarrow N$, $\neg P \Rightarrow \neg S$, $R \Rightarrow \neg N$
   Prove: $R \Rightarrow \neg S$
6. *Premises:* $P \Rightarrow Q$, $\neg P \wedge R \Rightarrow S$, $\neg Q$
   Prove: $R \vee P \Rightarrow S$
7. *Premises:* $P \Rightarrow Q$, $\neg P \wedge R \Rightarrow S$, $\neg Q$
   Prove: $R \Rightarrow S$

In Exercises 8 and 9, use proof by contradiction.
8. *Premises:* $P \Rightarrow \neg N$, $\neg P \Rightarrow \neg S$, S, $R \Rightarrow N$
   Prove: $\neg R$
9. *Premises:* $P \vee R$, $P \Rightarrow S$, $R \Rightarrow S$
   Prove: S

In Exercises 10–13, decide which conclusions can be proved from the given premises. Give a proof for those that can be proved and a counterexample for those that cannot.

Introduction to Logic   Chap. 1

10. *Premises:* $p \Rightarrow m$, $m \Rightarrow t$, m
    (a) *Conclusion:* $p \Rightarrow t$
    (b) *Conclusion:* $m \Rightarrow p$

11. *Premises:* $p \Rightarrow q$, $\neg p \wedge r \Rightarrow s$, $\neg q$, $(p \Rightarrow q) \Leftrightarrow (\neg q \Rightarrow \neg p)$
    (a) *Conclusion:* $r \Rightarrow s$
    (b) *Conclusion:* $\neg s \wedge \neg r$

12. *Premises:* $n \Rightarrow \neg j$, $\neg j \Rightarrow \neg d$, $\neg d \Rightarrow p$, $r \Rightarrow n$
    (a) *Conclusion:* $r \Rightarrow p$
    (b) *Conclusion:* $\neg d \vee p$

13. *Premises:* $n \Rightarrow \neg j$, $\neg j \Rightarrow \neg d$, $\neg d \Rightarrow p$
    (a) *Conclusion:* $\neg d \Rightarrow r$
    (b) *Conclusion:* $n \Rightarrow p$

14. Show that neither of the following statements logically follows from the premises $p \vee q$, $\neg r \Rightarrow s$, and $\neg s \Rightarrow p$.
    (a) $\neg p$
    (b) p

15. A set of premises is **inconsistent** if a statement of the form $P \wedge \neg P$ logically follows from the set of premises. It is often said that any conclusion logically follows from an inconsistent set of premises. Show that any statement logically follows from an inconsistent set of premises.

1.5 DERIVING TAUTOLOGIES

The following short list of tautologies is useful for reference:

T_1	$P \Leftrightarrow \neg\neg P$	law of double negation
T_2	$\neg P \vee P$	law of excluded middle
T_3	$Q \Rightarrow (P \Rightarrow Q \wedge P)$	law of conjunction
T_4	$P \wedge Q \Rightarrow Q$	law of simplification
T_5	$P \Rightarrow P \vee Q$	law of addition
T_6a	$P \vee Q \Leftrightarrow Q \vee P$	commutative law
T_6b	$P \wedge Q \Leftrightarrow Q \wedge P$	commutative law
T_7a	$P \vee (Q \vee R) \Leftrightarrow (P \vee Q) \vee R$	associative law
T_7b	$P \wedge (Q \wedge R) \Leftrightarrow (P \wedge Q) \wedge R$	associative law
T_8a	$P \vee (Q \wedge R) \Leftrightarrow (P \vee Q) \wedge (P \vee R)$	distributive law
T_8b	$P \wedge (Q \vee R) \Leftrightarrow (P \wedge Q) \vee (P \wedge R)$	distributive law
T_9	$P \wedge (P \Rightarrow Q) \Rightarrow Q$	modus ponens
T_{10}	$(P \Rightarrow Q) \Leftrightarrow (\neg P \vee Q)$	or-form law
$T_{11}a$	$\neg P \wedge \neg Q \Leftrightarrow \neg(P \vee Q)$	De Morgan's law
$T_{11}b$	$\neg(P \wedge Q) \Leftrightarrow \neg P \vee \neg Q$	De Morgan's law
T_{12}	$P \Rightarrow P$	implication form of T_2

T_{13}	$(P \Rightarrow Q) \Leftrightarrow (\neg Q \Rightarrow \neg P)$	contrapositive law
T_{14}	$(P \Rightarrow Q) \wedge (Q \Rightarrow R) \Rightarrow (P \Rightarrow R)$	transitivity of the implication
T_{15}	$(P \Leftrightarrow Q) \Leftrightarrow (Q \Leftrightarrow P)$	symmetry of biconditional
$\mathbf{T_{16}}$	$(P \Leftrightarrow Q) \Leftrightarrow (P \Rightarrow Q) \wedge (Q \Rightarrow P)$	decomposition of biconditional

All the tautologies in this list can be verified using truth tables. It is also possible, as well as more challenging and satisfying, to use proof techniques to derive tautologies from a given list of known tautologies.

In this section, we add to the method of proving a statement by allowing ourselves to use as a step in a proof any of the boldface-numbered tautologies, or any tautology that has been previously proved, even if it does not occur as a premise. For such a step, we list as a reason the number of the tautology in the former case and the example number (or exercise number, etc.) in the latter case.

With the foregoing modification, the boldface-numbered tautologies can be used to derive any other tautology. (The proof of this is quite involved and beyond the scope of the text.) In other words, the boldface-numbered tautologies can be considered axioms for all the tautologies. We can either verify them using truth tables or simply assume that they are true. In either case, we can then prove any other tautology.

Tautologies may be used with substitution. For example, T_{10} with P substituted for Q reads $(\neg P \vee P) \Leftrightarrow (P \Rightarrow P)$.

Example 1

Prove $P \Rightarrow P$ (T_{12}). Note that the premises can include any boldface-numbered tautology.

Solution

1. $(\neg P \vee P) \Leftrightarrow (P \Rightarrow P)$ (T_{10})
2. $\neg P \vee P$ (T_2)
3. $P \Rightarrow P$ (1, 2, Sub)

Chain of Biconditionals

Often in a derivation, we have the situation shown in the following example.

Example 2

Premises: $P \Leftrightarrow Q$, $Q \Leftrightarrow R$, $R \Leftrightarrow S$. Prove $P \Leftrightarrow S$.

Solution

1. $P \Leftrightarrow Q$ (Pr)
2. $Q \Leftrightarrow R$ (Pr)
3. $R \Leftrightarrow S$ (Pr)
4. $P \Leftrightarrow R$ (1, 2, Sub)
5. $P \Leftrightarrow S$ (3, 4, Sub)

Recall that in Sub one of the premises is a biconditional and the other is any statement. In Step 4, we used two biconditionals as premises for Sub. $P \Leftrightarrow R$ can be derived by substituting P for Q in $Q \Leftrightarrow R$. In this case, $P \Leftrightarrow Q$ is the biconditional. However, $P \Leftrightarrow R$ can also be derived by substituting Q for R in $P \Leftrightarrow Q$. In this case, $Q \Leftrightarrow R$ is the biconditional. Similar remarks hold for Step 5.

We further abbreviate the argument in Example 2 as follows:

1. $P \Leftrightarrow Q$
2. $ \Leftrightarrow R$
3. $ \Leftrightarrow S$
4. $P \Leftrightarrow S$ $\qquad (1, 2, 3)$

This method of can be generalized to any number of premises.

Another shortcut is not to list a boldface-numbered tautology, or any tautology that has already been derived, as a step in a derivation. Instead, we just refer to it.

Example 3

Prove $\neg(P \wedge Q) \Leftrightarrow \neg P \vee \neg Q$ (that is, $T_{11}b$).

Solution

1. $\neg P \vee \neg Q \Leftrightarrow \neg\neg(\neg P \vee \neg Q)$ $\qquad (T_1)$
2. $ \Leftrightarrow \neg(\neg\neg P \wedge \neg\neg Q)$ $\qquad (T_{11}a, 1, Sub)$
3. $ \Leftrightarrow \neg(P \wedge Q)$ $\qquad (T_1, 2, Sub)$
4. $\neg P \vee \neg Q \Leftrightarrow \neg(P \wedge Q)$ $\qquad (1, 2, 3)$

In Step 2, $\neg(\neg P \vee \neg Q) \Leftrightarrow (\neg\neg P \wedge \neg\neg Q)$ is an instance of $T_{11}a$, a boldface-numbered tautology, so we need not list it in the derivation. In Step 3, we use Sub to replace $\neg\neg P$ by P two times. Finally, in Step 4, we use a chain of biconditionals.

Practice Problem 1. Prove T_6a.

Chain of Implications

Suppose we have as premises $P_1 \Rightarrow P_2$, $P_2 \Rightarrow P_3$, $P_3 \Rightarrow P_4, ..., P_{n-1} \Rightarrow P_n$, and we wish to prove $P_1 \Rightarrow P_n$. Then we may proceed as follows:

1. $P_1 \Rightarrow P_2$
2. $ \Rightarrow P_3$
3. $ \Rightarrow P_4$
   $\vdots$
n-1. $ \Rightarrow P_n$
n. $P_1 \Rightarrow P_n$ $\qquad (1, 2, 3, ..., n-1)$

Example 4

Use a chain of implications to prove $(P \wedge Q) \wedge R \Rightarrow P$.

Solution

1. $(P \wedge Q) \wedge R \Rightarrow R \wedge (P \wedge Q)$ (T$_6$b)
2. $\Rightarrow P \wedge Q$ (T$_4$)
3. $\Rightarrow Q \wedge P$ (T$_6$b)
4. $\Rightarrow P$ (T$_4$)
5. $(P \wedge Q) \wedge R \Rightarrow P$ (1, 2, 3, 4)

Practice Problem 2. Use a chain of implications to prove $P \wedge \neg(P \wedge Q) \Rightarrow \neg Q$.

Suppose now that we have $P \Leftrightarrow Q$ as a premise, but we would like to use only $P \Rightarrow Q$ in a derivation. In this case, we may use $P \Rightarrow Q$ as a step without any further justification. The proof follows:

1. $P \Leftrightarrow Q$ (Pr)
2. $(P \Rightarrow Q) \wedge (Q \Rightarrow P)$ (1, T$_{16}$, Sub)
3. $(P \Rightarrow Q) \wedge (Q \Rightarrow P) \Rightarrow (Q \Rightarrow P) \wedge (P \Rightarrow Q)$ (T$_6$b)
4. $\Rightarrow (P \Rightarrow Q)$ (T$_4$)
5. $(P \Rightarrow Q) \wedge (Q \Rightarrow P) \Rightarrow (P \Rightarrow Q)$ (3, 4)
6. $P \Rightarrow Q$ (2, 5, MP)

Similarly, if we have $P \Leftrightarrow Q$ as a premise, we may use $Q \Rightarrow P$ (see Exercise 14).

In Exercise 3, you are asked to derive the contrapositive law T$_{13}$. The tautology derived in the next example is the biconditional version of the contrapositive law, and its derivation makes use of T$_{13}$. So, when deriving T$_{13}$, you may not use the result of Example 5.

Example 5

Prove $(S \Leftrightarrow R) \Leftrightarrow (\neg S \Leftrightarrow \neg R)$.

Solution

1. $(S \Leftrightarrow R) \Leftrightarrow [(S \Rightarrow R) \wedge (R \Rightarrow S)]$ (T$_{16}$)
2. $\Leftrightarrow [(\neg R \Rightarrow \neg S) \wedge (R \Rightarrow S)]$ (T$_{13}$, 1, Sub)
3. $\Leftrightarrow [(\neg R \Rightarrow \neg S) \wedge (\neg S \Rightarrow \neg R)]$ (T$_{13}$, 2, Sub)
4. $\Leftrightarrow [(\neg S \Rightarrow \neg R) \wedge (\neg R \Rightarrow \neg S)]$ (T$_6$b, 3, Sub)
5. $\Leftrightarrow (\neg S \Leftrightarrow \neg R)$ (T$_{16}$, 4, Sub)
6. $(S \Leftrightarrow R) \Leftrightarrow (\neg S \Leftrightarrow \neg R)$ (1, 2, 3, 4, 5)

Example 6

Prove $\mathcal{I} \wedge P \Leftrightarrow P$. (In other words, derive $(R \vee \neg R) \wedge P \Leftrightarrow P$.)

Solution

1. $(R \vee \neg R) \wedge P \Rightarrow P$ (T$_4$)
2. $(R \vee \neg R) \Rightarrow (P \Rightarrow (R \vee \neg R) \wedge P)$ (T$_3$)
3. $P \Rightarrow (R \vee \neg R) \wedge P$ (2, T$_2$, MP)

4. $[(R \vee \neg R) \wedge P \Rightarrow P] \wedge [P \Rightarrow (R \vee \neg R) \wedge P]$ (1, 3, Adj)

5. $(R \vee \neg R) \wedge P \Leftrightarrow P$ (T_{16}, 4, Sub)

Practice Problem 3. Use the results of Examples 5 and 6 to derive the biconditional $\mathcal{O} \vee Q \Leftrightarrow Q$. Use $R \wedge \neg R$ for the contradiction $\mathcal{O}$.

Example 7

Prove $(P \Rightarrow Q) \wedge (P \wedge Q \Rightarrow R) \Rightarrow (P \Rightarrow R)$.

Solution We first derive $(P \Rightarrow Q) \wedge (P \wedge Q \Rightarrow R) \Leftrightarrow (P \Rightarrow Q) \wedge (P \Rightarrow R)$. Then we can use substitution to reduce the problem to deriving

$$(P \Rightarrow Q) \wedge (P \Rightarrow R) \Rightarrow (P \Rightarrow R)$$

which is just T_4.

1. $(P \Rightarrow Q) \wedge (P \wedge Q \Rightarrow R)$ $\Leftrightarrow (\neg P \vee Q) \wedge (\neg (P \wedge Q) \vee R)$
2. $\Leftrightarrow (\neg P \vee Q) \wedge [(\neg P \vee \neg Q) \vee R]$
3. $\Leftrightarrow (\neg P \vee Q) \wedge [\neg P \vee (\neg Q \vee R)]$
4. $\Leftrightarrow \neg P \vee [Q \wedge (\neg Q \vee R)]$
5. $\Leftrightarrow \neg P \vee [(Q \wedge \neg Q) \vee (Q \wedge R)]$
6. $\Leftrightarrow \neg P \vee (Q \wedge R)$
7. $\Leftrightarrow (\neg P \vee Q) \wedge (\neg P \vee R)$
8. $\Leftrightarrow (P \Rightarrow Q) \wedge (P \Rightarrow R)$
9. $(P \Rightarrow Q) \wedge (P \wedge Q \Rightarrow R)$ $\Leftrightarrow (P \Rightarrow Q) \wedge (P \Rightarrow R)$
10. $(P \Rightarrow Q) \wedge (P \wedge Q \Rightarrow R)$ $\Rightarrow (P \Rightarrow Q) \wedge (P \Rightarrow R)$
11. $\Rightarrow (P \Rightarrow R)$
12. $(P \Rightarrow Q) \wedge (P \wedge Q \Rightarrow R)$ $\Rightarrow (P \Rightarrow R)$

Practice Problem 4. Give a reason for each step in the derivation of $(P \Rightarrow Q) \wedge (P \wedge Q \Rightarrow R) \Leftrightarrow (P \Rightarrow Q) \wedge (P \Rightarrow R)$ in Example 7.

After finishing Practice Problem 4, you will know that T_8b was used in the derivation of the tautology in Example 7. You are asked to derive T_8b in Exercise 2.

EXERCISE SET 1.5

In Exercises 1–11, prove the given tautologies from the boldface-numbered tautologies in the reference list. Any other tautology previously derived from the boldface-numbered ones, such as $T_{11}b$ and T_{12}, may be used in the derivations.

1. T_7b. (*Hint:* Use $T_{11}a$, T_7a, and $T_{11}b$.)
2. T_8b. (*Hint:* Use $T_{11}a$, T_8a, and $T_{11}b$.)
3. T_{13}. (Do not use Example 5.)
4. T_{14}.

5.   **(a)** $\neg P \Rightarrow (P \Rightarrow Q)$
      (b) $Q \Rightarrow (P \Rightarrow Q)$

6.   **(a)** $(\neg Q \Rightarrow R \wedge \neg R) \Rightarrow Q$
      (b) $(P \wedge \neg Q \Rightarrow R \wedge \neg R) \Leftrightarrow (P \Rightarrow Q)$

7. $[P \Rightarrow (R \Rightarrow Q)] \Leftrightarrow [P \wedge R \Rightarrow Q]$

8. $[(P \wedge R \Rightarrow Q) \wedge R] \Rightarrow (P \Rightarrow Q)$

9. $(P \Rightarrow Q) \wedge \neg Q \Rightarrow \neg P$ (*Hint:* Use T$_{13}$.)

10. $(P \Rightarrow Q \vee R) \Leftrightarrow (P \wedge \neg Q \Rightarrow R)$

11. $(P \wedge Q) \vee (P \wedge \neg Q) \Leftrightarrow P$

12.   **(a)** Calculate the appropriate truth table to verify $\neg(P \Rightarrow Q) \equiv P \wedge \neg Q$.
      (b) Derive the tautology $\neg(P \Rightarrow Q) \Leftrightarrow P \wedge \neg Q$.

13.   **(a)** Calculate the appropriate truth table to verify that $P \wedge (\neg P \vee Q) \Rightarrow Q$ is a tautology.
      (b) Derive the tautology $P \wedge (\neg P \vee Q) \Rightarrow Q$.

14. If we have $P \Leftrightarrow Q$ as a premise, justify the assertion that we may use $Q \Rightarrow P$ as a premise.

KEY CONCEPTS

Hypothesis	Semantics
Conclusion	Truth table
Backwards/forwards method	Tautology
Statement	Logical equivalence
Statement connective	Counterexample
Prime statement, composite statement	Proof of statements from premises
Syntactics	Rules of inference
Main connective	Valid and invalid argument

REVIEW EXERCISES

In Exercises 1–4, outline the proof, whether it is given or requested, and fill in any details (for example, reasons for each step) that will make the proof easier to understand. Answer each of the following questions:

What is the goal of the proof?

What is the hypothesis?

What definitions are necessary?

What previously proven facts or laws of logic are used in the proof?

1. Let x be an integer. Prove: If x^3 is even, then x is even.

2. Let x and y be real numbers. Prove: If $(x + y)^2 = (x - y)^2$, then $x = 0$ or $y = 0$.

3. Let x, y, and z be real numbers. Prove: If $y^2 + z^2 = 2x(y + z - x)$, then $x = y = z$. *Proof:* Assume $y^2 + z^2 = 2x(y + z - x)$. Then $x^2 - 2xy + y^2 + x^2 - 2xz + z^2 = 0$. Hence, $(x - y)^2 + (x - z)^2 = 0$. Therefore, $x = y = z$.

4. Find, in a calculus book, the proof of the theorem, "If a function is differentiable, then it is continuous."

5. First supply all the missing parentheses for the following statement. Then use the syntax rules S1–S6 to verify that it is, in fact, a statement, and specify the main connective.

$$p \Rightarrow q \Leftrightarrow \neg r \vee p \Rightarrow q$$

6. Construct a truth table for the statement in Exercise 5 above.

In Exercises 7 and 8, determine whether the given conclusion can be derived from the given premises. Give a proof for those that can be derived and a counterexample for those that cannot.

7. (a) Premises: $\neg a \Rightarrow \neg b, b, a \Rightarrow c, a \Rightarrow c \Leftrightarrow \neg c \Rightarrow \neg a$. Conclusion: c.
    (b) Premises: $p \Rightarrow q, p \vee r \Rightarrow s$. Conclusion: $q \Rightarrow s$.

8. (a) Premises: $p \vee q, \neg q, p \Rightarrow r, s \wedge r \Rightarrow t, \neg q \wedge (p \vee q) \Rightarrow p$. Conclusion: $s \Rightarrow t$.
    (b) Premises: $a \wedge b \Rightarrow c, d \Rightarrow c, d \vee a$. Conclusion: c.

9. Use the methods of Section 1.5 to derive the tautology T_{15}.

10. Use the methods of Section 1.5 to derive the tautology $(P \vee Q) \wedge \neg P \Rightarrow Q$.

SUPPLEMENTARY EXERCISES

In discussing the logical equivalence of two statements in this chapter, it was assumed that the two statements were composed of the same constituent statements. We now show that $(P \Rightarrow Q) \wedge (R \vee \neg R) \equiv P \Rightarrow Q$, using the following truth table:

P	Q	R	$(P \Rightarrow Q) \wedge (R \vee \neg R)$				$P \Rightarrow Q$
T	T	T	T	T	T	F	T
T	T	F	T	T	T	T	T
T	F	T	F	F	T	F	F
T	F	F	F	F	T	T	F
F	T	T	T	T	T	F	T
F	T	F	T	T	T	T	T
F	F	T	T	T	T	F	T
F	F	F	T	T	T	T	T

Since the main columns of the two statements are identical, the statements are logically equivalent. Note that even though R does not occur in $P \Rightarrow Q$, we include it in the truth table. In general, when we determine whether two statements are logically equivalent, we include in the truth table all of the constituent statements that occur in each statement.

In Exercises 1–4, determine whether the following pairs of statements are logically equivalent.

1. (a) $(\neg R \wedge P) \vee (R \wedge P)$, P
    (b) $((\neg R \vee S) \wedge (R \vee S)) \vee T$, $S \vee T$

2. (a) $P \wedge R \Leftrightarrow P, R$
   (b) $P \Leftrightarrow \neg P, Q \Leftrightarrow \neg Q$
3. (a) $P \vee \neg P, Q \vee \neg Q$
   (b) Any two tautologies
4. (a) $P \wedge \neg P, Q \wedge \neg Q$
   (b) Any two contradictions

There are instances when the truth values "true" and "false" are not adequate. Consider the equation $1/x = 2$, where x refers to a real number. If we use 0.5 as a replacement for x, the resulting statement is true. If we use 3, the resulting statement is false. What is the truth value if we use the replacement 0? The problem here is that $1/x$ is undefined when $x = 0$. However, we can introduce a new truth value, UNDEFINED, denoted by U, and assign this value to the statement $1/0 = 2$. In doing so, we arrive at a three-valued logic. The following truth tables define the AND and NOT connectives in this three-valued logic.

AND $P \wedge Q$				NOT $\neg P$	
P \ Q	T	U	F	P	$\neg P$
T	T	U	F	T	F
U	U	U	F	U	U
F	F	F	F	F	T

Exercises 5–8 involve three-valued logic.

5. Assume that $P \vee Q$ has the same truth table as $\neg(\neg P \wedge \neg Q)$. Use the given truth tables to construct a truth table for $\vee$.
6. Assume that $P \Rightarrow Q$ has the same truth table as $\neg P \vee Q$. Use the results of Exercise 5 to construct a truth table for $\Rightarrow$.
7. Assume that $P \Leftrightarrow Q$ has the same truth table as $(P \Rightarrow Q) \wedge (Q \Rightarrow P)$. Use the results of Exercises 5 and 6 to construct a truth table for $\Leftrightarrow$.
8. Construct a truth table for $P \wedge (P \Rightarrow Q) \Rightarrow Q$.

All statements in propositional logic are logically equivalent to, and hence can be written in terms of, statements that use only the three connectives AND, OR, and NOT. Specifically, we can replace any statement of the form $P \Rightarrow Q$ by $\neg P \vee Q$ and $P \Leftrightarrow Q$, by $(\neg P \vee Q) \wedge (P \vee \neg Q)$. It is possible, and frequently useful, to write all statements using only one or two connectives. For example, all statements can be written in terms of the connectives OR and NOT. To see this, we need only write AND in terms of OR and NOT, which is done using one of De Morgan's laws and the law of double negation, to obtain $(P \wedge Q) \equiv \neg(\neg P \vee \neg Q)$. A set of statement connectives is **functionally complete** if all statements in propositional logic can be written in terms of the connectives in the set. The set of three basic connectives AND, OR, and NOT is functionally complete. We can show that any set A of connectives is functionally complete by showing that the three basic connectives can all be written in terms of the connectives in A. We have just shown that the set {NOT, OR} is functionally complete using such a method.

9. Use De Morgan's law to show that the set of connectives {NOT, AND} is functionally complete.

10. Define a new statement connective NOR ($\downarrow$) by the following truth table [P NOR Q stands for NOT(P OR Q)]:

P	Q	$P\downarrow Q$
T	T	F
T	F	F
F	T	F
F	F	T

Show that $\{\downarrow\}$ is functionally complete. (*Hint:* Show that $P \vee Q \equiv \neg(P \downarrow Q)$ and $\neg P \equiv P \downarrow P$. Then use De Morgan's law to obtain an expression for $P \wedge Q$).

11. Define a new statement connective NAND ($\uparrow$) by the following truth table (P NAND Q is an abbreviation for NOT BOTH P AND Q):

P	Q	$P\uparrow Q$
T	T	F
T	F	T
F	T	T
F	F	T

Show that $\{\uparrow\}$ is functionally complete.

SOLUTIONS TO PRACTICE PROBLEMS

Section 1.1

1. Since n is even, $n = 2k$ for some integer k. Therefore, $n^2 = (2k)^2 = 2(2k^2)$. Hence, n^2 is of the required form for it to be even.

2. *Prove:* If n is not odd, then n^2 is not odd. If n is not odd, then it is even. By Practice Problem 1, n^2 is even. Therefore, n^2 is not odd. It follows logically that if n^2 is odd, then n is odd.

Section 1.2

1. (1) p, q, r are statements.   (S1)
   (2) $\neg p$ is a statement.   (1 and S6)
   (3) $(q \wedge r)$ is a statement.   (1 and S2)
   (4) $\neg p \Rightarrow (q \wedge r)$ is a statement.   (2, 3, and S4)
   The main connective is $\Rightarrow$

2. (a) $p \Rightarrow (\neg q \vee r)$
   (b) $(s \wedge \neg q) \Rightarrow (t \vee \neg r)$

(c) $(p \Rightarrow q) \Leftrightarrow (r \Rightarrow s)$

(d) $((p \Rightarrow q) \vee (s \wedge q)) \Rightarrow r$

3. **(a)** $(p \wedge q) \Rightarrow ((r \vee q) \wedge \neg r)$, $(p \wedge q) \Rightarrow (r \vee (q \wedge \neg r))$

   (b) $(p \Rightarrow (q \vee r)) \Rightarrow q$, $p \Rightarrow ((q \vee r) \Rightarrow q)$

4. **(a)** T   **(e)** T (vacuously)

   (b) T   **(f)** T

   (c) T   **(g)** T

   (d) F

5.

P	Q	R	$\neg$	$(P \wedge Q)$	$\vee$	R	$\Rightarrow$	$\neg$	P	$\vee$	R
T	T	T	F	T	T		T		F		T
T	T	F	F	T	F		T		F		F
T	F	T	T	F	T		T		F		T
T	F	F	T	F	T		F		F		F
F	T	T	T	F	T		T		T		T
F	T	F	T	F	T		T		T		T
F	F	T	T	F	T		T		T		T
F	F	F	T	F	T		T		T		T
(1)	(2)	(3)	(5)	(4)	(6)		(9)	(7)		(8)	

Section 1.3

1.

P	Q	$\neg$	$(P \vee Q)$	$\neg P$	$\wedge$	$\neg Q$
T	T	F	T	F	F	F
T	F	F	T	F	F	T
F	T	F	T	T	F	F
F	F	T	F	T	T	T
		↑			↑	
		main			main	

2.

P	Q	$(P \Rightarrow Q)$	$\wedge$	$(Q \Rightarrow P)$	$P \Leftrightarrow Q$
T	T	T	T	T	T
T	F	F	F	T	F
F	T	T	F	F	F
F	F	T	T	T	T
			↑		↑
			main		main

3.

p	q	$(p \vee q)$	$\Rightarrow$	p	
T	T	T		T	
T	F	T		T	
F	T	T		F	← counterexample
F	F	F		T	

The counterexample is pF and qT.

Section 1.4

1.
1. $T \Rightarrow R$ (Pr)
2. T (Pr)
3. R (1, 2, MP)
4. $R \Rightarrow S$ (Pr)
5. S (3, 4, MP)

2.
1. $\neg(R \wedge T) \Rightarrow \neg R \vee \neg T$ (Pr)
2. $\neg(R \wedge T)$ (Pr)
3. $\neg R \vee \neg T$ (1, 2, MP)
4. $\neg R \vee \neg T \Rightarrow (\neg R \Rightarrow T)$ (Pr)
5. $\neg R \Rightarrow T$ (3, 4, MP)
6. $\neg R$ (Pr)
7. T (5, 6, MP)

3.
1. R (Example 3, AP)
2. $R \Rightarrow S \wedge T$ (Pr)
3. $S \wedge T$ (1, 2, MP)

4.
1. Q (Hyp)
2. $Q \Rightarrow (P \Rightarrow \neg R)$ (Pr)
3. $P \Rightarrow \neg R$ (1, 2, MP)
4. P (Pr)
5. $\neg R$ (3, 4, MP)
6. $Q \Rightarrow \neg R$ (1, 5, DPI)

5.
1. Q (Pr)
2. $Q \Rightarrow P$ (Pr)
3. P (1, 2, MP)
4. $P \wedge Q$ (1, 3, Adj)
5. $P \wedge Q \Rightarrow R \vee T$ (Pr)
6. $R \vee T$ (4, 5, MP)

6.
1. R (Pr)
2. $R \Rightarrow (Q \vee P)$ (Pr)
3. $Q \vee P \Leftrightarrow (\neg Q \Rightarrow P)$ (Pr)
4. $R \Rightarrow (\neg Q \Rightarrow P)$ (2, 3, Sub)
5. $\neg Q \Rightarrow P$ (1, 3, MP)
6. $\neg Q$ (Pr)
7. P (4, 5, MP)

7.
1. $\neg(P \lor Q)$ (Neg of concl)
2. $\neg(P \lor Q) \Leftrightarrow \neg P \land \neg Q$ (Pr)
3. $\neg P \land \neg Q$ (1, 2, Sub)
4. $\neg P \land \neg Q \Rightarrow \neg P$ (Pr)
5. $\neg P$ (3, 4, MP)
6. $\neg P \land \neg Q \Rightarrow \neg Q$ (Pr)
7. $\neg Q$ (3, 6, MP)
8. $\neg P \Rightarrow R$ (Pr)
9. R (5, 8, MP)
10. $\neg Q \Rightarrow \neg R$ (Pr)
11. $\neg R$ (7, 10, MP)
12. $\neg R \land R$ (9, 11, Adj)
13. $P \lor Q$ (1, 12, Contra)

8.

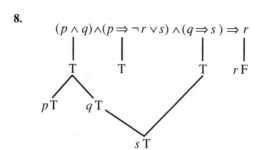

$$(p \land q) \land (p \Rightarrow \neg r \lor s) \land (q \Rightarrow s) \Rightarrow r$$

p	q	r	s	$(p \land q) \land (p \Rightarrow \neg r \lor s) \land (q \Rightarrow s) \Rightarrow r$
T	T	F	T	T T T T T T T F

The counterexample is pT, qT, rF, sT.

Section 1.5

1.
1. $P \lor Q \Leftrightarrow \neg\neg P \lor \neg\neg Q$ (T_1, Sub)
2. $\Leftrightarrow \neg(\neg P \land \neg Q)$ $(T_{11}b, \text{Sub})$
3. $\Leftrightarrow \neg(\neg Q \land \neg P)$ (T_6b, Sub)
4. $\Leftrightarrow \neg\neg Q \lor \neg\neg P$ $(T_{11}b, \text{Sub})$
5. $\Leftrightarrow Q \lor P$ (T_1, Sub)
6. $P \lor Q \Leftrightarrow Q \lor P$ (1, 2, 3, 4, 5)

2.
1. $P \land \neg(P \land Q) \Rightarrow P \land (\neg P \lor \neg Q)$ $(T_{11}b, \text{Sub})$
2. $\Rightarrow P \land (\neg\neg P \Rightarrow \neg Q)$ (T_{10}, Sub)
3. $\Rightarrow P \land (P \Rightarrow \neg Q)$ (T_1, Sub)
4. $\Rightarrow \neg Q$ (T_9)
5. $P \land \neg(P \land Q) \Rightarrow \neg Q$ (1, 2, 3, 4)

3.

1. $\neg((R \wedge \neg R) \vee Q) \Leftrightarrow \neg(R \wedge \neg R) \wedge \neg Q$ (T$_{11}$a, Sub)
2. $\Leftrightarrow (\neg R \vee \neg\neg R) \wedge \neg Q$ (T$_{11}$b, Sub)
3. $\Leftrightarrow \neg Q$ (Example 6)
4. $\neg((R \wedge \neg R) \vee Q) \Leftrightarrow \neg Q$ (1, 2, 3)
5. $((R \wedge \neg R) \vee Q) \Leftrightarrow Q$ (Example 5)

4.

Step	Reason
1	T$_{10}$, twice
2	T$_{11}$b
3	T$_7$a
4	T$_8$a
5	T$_8$b
6	Practice Problem 3
7	T$_8$a
8	T$_{10}$
9	1, 2, . . . , 8; chain of biconditionals
10	$\Rightarrow$ follows from $\Leftrightarrow$
11	T$_4$
12	10, 11; chain of implications

2

Methods of Proof

Several commonly used proof techniques are powerful analytical tools. In particular, one of the most useful proof techniques in mathematics is mathematical induction. There are many examples of proofs by mathematical induction in this chapter and throughout the book. Finally, in this chapter, we will discuss the predicate calculus and proofs involving quantifiers.

2.1 *PROOF TECHNIQUES*

Recall that a proof is a step-by-step demonstration showing that a given argument is valid. Mathematicians use logical rules of procedure based on known tautologies and validities to construct proofs. A few such rules will be made explicit and their use illustrated by examples. The goal is to understand how and when the rules are applied.

For easy reference, we reproduce several important laws of logic stated earlier in Section 1.3. Each of them will be useful for analyzing and producing proofs.

Laws of Logic for Proof Techniques

a. $\neg\neg P \equiv P$ [Rule of **double negation**]

b. $P \Rightarrow Q \equiv \neg P \vee Q$ [**Or-form** of an implication]

c. $P \Rightarrow Q \equiv \neg Q \Rightarrow \neg P$ [**Contrapositive** of an implication]

d. $\neg(P \vee Q) \equiv \neg P \wedge \neg Q$ $\Big\}$ [**De Morgan's laws**]

e. $\neg(P \wedge Q) \equiv \neg P \vee \neg Q$

f. $(P \wedge R \Rightarrow Q) \equiv (R \Rightarrow (P \Rightarrow Q))$ [Rule for direct proof]

g. $(P \wedge \neg Q \Rightarrow \mathcal{O}) \equiv (P \Rightarrow Q)$ [Rule for proof by contradiction]

i. $(P \vee R \Rightarrow Q) \equiv [(P \Rightarrow Q) \wedge (R \Rightarrow Q)]$ [Rule for proof by cases]

The rule for proof by cases was not stated earlier but it will be useful in what follows. We will also restate several of the rules of inference introduced in Section 1.4.

In the following examples, we use several facts pertaining to the ordering of the real numbers. The following is a short list of some of these:

A1. If $x < y$ and $y < z$, then $x < z$

A2. $x < y$, or $y < x$, or $x = y$

A3. If $x < y$, then $x + z < y + z$

A4. If $x < y$ and $z > 0$, then $xz < yz$

A5. $x \not< x$

A6. $n > 0$ if n is any positive integer.

We assume that the **positive integers** consist of 1, 2, 3,... Also, we may write $x > y$ for $y < x$, $x \leq y$ for $x < y$ or $x = y$, and $x \geq y$ for $y \leq x$.

PROOF TECHNIQUE   **Direct Proof of an Implication.**   To prove an implication $P \Rightarrow Q$, assume that P is true and use P along with other known true statements to deduce Q.

The direct proof technique is based on

$$(P \wedge R \Rightarrow Q) \equiv (R \Rightarrow (P \Rightarrow Q))$$

where R stands for the collection of previously assumed or proved statements.

The next example illustrates a direct proof of an implication.

Example 1

Prove: If $x < y$ and $u < v$, then $x + u < y + v$.

Proof

1. $x < y$	(Assumption)
2. $x + u < y + u$	(A3 and Step 1)
3. $u < v$	(Assumption)
4. $y + u < y + v$	(A3 and Step 3)
5. $x + u < y + v$	(A1 and Steps 2 and 4)

Let R stand for the reasons given in Steps 2, 4, and 5, P stand for $x < y$ and $u < v$, and Q stand for $x + u < y + v$. Then, the form of the proof is $(P \wedge R) \Rightarrow Q$. Hence, $P \Rightarrow Q$ follows logically from R.

Analysis of Step 2: A3 is of the form $P \Rightarrow Q$, where P is $x < y$ and Q is $x + z < y + z$. Step 1 is P. By modus ponens, if we have P and $P \Rightarrow Q$, then we have Q. So Step 2 follows.

Modus ponens is used so often that we make no further comment when it is used in a proof.

Practice Problem 1. *Prove*: If $x > 0$, $u > 0$, $x < y$, and $u < v$, then $xu < yv$.

Example 2

Prove: If $x < y$, then $-y < -x$.

Proof

1. $x < y$	(Assumption)
2. $x + (-y) + (-x) < y + (-y) + (-x)$	(A3 and Step 1)
3. $-y < -x$	(Algebra)

Analysis of Step 2: A3 says that we may add the same real quantity z to both sides of an inequality. In Step 2, that quantity is $(-y) + (-x)$.

The word "Algebra" in Step 3 refers to some algebraic facts pertaining to the real numbers that are used in the derivation of Step 3. In this case, we used the commutative law $(a + b = b + a)$, the additive inverse law $(a + (-a) = 0)$, and the additive identity law $(0 + a = a)$.

Example 3

Prove: If $z < 0$ and $x < y$, then $yz < xz$.

Proof

1. $x < y$	(Assumption)
2. $z < 0$	(Assumption)
3. $0 < -z$	(Example 2 and Step 2)
4. $x(-z) < y(-z)$	(A4 and Steps 1 and 3)
5. $-(xz) < -(yz)$	(Algebra)
6. $--(yz) < --(xz)$	(Example 2 and Step 5)
7. $yz < xz$	(Algebra)

Practice Problem 2. *Prove*: If $z < 0$, $v < 0$, $x < z$, and $u < v$, then $xu > zv$.

Many statements needing proof are biconditionals. Recall that $(P \Leftrightarrow Q) \equiv [(P \Rightarrow Q) \wedge (Q \Rightarrow P)]$.

In proofs of $P \Leftrightarrow Q$, the proof of $P \Rightarrow Q$, called the **sufficiency proof** (P is sufficient for Q), is labeled ($\Rightarrow$) while the proof of $Q \Rightarrow P$, called the **necessity proof** (P is necessary for Q), is labeled ($\Leftarrow$).

Example 4

Prove: $x < y$ if and only if $x + z < y + z$.

Proof ($\Rightarrow$) This is A3.
($\Leftarrow$) Assume $x + z < y + z$. Then, by A3, $x + z + (-z) < y + z + (-z)$. Since $z + (-z) = 0$, we obtain $x < y$.

In Example 4, we have proved that $x < y$ is a necessary and sufficient condition for $x + z < y + z$. Although the proof here is less formal than those of Examples 1, 2, and 3, it still contains all the information needed for understanding it. In practice problems that follow, you may use either of the two styles, but you are encouraged to use an informal style.

Practice Problem 3. Prove that $7x < 7y$ if and only if $x < y$. (*Hint*: You may use the fact that $7 > 0$ and $\frac{1}{7} > 0$.)

PROOF TECHNIQUE   Proof by Cases. To prove the statement $P \vee R \Rightarrow Q$, it is necessary and sufficient to consider two cases: (1) Prove that Q follows logically from P, and (2) prove Q also follows logically from R.

Proof by cases is based on the logical equivalence

$$(P \vee R \Rightarrow Q) \equiv [(P \Rightarrow Q) \wedge (R \Rightarrow Q)]$$

Example 5

Prove: If $x < y$ and $y \leq z$, then $x < z$.

Proof Assume $x < y$ and $y \leq z$. Since $y \leq z$ means $y < z$ or $y = z$, there are two cases.
Case 1. Assume $y < z$. Then $x < y$ and $y < z$, so by A1, $x < z$.
Case 2. Assume $y = z$. Then, since $x < y$ and $y = z$, it follows that $x < z$.

It is important that all cases are considered and that in each case we end with the desired conclusion.

Recall that the absolute value of a real number is defined by

$$|a| = \begin{cases} a & \text{if} \quad a \geq 0 \\ -a & \text{if} \quad a < 0 \end{cases}$$

Many proofs involving absolute values can best be done with cases, because the definition is in terms of two cases.

Example 6

Prove: $a \leq |a|$ for any real number a.

Proof Let a be an arbitrary real number. We do a proof by cases.
Case 1. Assume $a \geq 0$. Then $|a| = a$, and it follows that $a \leq |a|$.
Case 2. Assume $a < 0$. Then $|a| = -a$. Since $a < 0$ and $-0 = 0$, we obtain $-a > 0$ by Example 2. Hence, $|a| > a$ by A1, and we are done.

Practice Problem 4. Prove that $|3 \cdot y| = 3 \cdot |y|$ for any real number y.

Using proof by cases, the following can be shown (see Exercise Set 2.1):

(a) If $x \leq y$ and $y \leq z$, then $x \leq z$.
(b) If $x \leq y$, then $x + z \leq y + z$.
(c) If $z > 0$ and $x \leq y$, then $xz \leq yz$.
(d) If $x \leq y$ and $u \leq v$, then $x + u \leq y + v$.

In Exercises 1–12, supply a proof for each assertion.

1. If $x > 1$, then $x^2 > 1$.
2. If $x < -1$, then $x^2 > 1$.
3. If $0 < x$ and $x < y$, then $x^2 < y^2$.
4. If $x < 0$ and $y < 0$ and $x < y$, then $y^2 < x^2$.
5. If $x \leq y$ and $y \leq z$, then $x \leq z$.
6. If $x \leq y$, then $x + z \leq y + z$.
7. If $z > 0$ and $x \leq y$, then $xz \leq yz$.
8. If $z < 0$ and $x \leq y$, then $xz \geq yz$.
9. If $x \leq y$ and $u \leq v$, then $x + u \leq y + v$.
10. $-|a| \leq a$ for any real number a.
11. $|x| \leq b$ if and only if $-b \leq x \leq b$ for any real numbers b and x.
12. $-\big(|a| + |b|\big) \leq a + b \leq |a| + |b|$ for any real numbers a and b. (*Hint*: Use Example 6 and Exercise 10.)
13. Prove: $|a+b| \leq |a|+|b|$ for any real numbers a and b. This is called the **triangle inequality** for real numbers. (*Hint*: Use Exercises 11 and 12.)
14. Prove that $|x \cdot y| = |x| \cdot |y|$ for all real numbers x and y.

An integer z is an **even integer** (or z is **even**) if $z = 2 \cdot k$ for some integer k. In Exercises 15–17, supply a proof for each assertion.

15. If x is an even integer, then x^2 is even.
16. If x is an integer and y is an even integer, then $x \cdot y$ is even.
17. If x and y are even integers, then $x + y$ is even.

2.2 MORE PROOF TECHNIQUES

In this section, we give several examples of proofs that use the contrapositive of a statement. In addition, we discuss proof by contradiction. Proofs using these techniques are considered to be indirect proofs.

For some statements, it is easier to give an indirect proof. This can take several forms, the most useful of which is proof by contradiction.

PROOF TECHNIQUE **Proof by Contradiction.** To prove an implication $P \Rightarrow Q$, it is enough to assume that P is true and assume the negation of Q is true, and then deduce any contradiction.

Proof by contradiction is based on $(P \wedge \neg Q \Rightarrow \mathcal{O}) \equiv (P \Rightarrow Q)$, where $\mathcal{O}$ stands for any contradiction.

Example 1

Prove: If $x \leq y$ and $y \leq x$, then $x = y$.

Proof Assume $x \leq y$, $y \leq x$, and $x \neq y$ (the negation of the conclusion). By A2, $x = y$ or $x < y$ or $y < x$. Then, by the tautology $\neg P \wedge (P \vee Q) \Rightarrow Q$, we conclude that $x < y$ or $y < x$.

Case 1. Assume $x < y$. Then $x < y$ and $y \leq x$, and hence, by Example 5 in Section 2.1, $x < x$. This is contrary to A5.

Case 2. Assume $y < x$. Then $y < x$ and $x \leq y$, and hence, again by Example 5 in Section 2.1, $y < y$. Once more, this is contrary to A5.

Hence, the assumption that $x \leq y$ and $y \leq x$ and $x \neq y$ leads to a contradiction. Therefore, $x \leq y$ and $y \leq x$ implies $x = y$.

Example 2

Prove: $x \not< y$ if and only if $y \leq x$.

Proof ($\Rightarrow$) Assume $x \not< y$. Then, by A2, $x = y$ or $y < x$; that is, $y \leq x$.

($\Leftarrow$) This part is done by contradiction. Assume that $y \leq x$ and $x < y$ (the negation of the conclusion). By Example 5 in Section 2.1, it follows that $x < x$. This contradicts A5.

Practice Problem 1. *Prove*: If $x > 0$ and $x < y$, then $\frac{1}{x} > \frac{1}{y}$.

By A2, $x < y$ or $y < x$ or $x = y$. In general, a statement of the form $p \vee q \vee r$ is true when any (possibly all) of the statements p, q, or r is true. However, in the case of A2, we can prove that exactly one of $x < y$, $y < x$, and $x = y$ is true.

Example 3

Prove: For any real numbers x and y, exactly one of $x < y$, $y < x$, and $x = y$ is true.

Proof By A2, at least one of $x < y$, $y < x$, and $x = y$ is true. Hence, we need only prove:

a. If $x < y$, then $y \not< x$ and $x \neq y$.
b. If $y < x$, then $x \not< y$ and $x \neq y$.
c. If $x = y$, then $x \not< y$ and $y \not< x$.

We prove a and c. The proof of b is similar to that of a.

a. Assume $x < y$, and assume $y < x$ or $x = y$, the negation of the conclusion. We have two cases.

Case 1. $y < x$. Then, by A1, $y < y$, since we have assumed that $x < y$. But this contradicts A5.

Case 2. $x = y$. Then, since $x < y$, we have $y < y$, which again contradicts A5.

Hence, in each case we obtain a contradiction. Therefore, we must have $y \not< x$ and $x \neq y$.

c. Assume $x = y$, and assume $x < y$ or $y < x$, the negation of the conclusion. We have two cases.

Case 1. $x < y$. Then $x < x$, since $y = x$. But this contradicts A5.

Case 2. $y < x$. Then $x < x$ since $y = x$, again contradicting A5.

The contradiction in each case yields the desired conclusion.

Using A2, we are able to conclude that if $x \not< y$, then $y < x$ or $y = x$. By Example 3, we can also conclude that if $x < y$, then $y \not< x$ and $y \neq x$.

Before considering the next example, recall that for any nonnegative real number x, $\sqrt{x}$ is defined and $\sqrt{x} \geq 0$. Furthermore, if $x > 0$, then $\sqrt{x} > 0$.

Example 4

Prove: If $x > 0$, $y > 0$, and $x < y$, then $\sqrt{x} < \sqrt{y}$.

Proof Assume $x > 0$, $y > 0$, $x < y$, and $\sqrt{x} \geq \sqrt{y}$.
Case 1. Assume $\sqrt{x} = \sqrt{y}$. Then $x = (\sqrt{x})^2 = (\sqrt{y})^2 = y$, and hence, by Example 3, we cannot have $x < y$.
Case 2. Assume $\sqrt{x} > \sqrt{y}$. Since $\sqrt{x} > 0$, by A4, we obtain $\sqrt{x}\sqrt{x} > \sqrt{x}\sqrt{y}$. Also, by A4, $\sqrt{x}\sqrt{y} > \sqrt{y}\sqrt{y}$. Now, by A1, $\sqrt{x}\sqrt{x} > \sqrt{y}\sqrt{y}$. Hence, by algebra, $x > y$. But by Example 3, this is contrary to the assumption that $x < y$.

PROOF TECHNIQUE Proof by Contraposition. To prove an implication $P \Rightarrow Q$, it is sufficient to prove $\neg Q \Rightarrow \neg P$.

Proof by contraposition is based on the logical equivalence $(P \Rightarrow Q) \equiv (\neg Q \Rightarrow \neg P)$.

Example 5

Let m and n be nonnegative integers. *Prove*: If $m + n > 50$, then $m > 25$ or $n > 25$.

Proof The contrapositive of this statement is $\neg(m > 25 \text{ or } n > 25) \Rightarrow \neg(m + n > 50)$, which, by De Morgan's law, is equivalent to $(m \leq 25 \text{ and } n \leq 25) \Rightarrow m + n \leq 50$. But this follows from Exercise 9 in Exercise Set 2.1. (Note that $\neg(m > 25)$ is equivalent to $m \leq 25$.)

Practice Problem 2. Let m and n be nonnegative integers. *Prove*: If $m \cdot n < 100$, then $m < 10$ or $n < 10$.

Proofs by contraposition and by contradiction for an implication $P \Rightarrow Q$ are similar in the sense that both include the assumption $\neg Q$. The choice of one over the other is frequently a matter of taste. A proof by contradiction is often easier, since more is assumed true; you are able to assume both $\neg Q$ and P. On the other hand, a proof by contradiction is likely to be less elegant than a proof by contraposition. In any case, for elegance and clarity, it is better to choose a direct proof over an indirect proof whenever possible.

EXERCISE SET 2.2

Supply a proof for each assertion.
1. If $x > 0$, then $1/x > 0$.
2. If $x < y$, then $x^3 < y^3$.
3. If $x < y$, then $x^{1/3} < y^{1/3}$.
4. If $x^2 < x$, then $x < 1$.

5. Let m and n be nonnegative integers. If $m \cdot n = 100$ and $m < 10$, then $n > 10$.

6. Let m and n be integers. If $m + n = 100$ and $m < 0$, then $n > 100$.

7. If $x > 0$ and $y > 0$, then $x/y > 0$.

8. If $x > 0$ and $y < 0$, then $x/y < 0$.

9. If $x \geq 0$ and $y < 0$, then $x/y \leq 0$.

10. If x, y, and $z > 0$ and $xy > z$, then $x > \sqrt{z}$ or $y > \sqrt{z}$.

See Exercise Set 2.1 for the definition of an even integer. An integer z is an **odd integer** (or z is **odd**) if $z = 2 \cdot j + 1$ for some integer j. If an integer is not odd, then it is even, and vice versa.

11. If the integer $x \cdot y$ is odd, then x and y are both odd integers.

12. If the integer $x + y$ is even and x is odd, then y is odd.

13. If the integer $x + y$ is odd, then x is odd or y is odd.

14. The integer x is even if and only if x^2 is even.

15. The integer $x \cdot y$ is even if and only if at least one of the integers x and y is even.

16. The integer $x + y$ is even if and only if x and y are both odd or x and y are both even.

2.3 INTRODUCTION TO MATHEMATICAL INDUCTION

If consecutive odd integers, starting with 1, are added, a nice pattern emerges, namely,

$$n = 1: \quad 1 = 1$$

$$n = 2: \quad 1 + 3 = 4$$

$$n = 3: \quad 1 + 3 + 5 = 9$$

$$n = 4: \quad 1 + 3 + 5 + 7 = 16$$

where n is the number of odd numbers to be added.

It appears, then, that the sum of the first n odd integers is always equal to the square of n. But how is such a statement proved? Verifying an infinite sequence of statements, statement by statement, is out of the question. Mathematical induction is what is needed in such cases.

Principle of Mathematical Induction. To prove an infinite sequence of statements $p(1)$, $p(2)$, $p(3)$, ... (that is, $p(n)$ for all positive integers n), apply the following steps:

Basis Step: Prove $p(1)$.

Induction Step: Prove the implication $p(k) \Rightarrow p(k+1)$ for $k = 1, 2, 3, \ldots$. The statement $p(k)$ in the induction step is called the **induction hypothesis (IH)**.

Example 1

Prove that the sum of the first n odd integers is equal to the square of n for $n = 1, 2, 3, \ldots$.

Proof We use mathematical induction. Let $p(n)$ be the statement that the sum of the first n odd integers is n^2.

Basis step: The sum of the first (one) odd integer is 1, which is 1^2. Therefore, $p(1)$ is true.
Induction Step: We prove that $1+3+5+\ldots+(2k-1) = k^2$ implies $1+3+\ldots+(2k-1)+(2k+1) = (k+1)^2$, that is, that $p(k) \Rightarrow p(k+1)$. Assume that $1+3+\ldots+(2k-1) = k^2$, that is, $p(k)$. (IH) Then

$$1+3+\ldots+(2k-1)+(2(k+1)-1) = [1+3+\ldots+(2k-1)]+(2k+1)$$

$$= k^2 + (2k+1) \qquad \text{(by IH)}$$

$$= (k+1)^2.$$

Hence, we have $p(k+1)$, and by mathematical induction, $1+3+\ldots+(2n-1) = n^2$, for all positive integers n.

It may look like nothing is gained by invoking mathematical induction since the induction step is also an infinite sequence of statements. But in practice, the induction hypothesis $p(k)$ is assumed for an arbitrary integer k, and a proof of $p(k+1)$ is given. Because the integer k is arbitrary and the proof thus holds for any k, the induction step is only done once. So mathematical induction reduces the proof of an infinite number of statements to two proofs.

Practice Problem 1. Let $p(n)$ be $1 + 2 + 2^2 + \ldots + 2^{n-1} = 2^n - 1$ for $n = 1, 2, 3, \ldots$.

   a. State $p(1)$, $p(2)$, and $p(3)$.
   b. Prove $p(n)$ for $n = 1, 2, 3, \ldots$. (*Hint:* The induction hypothesis is $1 + 2 + 2^2 + \ldots + 2^{k-1} = 2^k - 1$. For the induction step, add 2^k to both sides of the equality.)

An Informal Justification of Mathematical Induction

The principle of mathematical induction is based on the well-ordering principle for the natural numbers, which says that if there is a natural number n such that $q(n)$, then there is a least natural number m such that $q(m)$. The well-ordering principle is an intuitively appealing property of the natural numbers. We will discuss it formally in the Supplementary Exercises; our purpose here is to show informally that the principle of mathematical induction follows from the well-ordering property.

To justify the principle of mathematical induction, we assume that $p(1)$ holds and $p(k+1)$ holds whenever $p(k)$ holds. We then seek to show that $p(n)$ holds for all positive integers n. Suppose not. Then there must be a positive integer n such that $p(n)$ does not hold. Let m be the least positive integer such that $\neg p(m)$. Can $m = 1$? If $m = 1$, then $p(1)$ does not hold. However, we have assumed that $p(1)$ holds. So $m \neq 1$. Consequently, m must be of the form $k+1$ for some positive integer k. Hence, $k+1$ is the least positive integer such that $p(k+1)$ does not hold. Now, since $k < k+1$, $p(k)$ holds. But by assumption, $p(k+1)$ holds if $p(k)$ holds. Hence, $p(k+1)$ holds, and we have a contradiction. Therefore, $p(n)$ holds for all positive integers n.

Suppose we wish to find a formula for the sum $1 + 2 + 3 + \ldots + n$ of the first n positive integers. Mathematical induction does not yield this formula. Instead, we must first try to find a plausible formula and then use mathematical induction to verify that it is, in fact, correct. We can try to find a formula by listing the first few sums, as was done before in the case of the sum of the first n odd integers:

$1 = 1$

$1 + 2 = 3$

$1 + 2 + 3 = 6$

$1 + 2 + 3 + 4 = 10$

No pattern appears obvious, so let us try to find a formula another way. By looking down the columns in Figure 2-1, we see that the number of black dots is $1 + 2 + 3 + 4 + 5 + 6 + 7 + 8 = 36$. We see that this is also the same as the number of white dots.

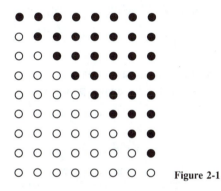

Figure 2-1

To count the total number of both dots, we can simply compute $8 \cdot 9$, since the rectangle is 8 dots by 9 dots. However, the number of black dots is half this amount, that is, $(8 \cdot 9)/2$, which is also 36. Notice that the number of rows is one more than the number of columns, so that the number we are after is $8 \cdot (8 + 1)/2$. This yields a possible formula for the sum of the first n positive integers. In the next example, we show by mathematical induction that

$$1 + 2 + \ldots + n = \frac{n(n + 1)}{2}$$

Example 2

Let $p(n)$ be the statement that the sum of the first n positive integers is $n(n + 1)/2$. Then $p(1)$ means that the sum of the first (one) positive integer is $1(1+1)/2$. This is easily seen to be true since the sum of the first (one) positive integer is, of course, 1, and $1(1 + 1)/2 = 1$. Now, let us check $p(2)$. The sum of the first two positive integers is $1 + 2$, which is 3, and $2(2 + 1)/2$ is also 3. It is easy to see that $p(3)$, $p(4)$, etc., are true, but this is not enough to show that $p(n)$ is true for *all* positive integers n. To use mathematical induction, we must show the basis step $p(1)$, which has already been shown above, and the induction step

$p(k) \Rightarrow p(k+1)$. To do the latter, assume $p(k)$; that is, assume $1+2+3+...+k = (k+1)k/2$. Then prove $p(k+1)$; that is, prove $1+2+3+...+k+(k+1) = [((k+1)+1)(k+1)]/2$:

$$1+2+3+...+k+(k+1) = [1+2+3+...+k]+(k+1)$$
$$= (k+1)k/2 + (k+1) \qquad \text{(By IH)}$$
$$= [(k+1)k + 2(k+1)]/2$$
$$= (k^2 + 3k + 2)/2$$
$$= [(k+2)(k+1)]/2$$
$$= [((k+1)+1)(k+1)]/2$$

So we have $p(k) \Rightarrow p(k+1)$ and, hence, by mathematical induction, $p(n)$ for all positive integers n.

Practice Problem 2. Use mathematical induction to prove that

$$1^2 + 2^2 + 3^2 + ... + n^2 = \frac{n(n+1)(2n+1)}{6}$$

Often the induction step is straightforward, as in the preceding examples. However, the induction step in the following proof is more involved and requires a needs assessment.

Example 3

Prove that $2^{n-1} \le 2^n - 1$ for $n = 1, 2, 3, \ldots$.

Proof We use mathematical induction again.
Basis step: $2^{1-1} = 2^0 = 1 \le 2 - 1 = 2^1 - 1$. Before doing the induction step, we do a needs assessment (see Section 1.1). We are given $2^{k-1} \le 2^k - 1$ (IH); we want to show that $2^{(k+1)-1} \le 2^{k+1} - 1$; that is, that $2^k \le 2^{k+1} - 1$. The idea is to somehow break what we want to show into parts so that we can recognize those parts which look like expressions in the induction hypothesis. We know that $2^k = 2(2^{k-1})$. So we want to show that $2(2^{k-1}) \le 2^{k+1} - 1$. We recognize 2^{k-1} as a part of the induction hypothesis. So we multiply both sides of the inequality assumed in the induction hypothesis by 2 to obtain $2(2^{k-1}) \le 2(2^k - 1)$, that is, $2^k \le 2^{k+1} - 2$. Our problem would be solved if we knew that $2^{k+1} - 2 \le 2^{k+1} - 1$, that is, that $-2 \le -1$. But this is true, and we can now use this information to obtain a proof of the induction step.
Induction step: The induction hypothesis is $2^{k-1} \le 2^k - 1$. Multiplying both sides by 2 we obtain $2^k \le 2^{k+1} - 2$. Since $-2 \le -1$, we have $2^{k+1} - 2 \le 2^{k+1} - 1$, and it follows that $2^k \le 2^{k+1} - 1$. Hence, by mathematical induction, $2^{n-1} \le 2^n - 1$ for $n = 1, 2, 3, \ldots$.

A proof by induction does not have to start at 1. If the theorem needing proof is of the form $p(n)$ for $n = b, b+1, \ldots$ then the basis step is to prove $p(b)$ and the induction hypothesis is for $k > b$. In the previous examples, b was always equal to 1, but b may be any integer. For example, one exercise asks for a proof that $n^2 < 2^n$ for $n = 5, 6, \ldots$. In this case, the basis step is $n = 5$.

By substituting for n, we see that $2n + 1 \leq 2^n$ is false for $n = 1, 2$. However, $2n + 1 \leq 2^n$ is true for $n = 0, 3, 4, 5, \ldots$. To verify this, we only need substitute $n = 0$ to check one case and then finish the following problem using mathematical induction with basis step $n = 3$.

Practice Problem 3. *Prove*: $2n + 1 \leq 2^n$ for $n = 3, 4, \ldots$.

The following example has basis step $n = 3$. A **convex n-gon** is an n-sided polygon with the property that the straight-line segment between any two points of the polygon is entirely within the polygon. A triangle is thus a convex 3-gon. Since a polygon must have at least three sides, a triangle is the convex n-gon with the least number of sides.

Example 4

Prove that the sum of the interior angles, in radians, of a convex n-gon is $(n - 2)\pi$. From Euclidean geometry, the interior angles of a triangle sum to π. So we have the basis step, namely, the sum of the interior angles of a convex 3-gon (that is, a triangle) is $(3 - 2)\pi$.

For the induction hypothesis, we assume that the sum of the interior angles of any convex k-gon is $(k - 2)\pi$, where $k \geq 3$. Now, an arbitrary convex $(k + 1)$-gon can be decomposed into a convex k-gon and a triangle with a line drawn between two corners having only one corner between them (see Figure 2-2). By examining the figure, it is not difficult to see that the sum of the interior angles of the $(k + 1)$-gon is equal to the sum of the interior angles of the convex k-gon plus the sum of the interior angles of the triangle. Hence, the sum of the interior angles of a convex $(k+1)$-gon is $(k-2)\pi + \pi = ((k+1) - 2)\pi$.

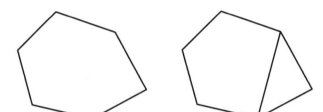

Figure 2-2 Convex$(k + 1)$-gon=convex k-gon+triangle

The following kind of definition is useful for proofs by mathematical induction.

Example 5

Define **n factorial,** written **n!.**

Solution   Basis step: $(n = 0)$ $0! = 1$.
Recursion step: $(k + 1)! = (k + 1) \cdot k!$ for $k = 0, 1, 2, \ldots$.
  Using this definition, we see that

$$1! = 1 \cdot 0! = 1 \cdot 1 = 1$$

$$2! = 2 \cdot 1! = 2 \cdot 1 = 2$$

$$3! = 3 \cdot 2! = 3 \cdot 2 \cdot 1 = 6$$

$$\vdots$$

$$n! = n \cdot (n-1)! = n \cdot (n-1) \cdots 1$$

The foregoing definition is unusual. In the recursion step, the right side of the definition contains a factorial. But this is what we want to define, and it is not usually proper to use the item to be defined in its definition. In this case, however, it is justified: On the basis of the definition, we can compute $1!$, $2!$, $3!$, and in fact $n!$ for any given natural number n. A definition of this kind is called a **definition by recursion.** Note the similarity between mathematical induction and definition by recursion.

Following are two more definitions by recursion:

Definition.   a. $\displaystyle\sum_{i=1}^{n} a_i$:

$$\sum_{i=1}^{1} a_i = a_1$$

$$\sum_{i=1}^{k+1} a_i = \sum_{i=1}^{k} a_i + a_{k+1}$$

b. $\displaystyle\prod_{i=1}^{n} a_i$:

$$\prod_{i=1}^{1} a_i = a_1$$

$$\prod_{i=1}^{k+1} a_i = \prod_{i=1}^{k} a_i \cdot a_{k+1}$$

Practice Problem 4.   Verify the following for $n = 2, 3, 4$:

a. $\sum_{i=1}^{n} a_i = a_1 + a_2 + \ldots + a_n$

b. $\prod_{i=1}^{n} a_i = a_1 \cdot a_2 \cdot \ldots \cdot a_n$

Definitions by recursion work very well with mathematical induction because the recursion step breaks expressions into parts which are easily seen to be parts of the induction hypothesis.

Example 6

Let us repeat Example 1 using the fact that $\sum_{i=1}^{n}(2i-1) = 1+3+5+\ldots+(2n-1)$. We must show that $\sum_{i=1}^{n}(2i-1) = n^2$.

Proof *Basis Step*: $\sum_{i=1}^{1}(2i-1) = 2(1)-1 = 1 = 1^2$.
Induction step: Assume that $\sum_{i=1}^{k}(2i-1) = k^2$. Then

$$\sum_{i=1}^{k+1}(2i-1) = \sum_{i=1}^{k}(2i-1) + (2(k+1)-1)$$

$$= k^2 + (2k+1) \quad \text{(By IH)}$$

$$= (k+1)^2$$

Therefore, by mathematical induction, $\sum_{i=1}^{n}(2i-1) = n^2 \qquad$ for $\quad n = 1,2,3,\ldots.$

EXERCISE SET 2.3

In Exercises 1–8, use mathematical induction to prove each assertion.

1.
$$1 + \frac{1}{2} + \frac{1}{4} + \ldots + \frac{1}{2^n} = 2 - \frac{1}{2^n}$$

for $n = 0,1,2,\ldots.$

2.
$$1 + 8 + 27 + \ldots + n^3 = \left[\frac{n(n+1)}{2}\right]^2$$

for $n = 1,2,3,\ldots.$

3.
$$1 + r + r^2 + \ldots + r^n = \frac{1 - r^{n+1}}{1 - r}$$

for $n = 0,1,2,3,\ldots$, and $r \neq 1$.

4. $n^2 \leq 2^n$ for $n = 4,5,6,\ldots.$ (*Hint*: You may use the result of Practice Problem 3.)
5. $1/1^2 + 1/2^2 + \ldots + 1/n^2 \leq 2n/(n+1)$ for $n = 1,2,3,\ldots.$
6. $1/\sqrt{1} + 1/\sqrt{2} + \ldots + 1/\sqrt{n} > \sqrt{n}$ for $n = 2,3,4,\ldots.$
7. $\sum_{i=1}^{n}(2i-1)^3 = n^2(2n^2-1)$ for $n = 1,2,3,\ldots.$
8. $\sum_{i=1}^{n} i \cdot i! = (n+1)! - 1$ for $n = 1,2,3,\ldots.$
9. Use mathematical induction to prove that in Euclidean geometry, given a line of unit length, it is possible to construct, with straightedge and compass, a line of length $\sqrt{n}$ for $n = 1,2,3,\ldots.$

In Exercises 1, 2, 3, 7, and 8, the right side of each equation is the **closed-form expression** of the left side. In Exercises 10 and 11 compute values of the left side for several integers and guess the closed-form expression $f(n)$. Then use mathematical induction to prove your conjecture correct.

10. $(1 - 1/4) \cdot (1 - 1/9) \cdots (1 - 1/n^2) = f(n)$ for $n = 2, 3, 4, \ldots$.

11.

$$\sum_{i=1}^{n} \frac{1}{i(i+1)} = f(n)$$

for $n = 1, 2, 3, \ldots$.

12. Give a definition by recursion for 2^n.

Exercises 13–18 contain results that are useful in calculus. Some of Exercises 15–18 make use of Exercises 13 and 14. Let $D^n(f(x))$ stand for the nth derivative of the function f at x.

13. Give a definition by recursion for $D^n(f(x))$.

14. *Prove:* $D^n(D(f(x))) = D(D^n(f(x)))$.

15. Let c be any constant. Prove that $D^n(cf(x)) = cD^n(f(x))$ for $n = 1, 2, \ldots$.

16. *Prove:* $D^n(x^n) = n!$ for $n = 1, 2, \ldots$.

17. *Prove:* $D^n(e^{bx}) = b^n e^{bx}$ for $n = 1, 2, \ldots$. You may assume the chain rule and that $D(e^x) = e^x$.

18. Prove that $D^n(\ln(x)) = (-1)^{n+1}(n-1)! x^{-n}$ for $n = 1, 2, \ldots$, and $x > 0$. You may assume that $D(\ln(x)) = 1/x$ for $x > 0$.

19. Find the error in the following incorrect argument that $7^n = 1$ for $n = 0, 1, 2, \ldots$.

Argument: We have $7^0 = 1$ by definition. Assume $7^j = 1$ for all $0 \le j \le k$. We prove that $7^j = 1$ for all j, $0 \le j \le k + 1$. To do this, we need only prove that $7^{k+1} = 1$. But $7^{k+1} = (7^k \cdot 7^k)/7^{k-1} = (1 \cdot 1)/1 = 1$. By mathematical induction, $7^m = 1$ for all m, $0 \le m \le n$, where $n = 0, 1, 2, \ldots$. The result follows.

2.4 PREDICATES AND QUANTIFIERS

The elementary logic that we studied in Section 1.2, is commonly called the propositional calculus. The propositional calculus, which involves only statements, is inadequate for many purposes.

Statements are sentences that have a truth value, in other words, that are either true or false, but not both. For example, the sentence "$-1 < 0$" is a statement, which happens to be true in the context of the real numbers. Many other sentences used in mathematics, however, are not statements. Sentences like "$x < 5$" and "$\sin x = 1$" are neither true nor false, because they contain a variable that denotes no particular object. A **variable** is used to symbolize an arbitrary element from a given universal set U. That is, elements from U can be substituted for x in a sentence like "$x < 5$." A sentence of the form $p(x)$, in which a substitution is possible and which becomes a statement after a substitution is made, is called a **predicate** (in one variable). If a variable occurs in several places in a predicate, as in "$x^5 + x^2 - 1 = 0$," then the same substitution must be made in each place the variable occurs. Predicates frequently have more than one variable. The sentence "$x^2 + y^2 = 1$" has two variables, while "$r < x \Rightarrow 1/x < \epsilon$" has three variables.

The **predicate calculus** involves statements, predicates, and quantifiers (to be introduced shortly). Just as with the propositional calculus, we will learn many useful rules of logic and examine many applications. But first, a brief discussion of sets is in order. A more detailed study will be done in Chapter 3.

A set is a collection of objects called *elements* of the set. If A denotes a set, and x denotes an element in the set, then we write $x \in A$. If x is not an element of A, then we write $x \notin A$.

If a set A has only the elements a, b, and c, then we write $A = \{a, b, c\}$. Similarly, $\{1, 2, 3, 4\}$ denotes the set with the elements 1, 2, 3, and 4. In this case, $2 \in \{1, 2, 3, 4\}$ whereas $5 \notin \{1, 2, 3, 4\}$. The set that does not have any elements is denoted by $\emptyset$. So for any x, we always have $x \notin \emptyset$. We always assume that the elements of a set come from some universal set U.

If $p(x)$ is a predicate, then the set of elements from a given universe U that make $p(x)$ true is called the **solution set for p(x) in U** and is written $\{x : p(x)\}$. For example, let the universe U be the set of real numbers. Then the solution set for $x > 4$ in the universe of real numbers is $\{x : x > 4\}$ (the set of real numbers greater than 4). The solution set for $4 < x < 9$ in the universe of integers is $\{x : 4 < x < 9\}$ (the set of integers between 4 and 9). This set may also be denoted by $\{5, 6, 7, 8\}$.

Certain sets occur so often that we use a standard notation for them:

$\mathcal{R}$ is the set of real numbers;

$\mathcal{N}$ is the set $\{0, 1, 2, 3, \ldots\}$ of nonnegative integers;

$\mathcal{Z}$ is the set $\{\ldots, -3, -2, -1, 0, 1, 2, 3, \ldots\}$ of integers.

We will always assume that the universal set U is nonempty.

Example 1

In this example, the universe is $\mathcal{R}$.

Predicate	Substitutions that make the predicate a	
	true statement	false statement
a. $x \cdot 0 = 0$	any	none
b. $x \cdot 5 = 0$	$x = 0$	$x \neq 0$
c. $\sin x = 2$	none	any
d. $x = 0 \Rightarrow x \cdot 5 = 0$	any	none
e. $x < 5$	$x = 3.1, 4$, etc.	$x = 5, 6.3$, etc.

Part d might be puzzling, so let us consider it a little more closely. The substitution of 0 for x yields the statement $(0 = 0) \Rightarrow (0 \cdot 5 = 0)$, which is certainly true. The substitution of 3 for x gives the statement $(3 = 0) \Rightarrow (3 \cdot 5 = 0)$, which is also (vacuously) true since $3 = 0$ is false.

Both sentences in the following example are statements and not predicates.

Example 2

The "x" which seems to be a variable in each of the following statements is not for substitution (try it yourself) and is frequently called a *dummy* or *bound* variable:

a. The solution set for $x^2 - 3x + 2 = 0$ is $\{1, 2\}$; that is, $\{x : x^2 - 3x + 2 = 0\} = \{1, 2\}$.

b. $\int_0^1 x^2 \, dx = \frac{1}{3}$.

Following are some examples of solution sets in various universes.

Example 3

Predicate	Universal Set	Solution Set
a. $x < 5$	$\mathcal{R}$	$\{x : x \in \mathcal{R}$ and $x < 5\}$
b. $x < 5$	$\mathcal{N}$	$\{0, 1, 2, 3, 4\}$
c. $\sin x = 0$	$\mathcal{R}$	$\{n\pi : n \in \mathcal{Z}\}$
d. $\sin x = 0$	$\mathcal{Z}$	$\{0\}$
e. $\sin x = 2$	$\mathcal{R}$	$\emptyset$
f. $x \cdot 0 = 0$	$\mathcal{R}$	$\mathcal{R}$
g. $\sin^2 x + \cos^2 x = 1$	$\mathcal{R}$	$\mathcal{R}$

A solution set thus depends on both the predicate and the given universal set.

Practice Problem 1. Find the solution sets for the following predicates in the given universes.

Predicate	Universal Set
a. $x + 4 = 1$	$\mathcal{Z}$
b. $x + 4 = 1$	$\mathcal{N}$
c. $x^2 = 2$	$\mathcal{R}$
d. $2x^2 - x - 1 = 0$	$\mathcal{R}$
e. $2x^2 - x - 1 = 0$	$\mathcal{N}$
f. $x^2 + 2x + 1 = (x + 1)^2$	$\mathcal{R}$

The predicates in parts f and g of Example 3 have the special property that their solution set is U. That is, each is true for every possible substitution from U. Asserting a predicate is true for all substitutions is so important that we use a special symbol called the universal quantifier for this purpose.

Universal Quantifier. $\forall x$ means "for all x" or "for any x."

Example 4

Assume that the universe is $\mathcal{R}$. Then:

a. $\forall x (x \cdot 0 = 0)$ means "any real number multiplied by 0 is equal to 0."

b. $\forall x (\sin^2 x + \cos^2 x = 1)$ means "the sum of the squares of the sine and cosine of any real number is equal to 1." This kind of assertion is made in every trigonometric identity.

The assertion that the solution set for a predicate is nonempty is also important. The symbol used for this assertion is the existential quantifier.

Existential Quantifier. $\exists x$ means "there exists an x such that" or "there is an x such that."

Example 5

Assume that the universe is $\mathcal{R}$. Then:
a. $\exists x(x \cdot 5 = 7)$ means $x \cdot 5 = 7$ has a solution. In this case, there is exactly one solution, namely, $\frac{7}{5}$.
b. $\exists x(\sin x = 0)$ means $\sin x = 0$ has at least one solution. In this case, there are an infinite number of solutions, given by $x = n\pi$ radians, for any integer n.

When the universal set U is given and $p(x)$ is a predicate in one variable, then both $\forall x p(x)$ and $\exists x p(x)$ are statements. Of course, these statements are not always true. By definition:

$\forall x p(x)$ is true if $p(a)$ is true for all substitutions a from U;

$\exists x p(x)$ is true if there is at least one substitution a from U for which $p(a)$ is true.

Example 6

In the following, let $\mathcal{R}$ be the universe.
a. $\forall x(x^2 + 2x + 5 > 0)$ is true.   c. $\exists x(x^2 + 2x + 5 = 0)$ is false.
b. $\forall x(\sin x = 1)$ is false.   d. $\exists x(\sin x \neq 1)$ is true.

Practice Problem 2. Which of the following statements are true, and which are false, for the given universes?

Statement	Universe
a. $\exists x(x + 3 = 1)$	$\mathcal{Z}$
b. $\exists x(x + 3 = 1)$	$\mathcal{N}$
c. $\forall x(\sin x < 1)$	$\mathcal{R}$
d. $\forall x(x < 1 \Rightarrow x + 3 < 4)$	$\mathcal{R}$

Earlier in this section, we assumed that $U \neq \emptyset$. One reason for this assumption is that the implication $\forall x q(x) \Rightarrow \exists x q(x)$ is always true for $U \neq \emptyset$, but is false when U is empty. To see the latter, let $U = \emptyset$ and $q(x)$ be $x \neq x$. Then $\forall x(x \neq x)$ is (vacuously) true and $\exists x(x \neq x)$ is false.

Definition. A statement in the predicate calculus that is true in all universes is called a **validity.**

Example 7

Let $q(x)$ be any predicate. Then $\forall x q(x) \Rightarrow \exists x q(x)$ is a validity.

Proof Assume $\forall x q(x)$ is true in some universe U. Because U is nonempty, we have an $a \in U$. Since $\forall x q(x)$ is true in U, $q(b)$ is true for all members b in U. In particular, $q(a)$ is true. Hence, there is an element in U, namely, a, such that $q(a)$. Therefore, $\exists x q(x)$ is true in U.

Later in this section we will give more example of validities.

The universal quantifier can be considered the generalized "and," the existential quantifier can be considered the generalized "or." To see this, let U be the two-element set $\{0, 1\}$. Then $\forall x p(x)$ is true if and only if $p(0) \wedge p(1)$ is true, and $\exists x p(x)$ is true if and only if $p(0) \vee p(1)$ is true.

One of the important uses of quantified statements is writing precise definitions. For many definitions, more than one quantifier must be used.

Example 8

Assume that the universe is $\mathcal{R}$.

a. A real number u is an **upper bound** for a set A of real numbers if $\forall x (x \in A \Rightarrow x \leq u)$ is true, that is, if every element of A is less than or equal to u.

b. A set A is **bounded above** if $\exists u \forall x (x \in A \Rightarrow x \leq u)$ is true, that is, if there is a real number u that is an upper bound of A.

c. A real number u is a **greatest element** of a set A of real numbers if $u \in A$ and u is an upper bound of A.

Practice Problem 3. Express the definition of a greatest element by means of a quantified statement.

The statement $\exists u \forall x (x \in A \Rightarrow x \leq u)$ in Example 8b has two quantifiers and the general form $\exists u \forall x p(x, u)$. The statement reads, "There exists an element u such that for all $x, p(x, u)$." As we will soon see, the order of quantifiers is important.

Quantifiers are very useful for stating and understanding the axioms for an algebraic structure such as $\mathcal{Z}$. The following example gives four of these axioms.

Example 9

The axioms for $\mathcal{Z}$ need more than one quantifier.

a. $x + y = y + x$ is a predicate in two variables; $\forall y (x + y = y + x)$ is a predicate in one variable x; the commutative law for $\mathcal{Z}$ is $\forall x \forall y (x + y = y + x)$.

b. *Identity laws:* $\exists z \forall x (x + z = x)$ and $\exists u \forall x (x \cdot u = x)$.

c. *Additive inverse law:* $\forall x \exists y (x + y = 0)$.

For the commutative law, it does not matter whether we write $\forall x \forall y$ or $\forall y \forall x$. However, the order of the quantifiers is important when both kinds of quantifiers occur in a statement. The difference between the statements like $\exists y \forall x p(x, y)$ and $\forall x \exists y p(x, y)$ is crucial. The statement "Every person has a mother" is of the form $\forall x \exists y$ (y is the mother of x). The statement "Someone is the mother of everyone" is of the form $\exists y \forall x$

(y is the mother of x). If the universe is the set of all people who have ever been alive, the first statement is true but the second is false.

For a mathematical example, $\forall x \exists y (x + y = 0)$ is true but $\exists y \forall x (x + y = 0)$ is false in $\mathcal{Z}$.

Practice Problem 4. Convince yourself that $\forall x \exists y (x + y = 0)$ is true but $\exists y \forall x (x + y = 0)$ is false in $\mathcal{Z}$, and hence, that $\forall x \exists y (x + y = 0) \Rightarrow \exists y \forall x (x + y = 0)$ is not a validity.

On the other hand, $\exists y \forall x p(x, y) \Rightarrow \forall x \exists y p(x, y)$ is a validity. As an example, suppose that in a certain small town there lived a person who knew everyone in town including, for the sake of the example, herself. Let the universe be the people of the town and y be such a person. Then our supposition is of the form $\exists y \forall x$ (y knows x). It clearly follows that $\forall x \exists y$ (y knows x) is true, that is, that everyone is known by someone.

Example 10

Let $p(x, y)$ be any predicate. Then

Q1 $\exists y \forall x p(x, y) \Rightarrow \forall x \exists y p(x, y)$

is a validity.

Proof Assume $\exists y \forall x p(x, y)$ is true. Then, since $U \neq \emptyset$, there is an element $b \in U$ such that $\forall x p(x, b)$ is true. Hence $p(a, b)$ is true for all substitutions a from U. So $\exists y p(a, y)$ is true for all substitutions a from U. Therefore, $\forall x \exists y p(x, y)$ is true.

Plainly, most definitions, axioms, and theorems are quantified statements. To decide whether or not a particular quantified statement is true, it is sometimes helpful to look at its negation. Let us look at some nonmathematical examples first.

Let the universe be the set of all people alive now, and let $p(x)$ be the predicate "x is a logic student." $\forall x p(x)$ means that all people are logic students. To assert the negation of this is to assert that there is some person who is not a logic student, that is, $\exists x \neg p(x)$ is true. Thus, we see that

Q2 $\neg \forall x p(x) \Leftrightarrow \exists x \neg p(x)$

is true in this universe. In fact, it is true in any universe, and hence, it is a validity.

$\exists x p(x)$ means that some person is a logic student. To assert the negation of this is to assert that no person is a logic student. In other words, every person is not a logic student; that is, $\forall x \neg p(x)$ is true. This means that

Q3 $\neg \exists x p(x) \Leftrightarrow \forall x \neg p(x)$

is true in this universe. It is also a validity.

Let us now look at a mathematical example.

Example 11

Let the universe be $\mathcal{R}$, and let $p(x)$ stand for "x has a positive square"; in other words, x^2 is positive. $\neg \forall x p(x)$ means "it is not the case that all real numbers have positive squares." $\exists x \neg p(x)$ means "there is a real number which does not have a positive square." This is, of course, true, since $0^2 = 0$.

Practice Problem 5. In the context of Example 11, read $\neg \exists x p(x)$ and $\forall x \neg p(x)$ aloud in English to see why Q3 makes sense.

Rules Q2 and Q3 are generalized De Morgan's laws. We can see this most easily by considering $U = \{0, 1\}$. For this two-element universe, $\neg \forall x p(x)$ is true if and only if $\neg(p(0) \wedge p(1))$ is true, that is, if and only if $\neg p(0) \vee \neg p(1)$ is true if and only if $\exists x \neg p(x)$ is true. Thus, we see that Q2 is a generalized De Morgan's law.

Practice Problem 6. Write Q3 with a two-element universe, and convince yourself that Q3 is a generalized De Morgan's law.

Following are some other statements and their negations.

Example 12

Statement	Negation
a. $\exists x \exists p(x + p = 0)$	$\forall x \forall p(x + p \neq 0)$
b. $\forall x \exists y(x + y = 0)$	$\exists x \forall y(x + y \neq 0)$
c. $\exists y \forall x(x + y = 0)$	$\forall y \exists x(x + y \neq 0)$
d. $\forall x \forall y[(x < y) \Rightarrow (x^2 < y^2)]$	$\exists x \exists y[(x < y) \wedge \neg(x^2 < y^2)]$

Rules Q2 and Q3 are used in the following manner to derive the negation in part b:

$$\neg \forall x \exists y(x + y = 0) \Leftrightarrow \exists x \neg \exists y(x + y = 0) \quad \text{(I)}$$

$$\Leftrightarrow \exists x \forall y \neg(x + y = 0) \quad \text{(II)}$$

$$\Leftrightarrow \exists x \forall y(x + y \neq 0). \quad \text{(III)}$$

This sequence of steps means the following:

$\neg \forall x \exists y(x + y = 0) \Leftrightarrow \exists x \neg \exists y(x + y = 0)$ is a validity. (I)

$\exists x \neg \exists y(x + y = 0) \Leftrightarrow \exists x \forall y \neg(x + y = 0)$ is a validity. (II)

$\exists x \forall y \neg(x + y = 0) \Leftrightarrow \exists x \forall y(x + y \neq 0)$ is a validity. (III)

Hence, $\neg \forall x \exists y(x + y = 0) \Leftrightarrow \exists x \forall y(x + y \neq 0)$ is a validity.

Following is an explanation of these steps:

(I) Let the predicate $p(x)$ be $\exists y(x + y = 0)$. Step (I) is an application of Q2.

(II) Let $q(x, y)$ be the predicate $x + y = 0$. Then $\neg \exists y(x + y = 0) \Leftrightarrow \forall y \neg(x + y = 0)$ is a validity. Now we replace $\neg \exists y(x + y = 0)$ by $\forall y \neg(x + y = 0)$ in $\exists x \neg \exists y(x + y = 0)$.

This rule for replacing predicates by equivalent predicates is similar to the substitution rule discussed in Section 1.4.

(III) $\neg(x + y = 0)$ is usually written as $x + y \neq 0$.

Practice Problem 7.   There are eight statements (four statements and their negations) in Example 12. Derive the negations in parts a and c. Assume that the universe is $\mathcal{R}$. Which four of the eight statements are true?

Restricted Quantifiers

It is sometimes useful to restrict the domain of a quantifier. This is usually accomplished by including a condition with the quantifier.

Definition.   Restricted Quantifiers.
a. $(\forall x \in B)(p(x))$ means $\forall x(x \in B \Rightarrow p(x))$.
b. $(\exists x \in B)(p(x))$ means $\exists x(x \in B \land p(x))$.

The next two examples illustrate the use of restricted quantifiers.

Example 13

Assume the universe is $\mathcal{R}$.
a. $(\forall x \in \mathcal{N})(p(x))$ means $\forall x(x \in \mathcal{N} \Rightarrow p(x))$.
b. $(\exists x > 0)(p(x))$ means $\exists x(x > 0 \land p(x))$.

Notice that restricting a universal quantifier is done using an "if . . ., then . . ." statement, while restricting an existential quantifier is done using an "and" statement.

Example 14

Assume the universe is $\mathcal{R}$.
a. $\mathcal{N}$ has no upper bound if and only if $\forall u(\exists x \in \mathcal{N})(u < x)$. With a standard quantifier we write $\forall u \exists x(x \in \mathcal{N} \land u < x)$.
b. We write "Every nonzero real number has a multiplicative inverse" with a restricted quantifier as $(\forall x \neq 0)(\exists y)(x \cdot y = 1)$. With a standard quantifier, we write $\forall x(x \neq 0 \Rightarrow \exists y(x \cdot y = 1))$.

Negating statements with restricted quantifiers is much the same as negating statements with standard quantifiers. We can prove that $\neg(\forall x \in S)p(x) \Leftrightarrow (\exists x \in S)\neg p(x)$ is a validity as follows:

1. $\neg(\forall x \in S)p(x)$ $\Leftrightarrow \neg\forall x(x \in S \Rightarrow p(x))$ (Definition)
2. $\Leftrightarrow \exists x\neg(x \in S \Rightarrow p(x))$ (Q2)
3. $\Leftrightarrow \exists x(x \in S \land \neg p(x))$ (Tautology)
4. $\Leftrightarrow (\exists x \in S)\neg p(x)$ (Definition)

Analysis:

Step 1. Actually, $\neg(\forall x \in S)p(x)$ is the same as $\neg \forall x(x \in S \Rightarrow p(x))$ by definition.

Step 2. Let $q(x)$ be $x \in S \Rightarrow p(x)$. Then this step is $\neg \forall xq(x) \Leftrightarrow \exists x \neg q(x)$.

Step 3. $\neg(r \Rightarrow t) \Leftrightarrow r \wedge \neg t$ is a tautology, so we can substitute $x \in S \wedge \neg p(x)$ for $\neg(x \in S \Rightarrow p(x))$.

Step 4. $\exists x(x \in S \wedge \neg p(x))$ is the same as $(\exists x \in S)\neg p(x)$ by definition.

Practice Problem 8. Show that $\neg(\exists x \in S)p(x) \Leftrightarrow (\forall x \in S)\neg p(x)$ is a validity.

The following validities are useful in mathematical proofs:

Q1 $\exists y \forall x p(x, y) \Rightarrow \forall x \exists y p(x, y)$

Q2 $\neg \forall x p(x) \Leftrightarrow \exists x \neg p(x)$

Q3 $\neg \exists x p(x) \Leftrightarrow \forall x \neg p(x)$

Q4 $\forall x(p(x) \wedge q(x)) \Leftrightarrow \forall x p(x) \wedge \forall x q(x)$

Q5 $\exists x(p(x) \vee q(x)) \Leftrightarrow \exists x p(x) \vee \exists x q(x)$

Q6 $\exists x(p(x) \wedge q(x)) \Rightarrow \exists x p(x) \wedge \exists x q(x)$

Q7 $\forall x p(x) \vee \forall x q(x) \Rightarrow \forall x(p(x) \vee q(x))$

EXERCISE SET 2.4

1. Find solution sets for each of the following predicates when $U = \mathcal{N}$ and when $U = \mathcal{R}$.
   (a) $x \cdot 5 = 3$
   (b) $x \cdot 3 < 7$
   (c) $\sin(x + \pi/2) = \cos x$
   (d) $\sin x = 1$

2. Find solution sets for each of the following predicates when $U = \mathcal{N}$ and when $U = \mathcal{R}$.
   (a) $|2 - x| < |x| + |2|$
   (b) $|2 - x| > |x| - |2|$
   (c) $\forall r(x \cdot r = r)$
   (d) $\exists r(x \cdot r = 1)$

3. Let $U = \mathcal{R}$. Determine whether each of the following statements is true or false.
   (a) $\forall x(x^2 + 2x - 3 = 0)$
   (b) $\exists x(x^2 + 2x - 3 = 0)$
   (c) $\forall x(\sin(x + \pi/2) = \cos x)$
   (d) $\exists x(\sin(x + \pi/2) = \cos x)$

4. Let $U = \mathcal{R}$. Determine whether each of the following statements is true or false.
   (a) $\forall x(x^2 - 2x + 5 > 0)$
   (b) $\exists x(x^2 - 2x + 5 = 0)$
   (c) $\forall x \exists r(x \cdot r = 1)$
   (d) $\forall x(\exists n \in \mathcal{N})(x < n)$

5. Write the negation, as in Example 12, of each statement in Exercise 3.

6. Write the negation, as in Example 12, of each statement in Exercise 4.

7. Write the negation, as in Example 12, of each of the following. Determine whether the resulting statement is true or false when $U = \mathcal{R}$.
   (a) $\forall x \exists m (x^2 < m)$
   (b) $\exists m \forall x (x^2 < m)$

8. In Example 12d, apply rules Q2 and Q3 to the statement on the left to obtain the statement on the right.

9. Write the negation, as in Example 12, of each of the following. Determine whether the resulting statement is true or false when $U = \mathcal{R}$.
   (a) $\forall z (\exists \epsilon > 0)(0 \le z < \epsilon \Rightarrow z = 0)$
   (b) $\exists m \forall x \big(x/(|x| + 1) < m\big)$

10. Write the negation, as in Example 12, of each of the following. Determine whether the resulting statement is true or false when $U = \mathcal{R}$.
    (a) $\exists m (\forall n \in \mathcal{N})(m < n \lor n < m)$
    (b) $\forall m (\exists k > 0)(1/(m^2 + 1) < k)$

11. (a) Use quantifiers to write the definition for a **lower bound** of a set B of real numbers.
    (b) The **least element** of a set of real numbers is an element of the set that is also a lower bound of the set. Write the definition for the least element of a set B of real numbers using quantifiers and no words.
    (c) Use quantifiers and no words to write as simply as possible: "$\{x : 0 < x \le 5\}$ has no least element."
    (d) Prove that there is no more than one least element of a set.

12. Let U = set of all animals alive today, $g(x)$: "x is a gibbon," $c(x)$: "x is carnivorous."
    (a) Express "All gibbons are carnivorous" in symbolic form. Use Q2 on the negation of this statement to derive the symbolic form for "Some gibbons are not carnivorous."
    (b) Express "Some gibbons are carnivorous" in symbolic form. Use Q3 on the negation of this statement to derive the symbolic form for "No gibbon is carnivorous."

13. State the principle of mathematical induction (see Section 2.3) using quantifiers.

14. Prove that Q4, $\forall x (p(x) \land q(x)) \Leftrightarrow \forall x p(x) \land \forall x q(x)$, is a validity.

15. Prove that Q5, $\exists x (p(x) \lor q(x)) \Leftrightarrow \exists x p(x) \lor \exists x q(x)$, is a validity.

2.5 COUNTEREXAMPLES, PROOFS, AND CONJECTURES

In a letter to Leonhard Euler in 1742, C. Goldbach (1690–1764) noted that it seemed that even integers greater than 2 could all be written as the sum of two primes. For example, $4 = 2 + 2$, $6 = 3 + 3$, $8 = 3 + 5$, $10 = 3 + 7 = 5 + 5$, etc. The **Goldbach conjecture** asserts that every even integer greater than 2 can be written as the sum of two prime numbers. An assertion that is still to be proved or refuted is called a **conjecture.** In order to refute a conjecture, it suffices to find a counterexample; for the Goldbach conjecture, an even positive integer that cannot be written as the sum of two primes would be a counterexample. Up to now, the Goldbach conjecture has been neither proved nor refuted.

Counterexamples

Example 1

Let the universe be $\mathcal{Z}$. Test the conjecture

$$\forall x(2x < 3 + 4x \Rightarrow x < -3)$$

Solution   To test this conjecture, we look at the negation, $\exists x(2x < 3 + 4x \land x \geq -3)$. We see that if $x = -1$, then $(2x < 3 + 4x \land x \geq -3)$ is true, and hence, $\exists x(2x < 3 + 4x \land x \geq -3)$ is true. Therefore, $\forall x(2x < 3 + 4x \Rightarrow x < -3)$ is false. The substitution of -1 for x is called a counterexample to the conjecture.

Whenever we make a conjecture of the form $\forall x p(x)$, and its negation $\exists x \neg p(x)$ is true, we say that a substitution for x that makes $\neg p(x)$ true is a **counterexample** to the conjecture. When faced with a conjecture, it is useful to write it in quantified form and then write its negation. By doing this, we can frequently tell whether the conjecture is true and find a counterexample if it is false. In a similar way, a conjecture of the form $\forall x \forall y p(x, y)$ is shown to be false by showing that $\exists x \exists y \neg p(x, y)$ is true. A substitution for x and y that makes $p(x, y)$ false is called a counterexample to the conjecture $\forall x \forall y p(x, y)$.

Example 2

Let the universe be $\mathcal{R}$. Test the conjecture

$$\forall x \forall y(x < y \Rightarrow x^2 < y^2)$$

Solution   We derive the negation of $\forall x \forall y(x < y \Rightarrow x^2 < y^2)$ and give a counterexample. The negation is $\exists x \exists y(x < y \land y^2 \leq x^2)$, since

$$\neg \forall x \forall y(x < y \Rightarrow x^2 < y^2) \Leftrightarrow \exists x \exists y \neg(x < y \Rightarrow x^2 < y^2)$$

$$\Leftrightarrow \exists x \exists y(x < y \land y^2 \leq x^2)$$

In this case, the negation is true because $x = -3$ and $y = 2$ makes

$$(x < y) \land (y^2 \leq x^2)$$

true. Hence, $x = -3$ and $y = 2$ is a counterexample to

$$\forall x \forall y(x < y \Rightarrow x^2 < y^2).$$

Example 3

Let $U = \mathcal{N}$, $p(x)$ stand for "x is a prime number," and $o(x)$ stand for "x is an odd number."

	Statement	Negation
a.	All prime numbers are odd.	Some prime numbers are not odd.
	$\forall x(p(x) \Rightarrow o(x))$	$\exists x(p(x) \land \neg o(x))$
b.	Some prime numbers are odd.	No prime numbers are odd.
	$\exists x(p(x) \land o(x))$	$\forall x(p(x) \Rightarrow \neg o(x))$

a. Derivation of the negation:

$$\neg \forall x (p(x) \Rightarrow o(x)) \Leftrightarrow \exists x \neg (p(x) \Rightarrow o(x))$$

$$\Leftrightarrow \exists x (p(x) \wedge \neg o(x))$$

Note that the substitution of 2 for x makes $p(x) \wedge \neg o(x)$ true. So 2 is a counterexample to "All prime numbers are odd."

b. Derivation of the negation:

$$\neg \exists x (p(x) \wedge o(x)) \Leftrightarrow \forall x \neg (p(x) \wedge o(x))$$

$$\Leftrightarrow \forall x (\neg p(x) \vee \neg o(x))$$

$$\Leftrightarrow \forall x (p(x) \Rightarrow \neg o(x))$$

Some prime numbers are odd, since 3 is both prime and odd.

Practice Problem 1.   Derive the negation of $\forall x \forall y$ (x is odd $\Rightarrow xy$ is odd). Show that this statement is false in $\mathcal{Z}$ by finding a counterexample.

Proofs Involving Existential Quantifiers

The chief way to prove a statement of the form $\exists x p(x)$ is to find an x and verify that it satisfies $p(x)$.

Example 4

Prove: There is an x such that $2x^3 - 3x^2 + 2x - 8 = 0$.

Proof   We will find an x that is a solution of the equation. Instead of using some of the techniques for solving such an equation, we just propose that $x = 2$ is a solution. Now we must verify that $x = 2$ works. In this case, it is a simple matter to verify that $2(2^3) - 3(2^2) + 2(2) - 8 = 0$.

Example 5

Prove: There is an x such that $8x + 1$ is a square of an integer.

Proof   Let $x = 3$. Then $8(3) + 1 = 25$, which is the square of an integer.

Practice Problem 2.   *Prove*: There is an integer x such that x^2 is a cube of an integer.

In the preceding examples, we obtained the number x by making an educated guess and then verified that it worked by a routine computation. In the next example, the proof requires a little more work.

Example 6

Prove: Between any two distinct real numbers there is a real number.

Proof Let x and y be two real numbers such that $x \neq y$, say $x < y$. We need to find a real number between x and y. The average of x and y should work. We let $z = (x + y)/2$. We now verify that z is between x and y. First we show that $x < z$. A needs assessment will help here.

We wish to show that $x < (x + y)/2$. We can do so if we show that $2x < x + y$, which is the same as $x < y$. We can now use this needs assessment to construct a proof.

We have assumed that $x < y$. Therefore, $x + x < x + y$; that is, $2x < x + y$. Hence, $x < (x + y)/2$.

We leave it as an exercise to verify that $z < y$.

Example 7

Prove that there is a real number $\delta > 0$ such that $1 < \sqrt{x} < 1.1$ whenever $1 < x < 1 + \delta$.

Proof We again use a needs assessment. We want to find a δ such that $1 < x < 1 + \delta$ implies $1 < \sqrt{x} < 1.1$. To find such a δ, let us examine the inequality $1 < \sqrt{x} < 1.1$. This is equivalent to $1 < \sqrt{x} < 1 + 0.1$. Squaring, we obtain $1 < x < 1 + 2(0.1) + (0.1)^2$. A candidate for δ is thus $2(0.1) + (0.1)^2$. So we let $\delta = 0.21$.

Now we prove that this δ works. Assume that $1 < x < 1 + \delta$. Therefore, $1 < x < 1.21$. Recall that $0 < a < b$ implies $\sqrt{a} < \sqrt{b}$. Hence, from $1 < x < 1 + \delta$, we obtain $1 < \sqrt{x} < \sqrt{1.21}$; that is, $1 < \sqrt{x} < 1.1$.

In Example 7, what we really proved is that there is a $\delta > 0$ such that *for all* x, if $1 < x < 1 + \delta$, then $1 < \sqrt{x} < 1.1$. That is, there is a hidden universal quantifier in this statement. The word "whenever" used in this context indicates that there is a universal quantifier involved.

Practice Problem 3. Prove that there is a real number $\delta > 0$ such that $2 < \sqrt{x} < 2.1$ whenever $4 < x < 4 + \delta$.

In the foregoing examples and practice problems, we illustrated one method of proving a statement of the form $\exists x p(x)$. However, sometimes this method cannot be used. Supplementary Exercises 11 and 12 require other techniques.

Proofs Involving Universal Quantifiers

The chief strategy for proving $\forall x p(x)$ is to pick an arbitrary x in the universe and then verify $p(x)$.

Example 8

Prove: For all real numbers x, $2(x + 3) + x = 3(x + 1) + 3$.

Proof Let x be any real number. Then $2(x + 3) + x = 2x + 6 + x = 3x + 6$. Also, $3(x + 1) + 3 = 3x + 3 + 3 = 3x + 6$. Hence, both sides of the equation are equal to $3x + 6$.

Practice Problem 4. *Prove*: For all real numbers x, $(x + 1)^2 + (x - 1)^2 = 2(x^2 + 1)$.

If we examine the proofs in Sections 2.1 and 2.2, we see that they are really proofs involving universal quantifiers. For example, Example 1 of Section 2.2 really should be read "for all real numbers x and y, if $x \leq y$ and $y \leq x$, then $x = y$." However, it is customary to omit universal quantifiers that precede a statement. So, for example, instead of "for all x and for all y, $x \leq y$ and $y \leq x$ implies $x = y$," we simply write "$x \leq y$ and $y \leq x$ implies $x = y$." In fact, all of the examples and practice problems of Sections 2.1 and 2.2 illustrate proofs involving universal quantifiers.

In Section 2.3, all of the proofs by mathematical induction are proofs involving universal quantifiers. Example 1 of Section 2.3 can be written "for all positive natural numbers n, $1 + 3 + 5 + ... + (2n - 1) = n^2$." All of the other statements proved by mathematical induction may be similarly written.

In order to refute a conjecture of the form $\exists x p(x)$, we must prove $\forall x \neg p(x)$. So to disprove a conjecture involving an existential quantifier, we must supply a proof involving a universal quantifier. Several examples are given in Exercises 1–3 of the Supplementary Exercises.

Conjectures

While doing research in mathematics, you are faced with the unknown. The situation is not like exercises in a textbook, where you are usually asked to prove a given statement. Instead, you encounter statements that may be provable or perhaps have counterexamples. At this stage, you have a conjecture, which must be studied until you indeed have a proof or a counterexample.

In the following two examples, you are not told whether to supply a proof or to give a counterexample for a given assertion; rather, you are asked to test a given conjecture. These examples and Practice Problem 5 involve real numbers.

Example 9

Conjecture: If $x < y$, then $x^2 < y^2$.

Solution   To test this conjecture, we can try different values substituted for x and y. First, we derive the negation. The negation is "there is an x and a y such that $x < y$ and $x^2 \not< y^2$." Consider $x = 2$ and $y = 3$. If $2 < 3$, then $2^2 < 3^2$ is certainly true. Also, $2 < 3$ and $2^2 \not< 3^2$ is false. So $x = 2$ and $y = 3$ satisfy the conjecture. Note that a choice such as $x = 3$ and $y = 2$ satisfies the conjecture vacuously. At this point, we might be tempted to try a proof of the conjecture. However, it is wise to try a wide variety of examples before attempting a proof.

Take $x = -2$ and $y = 1$. The negation $-2 < 1$ and $(-2)^2 \not< 1^2$ is true. Hence, $x = -2$ and $y = 1$ is a counterexample. Therefore, the conjecture is false.

Example 10

Conjecture: If $0 < y$, then $\sqrt{y} < y$.

Solution   We try several positive real numbers. For $y = 9$, $y = 4$, and $y = \pi$, the conjecture holds. In fact, if we substitute any $y > 1$, the conjecture holds. However, the conjecture is false, since $y = 1/9$ is a counterexample.

Practice Problem 5. Following are three conjectures. Supply a proof or a counterexample, as appropriate. Remember that the universal quantifiers before each statement are omitted.

    a. If $0 < x < y$, then $x^2 < y^3$.

    b. If $x < y$, then $x^3 < y^3$.

    c. If $x < y$, then $\sin x < \sin y$.

EXERCISE SET 2.5

1. Prove that there is a real number x such that $x^2 - 5x + 6 = 0$.
2. Prove that there is a real number x such that $x^3 + 5x + 6 = 0$.

Assume that x and y are integers with $x \neq 0$. We say that x **divides** y if there exists an integer k such that $y = k \cdot x$. When x divides y, we also say that y is a **multiple of** x. Note that y is even if and only if 2 divides y.

3. *Prove*:
    (a) 9 divides 486.
    (b) If 9 divides a and 9 divides b, then 9 divides $(a + b)$.

4. *Prove*:
    (a) b divides 0.
    (b) 1 divides b.

5. Prove that if d divides a, then d divides $a \cdot b$.
6. Prove that if d divides a and d divides b, then d divides $(a + b)$.
7. Prove that there is a real number $\delta > 0$ such that $2 < \sqrt{x} < 2.01$ whenever $4 < x < 4 + \delta$.
8. Prove that there is a real number $\delta > 0$ such that $3 < \sqrt{x} < 3.01$ whenever $9 < x < 9 + \delta$.
9. Let d be a nonnegative real number. Prove that there is a real number $\delta > 0$ such that $d < \sqrt{x} < d + .01$ whenever $d^2 < x < d^2 + \delta$.
10. Let c be a nonnegative real number. Prove that there is a real number $\delta > 0$ such that $\sqrt{c} < \sqrt{x} < \sqrt{c} + .01$ whenever $c < x < c + \delta$.
11. Prove that for any real number $\epsilon > 0$, there is a real number $\delta > 0$ such that $3 < \sqrt{x} < 3 + \epsilon$ whenever $9 < x < 9 + \delta$.
12. Let c be a nonnegative real number. Prove that for any real number $\epsilon > 0$, there is a real number $\delta > 0$ such that $\sqrt{c} < \sqrt{x} < \sqrt{c} + \epsilon$ whenever $c < x < c + \delta$.

In Exercises 13–18, find a counterexample or a proof for the given conjectures.

13. For all real numbers x, $x^{\frac{1}{3}} < x$.
14. If $x \cdot y$ is a nonnegative real number, then $x + 4y \geq 4\sqrt{xy}$.
15. If $x < y$ and $xy \neq 0$, then $1/y < 1/x$.
16. If x divides y, then $x \leq y$.
17. If d divides $a \cdot b$, then d divides a or d divides b.
18. $(1 + x)^n \geq 1 + n \cdot x$ for $n = 0, 1, 2, 3, \ldots$.

KEY CONCEPTS

Basic proof techniques: direct proof,
  proof by cases, indirect proof
Mathematical induction: basis step,
  induction step, induction hypothesis
Definition by recursion
Predicate
Universal quantifier

Existential quantifier
Use of quantifiers: definitions,
  theorems, counterexamples
Validity
Restricted quantifier
Conjecture
Counterexample

PROOFS TO EVALUATE

In each of the following exercises, a conjecture is given together with an argument proposed as a proof of the conjecture. Determine whether the conjecture is correct. If the conjecture is incorrect, find the flaw or flaws in the argument and supply a counterexample. If the conjecture is correct, determine whether the argument is valid. If the argument is invalid, find the flaws and correct them. If the argument is valid but not clear or complete, add clarifying comments.

1. *Conjecture*: If $x < 0$, then $1/x < 0$.

   Argument: Since $x \neq 0$, $x^2 > 0$. Assume $1/x < 0$. Then $x^2 \cdot (1/x) < x^2 \cdot 0$. Hence, $x < 0$. Therefore, $x < 0$ implies $1/x < 0$.

2. *Conjecture*: If $x < y$, then $x^3 < y^4$.

   Argument: First, note that $y^4 \geq 0$. We do a proof by cases. *Case 1*. Assume $x < 0$. Then $x^3 < 0$. Hence, $x^3 < y^4$. *Case 2*. Assume $x \geq 0$. Then $x^3 < y^3$, since $0 \leq x < y$. Hence, $x^3 < y^3 \cdot y = y^4$, since $x < y$.

3. *Conjecture*: If $x \cdot y$ is even, then x is even or y is even.

   Argument: We prove the contrapositive. Assume x is odd or y is odd. By Exercise 11 in Exercise Set 2.2, if $x \cdot y$ is odd, then x is odd and y is odd. But this contradicts our assumption. Therefore, $x \cdot y$ is even.

4. *Conjecture*: m is odd if and only if m^4 is odd.

   Argument: ($\Rightarrow$) Assume m is even and m^4 is odd. By Exercise 14 of Exercise Set 2.2, an integer x is even if and only if x^2 is even. Hence, m^2 is even. But since m^2 is even, $(m^2)^2 = m^4$ is even. Thus, we have a contradiction. ($\Leftarrow$) Assume m is odd and m^4 is even. Since m^4 is even, m^2 is even by Exercise 14 of Exercise Set 2.2. Similarly, m is even, and we have a contradiction.

5. *Conjecture*: $\pi + \sum_{i=1}^{n}(2i - 1) = n^2$, for $n = 1, 2, 3, \ldots$.

   Argument: For the induction hypothesis, we assume $\pi + \sum_{i=1}^{k}(2i - 1) = k^2$. Then $\pi + \sum_{i=1}^{k+1}(2i - 1) = \pi + \sum_{i=1}^{k}(2i - 1) + [2(k + 1) - 1] = \pi + k^2 + 2k + 1 = \pi + (k + 1)^2$.

6. *Conjecture*: All the people in any collection of n people have the same height, for $n = 1, 2, \ldots$.

   Argument: All people in any collection containing one person certainly have the same height.

Assume that all people in any collection containing k people have the same height. Let $A = \{p_1, p_2, \ldots, p_k, p_{k+1}\}$ be an arbitrary collection of $k + 1$ people. Then $B = \{p_1, p_2, \ldots, p_k\}$ and $C = \{p_2, \ldots, p_k, p_{k+1}\}$ are both collections containing k people. So all the people in B have the same height, and all the people in C have the same height. But p_2 is in both collections. Hence, all the people in A have the same height as p_2. Therefore, all the people in A have the same height.

By the principle of mathematical induction, all people in any collection of n people have the same height, for $n = 1, 2, \ldots$.

REVIEW EXERCISES

1. Assume that x, y, $z > 0$. Prove that if $xy < z$ and $\sqrt{z} < x$, then $y < \sqrt{z}$.
2. Prove that m is odd if and only if m^3 is odd.

The following applies to Exercises 3–6. In the time of Pythagoras, there were three means (averages): the arithmetic, the geometric, and the harmonic mean. For positive numbers x and y these means are defined, respectively, as

$$A = \frac{x + y}{2} \qquad G = \sqrt{x \cdot y} \qquad H = \frac{2xy}{x + y}$$

3. Assume that $x < y$. In Example 6 of Section 2.5, we proved that $x < A < y$. Now prove that:
   (a) $x < G < y$.
   (b) $x < H < y$.
4. *Prove*:
   (a) $A = x$ if and only if $x = y$.
   (b) $G = x$ if and only if $x = y$.
   (c) $H = x$ if and only if $x = y$.
5. *Prove:*
   (a) $A \geq G$.
   (b) $G \geq H$.
6. Prove that $H = G = A$ if and only if $x = y$.
7. Use mathematical induction to prove that

$$\sum_{i=1}^{n} \frac{1}{(i+1)^2 - 1} = \frac{3}{4} - \frac{1}{2(n+1)} - \frac{1}{2(n+2)} \qquad \text{for} \quad n = 1, 2, 3, \ldots.$$

8. Prove that

$$\sum_{i=1}^{2n} \frac{(-1)^{i+1}}{i} = \sum_{i=n+1}^{2n} \frac{1}{i} \qquad \text{for} \quad n = 1, 2, 3, \ldots.$$

9. Assume $x > -1$. Prove that $(1+x)^n \geq 1 + n \cdot x$ for $n = 0, 1, 2, 3, \ldots$.

10. Prove that

$$\prod_{k=0}^{n} \cos 2^k x = \frac{\sin 2^{n+1} x}{2^{n+1} \sin x} \qquad \text{for} \quad n = 0, 1, 2, \ldots$$

(*Hint:* Use $\sin 2\theta = 2 \sin \theta \cos \theta$.)

11. This exercise requires a basic acquaintance with complex numbers. Recall, for example, that $i^2 = -1$. Prove that $(\cos x + i \cdot \sin x)^n = \cos nx + i \cdot \sin nx$. (*Hint:* $\cos(\theta + \tau) = (\cos \theta)(\cos \tau) - (\sin \theta)(\sin \tau)$ and $\sin(\theta + \tau) = (\sin \theta)(\cos \tau) + (\sin \tau)(\cos \theta)$.)

12. Prove that $(n!)^4 \leq 2^{n(n+1)}$ for $n = 0, 1, 2, \ldots$. (*Hint:* You may wish to treat the cases $n \leq 2$ and $n \geq 3$ separately. For the induction step, use Exercise 4 of Section 2.3 in the form $(k+1)^2 \leq 2^{k+1}$, $k \neq 2$, with the induction hypothesis.)

13. Prove that $(2n)! < 2^{2n} (n!)^2$ for $n = 1, 2, \ldots$.

The following applies to Exercises 14–16. According to an ancient (perhaps apocryphal) fable, Brahman priests were given a platform with three diamond needles at the time the world was created. On the first needle were 64 gold disks, with each disk slightly larger than the one above it (see Figure 2–3). The priests were given the task of moving the disks to another needle while not violating certain rules. It was said that the world would end when the task was completed. Based on this story, a game called the **towers of Hanoi** was invented.

Figure 2–3   Towers of Hanoi

In this game, there are three posts with n disks stacked on the first post. Each disk is larger than the one above it. The object of the game is to move all the disks to one of the other posts subject to the conditions that (1) only one disk can be moved at a time and (2) no disk can be placed on top of a smaller disk. Let T_n be the minimum number of moves it takes to move the n disks from the first post to another post.

a. Find T_1, T_2, T_3, and T_4.

b. Obtain a recursion formula.

c. Guess a formula for T_n.

d. Use mathematical induction to prove the formula correct.

Review Exercises

Solution:

    a. With some experimentation, we see that $T_1 = 1$, $T_2 = 3$, $T_3 = 7$, and $T_4 = 15$.

    b. Suppose we have moved k disks to either the second or third post, using T_k moves. We can then move the $(k+1)$th disk to the unoccupied post. It then takes T_k moves to move the k disks on top of the $(k+1)$th disk. Hence, we obtain the recursion formula $T_{k+1} = 2T_k + 1$ for $k = 0, 1, 2, \ldots$.

    c. Based on the data in part a, we conjecture that $T_n = 2^n - 1$.

    d. The basis step is $T_1 = 1 = 2^1 - 1$. For the induction hypothesis, we assume $T_k = 2^k - 1$. Then, by the recursion formula and the induction hypothesis, $T_{k+1} = 2T_k + 1 = 2(2^k - 1) + 1 = 2^{k+1} - 1$.

    Assuming the fable concerning the towers of Hanoi is true, how long before the world ends? Suppose the priests take the minimum number of moves, $2^{64} - 1$. Recall that $2^{10} = 1,024 > 10^3$. Then $2^{64} = 2^4 \cdot 2^{60} = 16 \cdot (2^{10})^6 > 16 \cdot (10^3)^6 = 1.6 \cdot 10^{19}$. Furthermore, suppose they move a disk every second. Each year has about $3.2 \cdot 10^7$ seconds. Therefore, the task will take more than $5 \cdot 10^{11}$ years, in other words, 500 billion years. Since 500 billion years is approximately 50 times the estimated age of the universe, the world still has many years left.

    In Exercises 14–16 use the method discussed in the towers of Hanoi game.

  14. If all the people in a group of n people shake hands with one another, how many handshakes occur? Let H_n be the number of handshakes among n people.
    (a) Assume $H_1 = 0$. Find H_2, H_3, H_4, and H_5.
    (b) Find a recursion formula, $H_{k+1} = $_____, for $k = 1, 2, \ldots$. Justify your conjecture. (*Hint:* Note that $H_3 = 2 + H_2$.)
    (c) Conjecture a formula for H_n. Use mathematical induction and the recursion formula found in part b to prove your conjecture.

  15. Given n lines in the plane such that no two lines are parallel and no three lines intersect, let R_n be the number of regions into which the plane is divided by the n lines.
    (a) Assume $R_0 = 1$. Find R_1, R_2, R_3, and R_4.
    (b) Use geometrical reasoning to show that $R_{k+1} = R_k + (k+1)$ for $k = 0, 1, 2, 3, \ldots$.
    (c) Guess a formula for R_n.
    (d) Use mathematical induction and the recursion formula in part b to prove your formula correct.

  16. In an alternative version of the towers of Hanoi game, the rules are the same as in the original game, except that the object of the game is to move the disks to the third post under the additional condition that disks may only be moved to adjacent posts. Let S_n be the minimum number of moves it takes to move the n disks from the first post to the third post.
    (a) Find S_1, S_2, S_3, and S_4.
    (b) Find a recursion formula $S_{k+1} = $_____, for $k = 0, 1, 2, \ldots$.
    (c) Guess a formula for S_n.
    (d) Use mathematical induction to prove your formula correct.

  17. Prove that Q6, $\exists x(p(x) \wedge q(x)) \Rightarrow \exists x p(x) \wedge \exists x q(x)$, is a validity.

  18. Prove that Q7, $\forall x p(x) \vee \forall x q(x) \Rightarrow \forall x(p(x) \vee q(x))$, is a validity.

  19. Find a counterexample to the converse of Q6. A counterexample in this case must include a specified universe and predicates $p(x)$ and $q(x)$.

20. Find a counterexample to the converse of Q7. A counterexample in this case must include a specified universe and predicates $p(x)$ and $q(x)$.

SUPPLEMENTARY EXERCISES

From Section 2.5, we know that to show a conjecture of the form $\forall x p(x)$ is false, we use a counterexample. By contrast, to show a conjecture of the form $\exists x p(x)$ is false, we must prove $\forall x \neg p(x)$.

1. Show that the following conjecture is false: There is a real number x such that $x^2 - 2x + 2 = 0$. (*Hint*: Note that $x^2 - 2x + 2 = (x - 1)^2 + 1$.)

2. Show that the following conjecture is false: There is a real number x such that $x^2 + 6x + 10 = 0$.

3. Show that the following conjecture is false: There are positive integers m and n such that m is odd and $2 \cdot m^2 = n^2$. (*Hint*: Recall that if the square of an integer is even, the integer is also even.)

4. Prove that $\sqrt{2}$ is not a rational number. (*Hint*: Suppose that $\sqrt{2} = n/m$, where n and m are positive integers. Without loss of generality, assume that n and m are not both even. Use the result of Exercise 3.)

The *well-ordering principle* for the natural numbers states that every nonempty subset of $\mathcal{N}$ has a least element.

5. Prove that the well-ordering principle does not hold for the real numbers. (For example, the set $A = \{x : 0 < x < 2\}$ does not have a least element.)

Another version of the principle of mathematical induction is called the *principle of mathematical induction for sets*: Let A be a set of natural numbers. If $0 \in A$, and for all natural numbers k, $k \in A$ implies that $k + 1 \in A$, then $A = \mathcal{N}$.

6. Show that the principle of mathematical induction for sets implies the principle of mathematical induction with basis step $n = 0$ stated in Section 2.3. (*Hint*: Let $A = \{k : k \in \mathcal{N}$ and $p(k)\}$, and note that for all natural numbers k, $k \in A$ if and only if $p(k)$.)

7. Show that the well-ordering principle for natural numbers implies the principle of mathematical induction for sets. (*Hint*: First prove the principle of mathematical induction for sets by assuming that $0 \in A$ and for all $k \in \mathcal{N}$, $k \in A$ implies $k + 1 \in A$. Then let $B = \{n : n \notin A\}$, and use an indirect proof and the well-ordering principle to prove that $B = \emptyset$. You may use the fact that if m is a natural number and $m \neq 0$, then $m = k + 1$, for some natural number k.)

The *principle of strong mathematical induction* states that for A a set of natural numbers, if k is in A whenever all natural numbers less than k are in A, then $A = \mathcal{N}$. Examples using this form of induction can be found in Section 8.4.

8. Prove that the principle of strong mathematical induction implies the following: If $p(q)$ for all $q < k$ implies $p(k)$, then for all n, $p(n)$. (*Hint*: Let $A = \{k : k \in \mathcal{N}$ and $p(k)\}$.)

9. Prove that the principle of mathematical induction for sets implies the principle of strong mathematical induction. (*Hint*: Assume the principle of mathematical induction for sets. To prove strong mathematical induction, assume, in addition, that $q \in A$ for all $q < k$

implies $k \in A$. Let $B = \{n : n \in \mathcal{N}$ and for all $j < n, j \in A\}$. Now use the principle of mathematical induction for sets to show that $B = \mathcal{N}$. From this, it follows that $A = \mathcal{N}$.)

10. Prove that the principle of strong mathematical induction implies the well-ordering principle for natural numbers. (*Hint*: Assume that a set of natural numbers B has no least element. Let $A = \{n : n \notin B\}$. Use the principle of strong mathematical induction to show that $A = \mathcal{N}$. Hence, $B = \emptyset$.)

Recall that the chief way to prove a statement of the form $\exists x p(x)$ is to find an x and verify that it satisfies $p(x)$. It is also possible to prove such a statement without actually finding a specific element. To illustrate this idea, finish the proof in Exercise 11, and give a proof in Exercise 12.

11. Prove that every integer greater than one has a prime divisor. (A *prime* is a positive integer with exactly two positive divisors, itself and one.):

Proof: Assume n is any positive integer greater than one. Let $A = \{m : m > 1$ and m divides $n\}$.

Step 1. Prove that A has a least element u.

Step 2. Prove that u is a prime.

12. Prove that $x^5 + x^3 - x + 1 = 0$ has a real root. Use the following two theorems of algebra:

Theorem A. Every polynomial of degree n has n complex roots. A multiple root is counted once for each time it occurs.

Theorem B. The complex roots that are not real occur in pairs.

13. Prove that the number of primes is infinite. We outline the proof and ask that you complete each step. The proof is attributed to Euclid. It may be the most famous example of an indirect proof.

Proof: Suppose there is a finite number k of primes. We list these as $p_1, p_2, \ldots, p_k$. Let $q = 1 + p_1 \cdot p_2 \cdots p_k$. By the result of Exercise 11, q has a prime divisor p. Prove that p is not equal to p_i for any $i = 1, 2, \ldots, k$. This contradicts the assumption that the list of primes $p_1, p_2, \ldots, p_k$ includes all primes.

SOLUTIONS TO PRACTICE PROBLEMS

Section 2.1

1. *Proof:*

1. $x < y$	(Assumption)
2. $u > 0$	(Assumption)
3. $xu < yu$	(1, 2, A4)
4. $u < v$	(Assumption)
5. $0 < x$	(Assumption)
6. $0 < y$	(1, 5, A1)
7. $yu < yv$	(4, 6, A4)
8. $xu < yv$	(3, 7, A1)

2. *Proof:*   1. $u < v$ (Assumption)

2. $z < 0$ (Assumption)

3. $zu > zv$ (1, 2, Example 3)

4. $x < z$ (Assumption)

5. $v < 0$ (Assumption)

6. $u < 0$ (1, 2, A1)

7. $xu > zu$ (4, 6, Example 3)

8. $xu > zv$ (3, 7, A1)

3. *Proof:* ($\Rightarrow$) Assume $7x < 7y$. Since $1/7 > 0$, $(1/7)(7x) < (1/7)(7y)$ by A4. Hence $1 \cdot x < 1 \cdot y$. Therefore, $x < y$. ($\Leftarrow$) Assume $x < y$. Since $7 > 0$, $7x < 7y$ by A4.

4. *Proof:* Let y be an arbitrary real number. We use cases.

Case 1: $y \geq 0$. Therefore, $3 \cdot y \geq 0$. Hence, $|3 \cdot y| = 3y = 3 \cdot |y|$.

Case 2: $y < 0$. Therefore, $3 \cdot y < 0$. Hence, $|3 \cdot y| = -3y = 3(-y) = 3 \cdot |y|$.

Section 2.2

1. *Proof:* Assume $x > 0$, $y > 0$, and $x < y$. Assume the negation of the conclusion, in other words, $1/x \leq 1/y$.

Case 1. Assume $1/x < 1/y$. Since $x > 0$ and $y > 0$, by A4, we have $xy > 0$. Also, by A4, we obtain $xy(1/x) < xy(1/y)$. That is, $y < x$. Since $x < y$ and $y < x$, $x < x$, a contradiction.

Case 2. Assume $1/x = 1/y$. By algebra, $x = y$. By Example 2, $x \not< y$. But this contradicts $x < y$.

2. *Proof:* Assume $mn < 100$. Assume the negation of the conclusion. Therefore, $m \geq 10$ and $n \geq 10$. Hence, $mn \geq 100$. (Why?) This contradicts the assumption.

Section 2.3

1. (a) $p(1)$ is $2^0 = 1 = 2^1 - 1$; $p(2)$ is $1 + 2 = 2^2 - 1$; $p(3)$ is $1 + 2 + 2^2 = 2^3 - 1$.

(b) *Proof:*

Basis step ($n = 1$). The sum of the first (one) summand on the left is 1, and $1 = 2^1 - 1$.

Induction hypothesis. Assume $1 + 2 + 2^2 + \ldots + 2^{k-1} = 2^k - 1$. Then $1 + 2 + 2^2 + \ldots + 2^{k-1} + 2^k = (2^k - 1) + 2^k = 2 \cdot 2^k - 1 = 2^{k+1} - 1$.

2. *Proof:*

Basis step ($n = 1$).

$$1^2 = 1 = \frac{1(1+1)(2 \cdot 1 + 1)}{6}$$

Induction hypothesis. Assume

$$1^2 + 2^2 + \ldots + k^2 = \frac{k(k+1)(2k+1)}{6}.$$

Then

$$1^2 + 2^2 + \ldots + k^2 + (k+1)^2 = [1^2 + 2^2 + \ldots + k^2] + (k+1)^2$$

$$= \frac{k(k+1)(2k+1)}{6} + (k+1)^2$$

$$= \frac{(k+1)(k+2)(2k+3)}{6}.$$

3. *Proof:*

Basis step ($n = 3$). $2 \cdot 3 + 1 = 7 \leq 8 = 2^3$.

Induction hypothesis. Assume $2k + 1 \leq 2^k$. We do a needs assessment. We wish to show that $2(k+1) + 1 \leq 2^{k+1}$. That is, $2k + 1 + 2 \leq 2 \cdot 2^k$. Since $2 \cdot 2^k = 2^k + 2^k$, we wish to show that $2k + 1 + 2 \leq 2^k + 2^k$. So by the induction hypothesis, we need to show that $2 \leq 2^k$. But this is true for $k = 1, 2, 3, \ldots$. For the proof of the induction step, $2 \leq 2^k$ for $k = 1, 2, 3, \ldots$. Therefore, by the induction hypothesis, we have $2k + 1 + 2 \leq 2^k + 2^k$. Hence, $2k + 2 + 1 \leq 2 \cdot 2^k$, from which it follows that $2(k+1) + 1 \leq 2^{k+1}$.

4. We do the case $n = 3$:

(a) Assume that we have already shown that $\sum_{i=1}^{2} a_i = a_1 + a_2$. Then

$$\sum_{i=1}^{3} a_i = \sum_{i=1}^{2} a_i + a_3 = a_1 + a_2 + a_3$$

Section 2.4

1. (a) $\{-3\}$
   (b) $\emptyset$
   (c) $\{-\sqrt{2}, \sqrt{2}\}$
   (d) $\{-1/2, 1\}$
   (e) $\{1\}$
   (f) $\mathcal{R}$

2. (a) T
   (b) F
   (c) F, for example $\sin(\pi/2) = 1$
   (d) T

3. u is a greatest element for a set A of real numbers if $u \in A \land \forall x(x \in A \Rightarrow x \leq u)$ is true in $\mathcal{R}$.

4. The statement $\forall x \exists y(x + y = 0)$ is true because, for $U = \mathcal{Z}$, for all x, there is an additive inverse $-x$ such that $x + (-x) = 0$. The statement $\exists y \forall x(x + y = 0)$ is false because when y is chosen first, there exists an element x such that $x + y \neq 0$. For example, $x = 1 - y$ will make $x + y \neq 0$.

5. The statement $\neg \exists x(p(x))$ is read, "There is no element x such that $p(x)$ is true." $\forall x(\neg p(x))$ is read, "For every x, $p(x)$ is false."

6. Let $U = \{0, 1\}$. Now we write Q3: $\neg \exists x(p(x))$ is true if and only if $\neg(p(0) \lor p(1))$ is true. By De Morgan's law, $\neg(p(0) \lor p(1))$ is true if and only if $(\neg p(0)) \land (\neg p(1))$ is true. But $(\neg p(0)) \land (\neg p(1))$ is true if and only if $\forall x \neg p(x)$ is true.

7. **(a)**

$$\neg\exists x\exists p(x+p=0)\Leftrightarrow \forall x[\neg\exists p(x+p=0)]\Leftrightarrow \forall x\forall p(x+p\neq 0)$$

(c)

$$\neg\exists y\forall x(x+y=0)\Leftrightarrow \forall y[\neg\forall x(x+y=0)]\Leftrightarrow \forall y\exists x(x+y\neq 0)$$

Statements a and b are true. The negations of c and d are also true.

8. *Proof:*

$$\neg(\exists x\in S)p(x)\Leftrightarrow \neg\exists x(x\in S\wedge p(x))$$

$$\Leftrightarrow \forall x\neg(x\in S\wedge p(x))$$

$$\Leftrightarrow \forall x[\neg(x\in S)\vee \neg p(x))$$

$$\Leftrightarrow \forall x[x\in S\Rightarrow \neg p(x))$$

$$\Leftrightarrow (\forall x\in S)\neg p(x)$$

Section 2.5

1.

$$\neg\forall x\forall y(x\text{ is odd }\Rightarrow xy\text{ is odd })\Leftrightarrow \exists x\exists y\neg(x\text{ is odd }\Rightarrow xy\text{ is odd })$$

$$\Leftrightarrow \exists x\exists y(x\text{ is odd and }xy\text{ is not odd })$$

$$\Leftrightarrow \exists x\exists y(x\text{ is odd and }xy\text{ is even })$$

$x=3$ and $y=4$ is a counterexample.

2. *Proof:* Let $x=8$. Then $x^2=64=4^3$. Therefore, x^2 is the cube of 4.

3. *Needs assessment*: Examine $2<\sqrt{x}<2.1$. Squaring, we obtain $4<x<(2+0.1)^2$, that is, $4<x<4+0.4+0.01$. The needs assessment suggests that we let $\delta=0.41$. We now verify that this in fact works. *Proof:* Assume $4<x<4+0.41$. Therefore, $4<x<4.41$. Taking square roots, $2<\sqrt{x}<\sqrt{4.41}$. Hence, $2<\sqrt{x}<2.1$.

4. *Proof:* Let x be a real number. Then $(x+1)^2+(x-1)^2=x^2+2x+1+x^2-2x+1=2x^2+2=2(x^2+1)$.

5. **(a)** If $0<x<y$, then $x^2<y^3$.
Solution: This is false, since $x=\frac{1}{4}$ and $y=\frac{1}{3}$ is a counterexample.
(b) If $x<y$, then $x^3<y^3$.
Solution: This is true. Its proof is given as Exercise 2 in Section 2.2.
(c) If $x<y$, then $\sin x<\sin y$.
Solution: This is false, since $x=0$ and $y=\pi$ is a counterexample.

3

Set Theory

In the preface to his classic text, *Naive Set Theory* (see references), Paul Halmos states: "Every mathematician agrees that every mathematician must know some set theory." This is so for two reasons: (1) Set theory provides a foundation for mathematics, because essentially every object of study in mathematics can be interpreted as a set. (2) In addition, much mathematics is done in the language of elementary set theory.

3.1 INTRODUCTION TO SETS

Intuitively, a **set** is a collection of objects. The objects are called **elements** or **members** of the set. For example, all the sequoia trees in California form a set, and each sequoia tree is an element or a member of this set.

There are several ways we can denote a set. One way is to list the elements separated by commas inside two braces, as shown in the following example. This is called the **list form** of a set.

Example 1

$A = \{a, b, c, d, e\}$ is the set containing the letters a, b, c, d, and e.

$L = \{a, b, c, ..., x, y, z\}$ is the set containing the letters of the English alphabet. The three dots between c and x are called an *ellipsis*. We use an ellipsis for brevity when it is clear from the context what elements are included in the set.

$D = \{0, 1, 2, ..., 9\}$ is the set of digits.

Following is a basic assumption about sets.

Extensionality. Two sets are equal if and only if they include the same elements.

As a consequence of extensionality, the set $\{a, b, c\}$ is equal to the set $\{c, a, b\}$, since they have the same elements. Order is thus unimportant when representing a set in list form. Also, the set $\{a, b, c\}$ is equal to the set $\{a, b, c, b\}$, since both sets have the same elements. Thus, repetition is permitted, and $\{a, b, c\}$, $\{c, a, b\}$, and $\{a, b, c, b\}$ denote the same set. However, we usually do not write $\{a, b, c, b\}$ to designate $\{a, b, c\}$.

Given a set A, $x \in A$ means x **is an element** or **is a member** of A. We write $x \notin A$ for x **is not an element** of A or x **is not a member** of A. For example, let $A = \{1, 2, 3\}$. Then $1 \in A$ but $4 \notin A$. When $x \in A$, we frequently say that x is in A.

Some sets are used frequently, so special letters are reserved for them. For example,

$\mathcal{N} = \{0, 1, 2, 3, ...\}$, the **natural numbers** or **nonnegative integers**.

$\mathcal{Z} = \{..., -3, -2, -1, 0, 1, 2, 3, ...\}$, the **integers**.

$\mathcal{Q} =$ the set of **rational numbers**, that is, quotients of integers.

$\mathcal{R} =$ the set of **real numbers**.

$\mathcal{R}^+ =$ the set of **positive real numbers**.

For now, we accept as a fact that for any property, there is a set with exactly those elements that satisfy the property. As we shall see in Example 7, there is a problem with this assertion, which will be subsequently rectified. If A and B are two sets whose elements satisfy the same property, then A and B have the same elements, and hence, by extensionality, $A = B$. Therefore, there is a unique set whose elements satisfy a given property.

By the previous paragraph, we may designate a set by stating the property or properties that are satisfied by the elements of the set. We use **set-builder** notation to denote the set of elements satisfying a given property. For example, $C = \{x : x \in \mathcal{N} \text{ and } x < 13\} = \{0, 1, 2, 3, 4, 5, 6, 7, 8, 9, 10, 11, 12\}$. We say, "$C$ is the set of all x such that x is in $\mathcal{N}$ and x is less than 13."

Example 2

The set $E = \{x : x \in \mathcal{R} \text{ and } 0 < x < 5\}$ describes the set of all real numbers greater than 0 and less than 5. This set has infinitely many elements, and it is not possible to list all of its elements.

Practice Problem 1. Let $F = \{x : x \in \mathcal{N} \text{ and } x < 20\}$ and $G = \{x : x \in \mathcal{Z} \text{ and } -3 < x < 6\}$.

(a) Give the list form of F.

(b) Give the list form of G.

Example 3

Let $A = \{1, 2, 3, 4, 5, 6\}$ and $B = \{x : x \text{ is a positive integer less than } 7\}$. Show that $A = B$.

Solution The list form of B is $\{1, 2, 3, 4, 5, 6\}$. Hence, B has the same elements as A. Therefore, $A = B$.

Practice Problem 2. Let $S = \{x : x \in \mathcal{R} \text{ and } x^2 = 100\}$ and $T = \{10, -10\}$. Show that $S = T$.

A set may have no elements. Consider the set $E = \{x : x \in \mathcal{R} \text{ and } x^2 = -1\}$. Since the square of every real number is nonnegative, E has no elements.

By extensionality, there can only be one set with no elements. (Why?)

Definition. The unique set with no elements is called the **empty set** or the **null set** and is denoted by $\emptyset$.

Note that $\{0\}$ is a set containing the integer 0 and is not the same as the empty set $\emptyset$. The set $\{0\}$ contains exactly one element. Such a set is called a **singleton set.** For example, $\{a\}$, $\{*\}$, $\{9\}$ are all singleton sets. Note that the singleton set $\{a\}$ is not the same as the element a; that is, $\{a\} \neq a$. As Halmos (*Naive Set Theory*) has pointed out: A box containing a hat is not the same as the hat alone.

Definition. Given two sets A and B, if every element of A is also an element of B, then A is a **subset** of B, written $A \subseteq B$. If $A \subseteq B$ and $A \neq B$, we write $A \subset B$ and say that A is a **proper subset** of B.

When we write $A \subseteq B$ in the language of predicate logic, we obtain $(\forall x \in A)x \in B$. The negation $\neg(\forall x \in A)x \in B$ is equivalent to $(\exists x \in A)x \notin B$; that is, there is an element of A that does not belong to B. In this case, we write $A \nsubseteq B$. Now, since $\emptyset$ contains no elements, it can never be true that there is an element of $\emptyset$ that does not belong to the set B. Hence, it is not the case that $\emptyset \nsubseteq B$. Therefore, $\emptyset \subseteq B$. In other words, $\emptyset$ is a subset of any set B.

Since every element of A is an element of A, $A \subseteq A$ for any set A.

Example 4

Let $A = \{1, 2, 3, 4, 5, 6, 7\}$, $B = \{2, 4, 6\}$, and $C = \{a, b, 2, d, 8\}$. Then $B \subseteq A$, $B \subset A$, $B \neq C$, and $A \nsubseteq B$.

Practice Problem 3. Let $A = \{b, f, p, c, a\}$, $B = \{p, a\}$, and $C = \{f, p, c\}$. Which of the symbols $(\subset, \subseteq, \nsubseteq)$ yields a true statement when placed in the blanks?

      a. $B\_\_\_A$ b. $B\_\_\_C$ c. $C\_\_\_A$.

Example 5

Let $S = \{a, b\}$. Find all the subsets of S.

Solution The subsets of S are $\emptyset$, $\{a\}$, $\{b\}$, and S. Note that $\emptyset$ and S are always subsets of S.

Definition.   Let S be a set. The **power set** of S is $P(S) = \{X : X \subseteq S\}$

That is, the power set of a set S is the set of all subsets of S.

Example 6

Let $S = \{a, b, c\}$. There are eight subsets of S, namely, $\emptyset$, $\{a\}$, $\{b\}$, $\{c\}$, $\{a, b\}$, $\{a, c\}$, $\{b, c\}$, and $\{a, b, c\}$. Thus, $P(S) = \big\{\emptyset, \{a\}, \{b\}, \{c\}, \{a, b\}, \{a, c\}, \{b, c\}, \{a, b, c\}\big\}$.

A Logical Paradox

In 1895, Georg Cantor (1845–1918) gave the following definition of a set: "By a set we shall understand any collection into a whole of definite distinguishable objects of our intuition or thought. The objects will be called members of the collection." According to Cantor's definition, the elements in a set are "objects of our intuition." But a set itself is an object of our intuition and can therefore be an element of a set. For example, let D be the set of all sets of single digit natural numbers. Then, $\{2, 3, 7\} \in D$, $\{3, 4, 5, 6\} \in D$, etc.

Following is a set with sets as elements, which has an extraordinary property. Let T be the set of all sets with at least two elements. Then $\{1, 4\} \in T$, $\{a, b\} \in T$, and $\{a, 1, 3\} \in T$. Hence, T has at least two elements, and so, $T \in T$. In other words, the set T is an element of itself.

In 1902, the philosopher and mathematician Bertrand Russell (1872–1970) constructed an object of his intuition satisfying Cantor's definition of a set, but that led to a logical contradiction. This situation, called **Russell's paradox,** is illustrated in the following example.

Example 7

Russell's Paradox

For a set to be a member of itself is not the ordinary state of affairs. Bertrand Russell called a set that was not a member of itself an **ordinary set.** Let R be the set of all ordinary sets. Hence, $R = \{X : X \notin X\}$. Now, given a set and any object, it must be possible to determine whether the object belongs to the set. In particular, does R belong to itself? By the law of excluded middle, either $R \in R$ or $R \notin R$.

Case 1. Suppose $R \notin R$. Then R satisfies the property $X \notin X$. Hence, $R \in R$. But this is a contradiction.

Case 2. Suppose $R \in R$. Then R satisfies the property $X \notin X$; that is $R \notin R$, which is again a contradiction.

This exhausts all possible cases, and each case leads to a contradiction.

In the early 1900s, mathematicians were successfully showing how the theory of sets could be used as a foundation for much of mathematics. In particular, set theory was being used to construct the natural numbers and the real number system. That is why Russell's paradox was viewed as a catastrophe by the mathematical community. In fact, any result concerning sets that leads to a logical contradiction is unacceptable.

Russell's paradox forced mathematicians to examine carefully what was meant when the word "set" was used. Indeed, the paradox made clear that Cantor's definition and assumptions about sets were inadequate.

One solution to Russell's paradox is to require that elements of all sets being discussed also be elements of some given universal set U. That is, in any discussion of sets $A, B, \ldots$, the elements in $A, B, \ldots$ are assumed to belong to a fixed set, called the **universal set,** describing the context for discussion.

With the assumption of the existence of a universal set U, for a given predicate $p(x)$, a set is formed by collecting all elements $z \in U$ such that $p(z)$ is true. So sets are written $A = \{x : x \in U \text{ and } p(x)\}$, and $z \in A$ if and only if $z \in U$ *and $p(z)$*.

Set Construction. For any set U and any predicate $p(x)$, there is a set including exactly those elements from U that satisfy $p(x)$.

With the preceding convention, Russell's set of all ordinary sets must be written as $\{X : X \in U \text{ and } X \notin X\}$ for some universal set U. It is an exercise to show that the argument for Russell's paradox no longer leads to a logical contradiction.

The universal set is frequently called the *universe of discourse* and contains all the elements under discussion. Sometimes the universal set is unspecified and assumed to exist, while at other times the universal set must be defined specifically. We will usually denote the universal set by the capital letter U.

Set Operations

A *binary* operation on sets produces a set from two sets, and a *unary* operation produces a set from one set. We will discuss the unary operation, complement, and the binary operations, difference, intersection, and union.

Definition. Let A be a set and U be a universal set. The **complement** of A is $A' = \{x : x \in U \text{ and } x \notin A\}$.

Thus, the complement of a set is the set of all elements that are in the universe but not in the set.

Example 8

Let $U = \{0, 1, 2, 3, 4, 5, 6, 7, 8\}$ and $A = \{2, 4, 6, 8\}$. Then $A' = \{0, 1, 3, 5, 7\}$. For example, $1 \in A'$ since $1 \notin A$, and $2 \notin A'$ since $2 \in A$. Note that the choice of universal set is important. If, for example, we had chosen U to be the set of all digits, then A' would include 9 also.

Definition. Given two sets A and B, the **difference** set is $A - B = \{x : x \in A \text{ and } x \notin B\}$.

Thus, the difference set $A - B$ is the set of all elements that belong to A but not B. In general, $A - B$ and $B - A$ are not the same. Also, note that $U - B = B'$.

Example 9

Let $A = \{2, 4, 6, 8\}$ and $B = \{0, 1, 2\}$. Then $A - B = \{4, 6, 8\}$ and $B - A = \{0, 1\}$. So $6 \in A - B$ since $6 \in A$ and $6 \notin B$. However, $2 \notin A - B$ since $2 \in B$. Note that $A - B$ is defined whether or not B is a subset of A.

Practice Problem 4. Let $U = \{-5, -4, \ldots, 4, 5\}$, $A = \{-1, 0, 2, 4\}$, and $B = \{1, 2, 3, 4\}$. Find A', B', $A - B$, and $B - A$. Observe that $A - B \neq B - A$.

Definition. Given two sets A and B, the **intersection** of A and B is $A \cap B = \{x : x \in A \text{ and } x \in B\}$.

The intersection of two sets is the set of all elements that belong to both sets.

Example 10

Let $C = \{a, b, c, d, e, f\}$, $D = \{x, y, d, f\}$, $E = \{a, d, t\}$, and $F = \{s, t\}$. Then $C \cap D = \{d, f\}$, $C \cap E = \{a, d\}$, $E \cap F = \{t\}$, $D \cap F = \emptyset$, and $D \cap \emptyset = \emptyset$. Note that $d \in C \cap D$ since $d \in C$ and $d \in D$, but $x \notin C \cap D$ since $x \notin C$.

Definition. Two sets A and B are **disjoint** if $A \cap B = \emptyset$.

In other words, two sets are disjoint if they have no elements in common.
In Example 10, the sets D and F are disjoint, and the sets E and F are not disjoint.

Definition. Given two sets A and B, the **union** of A and B is $A \cup B = \{x : x \in A \text{ or } x \in B\}$.

That is, the union of two sets is the set of all elements that belong to one or the other set. In other words, an element is in $A \cup B$ if and only if it is in at least one of A and B.

Example 11

Let $A = \{1, 2, 3, 4, 5\}$ and $B = \{-3, -4, 3, 4\}$. Then $A \cup B = \{-3, -4, 1, 2, 3, 4, 5\}$ and $A \cap B = \{3, 4\}$. Thus, $1 \in A \cup B$ since $1 \in A$, $3 \in A \cup B$ since $3 \in B$, and $4 \in A \cup B$ since $4 \in A$.

Practice Problem 5. Let $A = \{a, b, c, d, e, f\}$ and $B = \{1, 2, d, f\}$. Find $A \cap \emptyset$, $B \cup \emptyset$, $A \cap B$, and $A \cup B$.

EXERCISE SET 3.1

1. Let $A = \{a, b, c, d, 2, 4, 6\}$. Answer true or false to the following statements.
   (a) $b \in A$
   (b) $7 \in A$

(c) $\emptyset \subseteq A$
   (d) $\{2, 4, 6, 8\} \subseteq A$
   (e) $A \in \{d\}$
   (f) $\{a, b, c, d\} \subseteq A$

2. Let $S = \{x : x \in \mathcal{N} \text{ and } x \leq 10\}$. Answer true or false to the following statements.
   (a) $5 \in S$
   (b) $13 \in S$
   (c) $S \subseteq \mathcal{N}$
   (d) $S \subseteq \mathcal{Z}$
   (e) $8 \notin S$
   (f) $4.5 \in S$

3. Let $A = \{x : x \in \mathcal{R} \text{ and } x^2 = 25\}$. List the elements of A.

4. Let $B = \{x : x \in \mathcal{R} \text{ and } x^2 - 6x + 8 = 0\}$. List the elements of B.

5. In each of the following, form a set whose elements are the letters of the given word.
   (a) NOON
   (b) MADAM
   (c) MATHEMATICS
   (d) RACECAR

6. Write the following sets in set-builder notation.
   (a) $A = \{2, 3, 4, 5, 6, 7\}$
   (b) $B = \{-3, 3\}$
   (c) $C = \{0, 1, 2, 3, 4, 5, 6, 7, 8\}$

7. For each of the following pairs of sets, determine whether the two sets are equal.
   (a) $\{1, 2, 3, 4\}, \{1, 3, 4, 4, 2\}$
   (b) $\{x : x \in \mathcal{R} \text{ and } x^2 - 3x - 18 = 0\}, \{3, -6\}$
   (c) $\{x : x \in \mathcal{R} \text{ and } |x| = 5\}, \{5\}$
   (d) $\{3, 4, 5, 6\}, \{x : x \in \mathcal{N} \text{ and } 2 < x < 7\}$

8. For each of the following pairs of sets, determine whether the two sets are equal.
   (a) $\{x : x \in \mathcal{N} \text{ and } 2 < x < 7\}, \{4, 5, 6, 7\}$
   (b) $\{x : x \in \mathcal{R} \text{ and } |x + 2| = -1\}, \{-3\}$
   (c) $\{x : x \in \mathcal{Z} \text{ and } -5 < x < 1\}, \{-4, -3, -2, -1, 0\}$
   (d) $\{2, 9\}, \{x : x \in \mathcal{N} \text{ and } x^2 - 11x + 18 = 0\}$

9. (a) List all subsets of the set $\{a\}$ and find $P(\{a\})$.
   (b) List all subsets of $\{a, b, c, d\}$ and find $P(\{a, b, c, d\})$.

10. Let $A = \{2, 6, 10\}$ and $B = \{6, 10, 33\}$.
    (a) Find $A \cup B$.
    (b) Find $A \cap B$.
    (c) Find $A \cup \emptyset$.
    (d) Find $B \cap \emptyset$.
    (e) Are A and B disjoint?

11. Let $C = \{a, b, c, d\}$ and $D = \{d, e, f\}$.
    (a) Find $C \cup D$.
    (b) Find $C \cap D$.
    (c) Find $D \cup \emptyset$.

(d) Find $\emptyset \cap C$.

(e) Are C and D disjoint?

12. Let $A = \{x : x \in \mathcal{R} \text{ and } x^2 - 3x + 2 = 0\}$ and $B = \{x : x \in \mathcal{R} \text{ and } x^2 - 2x + 1 = 0\}$.
    (a) Find $A \cap B$.
    (b) Find $A \cup B$.
    (c) Find $A - B$.
    (d) Find $B - A$.

13. Let $C = \{x : x \in \mathcal{R} \text{ and } x^2 + 1 = 0\}$ and $D = \{x : x \in \mathcal{R} \text{ and } x^2 - x - 6 = 0\}$.
    (a) Find $C \cup D$.
    (b) Find $C \cap D$.
    (c) Find $C - D$.
    (d) Find $D - C$.

14. Let $A = \{a, b, c\}$, $B = \{b, c\}$, and $C = \{a, e, f\}$. Determine which of the following statements are true.
    (a) $c \in A$ **(b)** $\{c\} \in A$ **(c)** $\{c\} \subseteq A$ **(d)** $B \subseteq A$
    (e) $A \subseteq B$ **(f)** $\emptyset \subseteq C$ **(g)** $\emptyset \in B$ **(h)** $A \cap C \subseteq B$

15. Let $A = \{1, 2, 3, 4\}$, $B = \{1, 4\}$, and $C = \{4, 7, 9, 10\}$. Determine which of the following statements are true.
    (a) $4 \in B$ **(b)** $\{4\} \in B$ **(c)** $\{4\} \subseteq B$ **(d)** $B \subseteq A$
    (e) $A \subseteq B$ **(f)** $\emptyset \in A$ **(g)** $A \cap B \subseteq C$

16. Let $A = \{a, b, c\}$ and $B = \{c, d\}$. Find $A - B$ and $B - A$, and verify that $A - B \neq B - A$.

17. Let $C = \{1, 2, 3, 4\}$ and $D = \{8\}$. Find $C - D$ and $D - C$, and verify that $C - D \neq D - C$.

18. Let $A = \{a, b, c, d, e\}$ and $U = \{a, b, c, d, e, 1, 2, 3\}$.
    (a) Find A'.
    (b) Find $A - U$.
    (c) Find $U - A'$.
    (d) Find $A \cap U$.
    (e) Find U'.

19. Let $C = \{1, 2, 3, 4, 5, 6\}$ and $U = \{1, 2, 3, 4, 5, 6, 7, 8, 9\}$.
    (a) Find C'.
    (b) Find U'.
    (c) Find $U - C'$.
    (d) Find $C' \cap U$.
    (e) Find $C \cup C'$.

20. Let $A = \{1, \{b\}, \{1, b\}\}$.
    (a) Place an $\in$ or $\subseteq$ in the blank to make each of the following statements true.
       1 _____ A; $\{1\}$ _____ A; $\{b\}$ _____ A; $\{\{b\}\}$ _____ A; $\{1, b\}$ _____ A
    (b) Find $P(A)$. Note that there are eight elements in $P(A)$.

21. Write each of the following sets as simply as possible.
    (a) $\emptyset \cup \{\emptyset\}$ **(b)** $\emptyset \cap \{\emptyset\}$ **(c)** $\{\emptyset\} - \emptyset$
    (d) $\{\{\emptyset\}, \emptyset\} - \{\emptyset\}$ **(e)** $P(\{\emptyset\})$

22. Let $S = \{X : X \in U \text{ and } X \notin X\}$ for some universal set U.
    (a) Show that $S \notin S$.
    (b) Show that $S \notin U$.

(c) Conclude that this does not lead to a paradox regarding S that is similar to Russell's paradox regarding the set of ordinary sets.

23. In 1918, Bertrand Russell gave another version of his 1902 paradox. He hypothesized a certain village in which a barber lived. The barber does not shave people who shave themselves. Moreover, he shaves all those people who do not shave themselves. Analyze this situation. In particular, who shaves the barber?

3.2 VENN DIAGRAMS AND CONJECTURES

It is frequently helpful to picture the possible combinations of sets resulting from the operations of intersection, union, difference, and complement. For this purpose, we use Venn diagrams, first employed by Leonhard Euler (1707–1783) and subsequently named after John Venn (1834–1923). A **Venn diagram** is a picture in which the universal set U is represented by points in a rectangle. Other sets are then represented by points in simple (usually circular) regions in the rectangle. Each circular region is labeled by a symbol designating the set for which it stands.

Example 1

Figure 3–1 shows Venn diagrams for the intersection, union, and difference of two sets A and B, and for the complement of the set A. The shaded area in each diagram shows where the elements are in each set. Of course, the shaded area can be empty. For example, when A and B are disjoint ($A \cap B = \emptyset$), the shaded area in the Venn diagram for $A \cap B$ is empty.

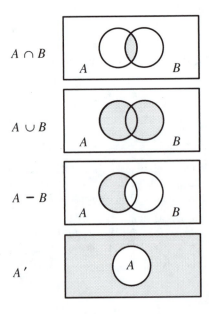

Figure 3–1

Example 2

The Venn diagrams in Figure 3–2 show $A \subseteq B$ and $A \cap B = \emptyset$.

$A \subseteq B$

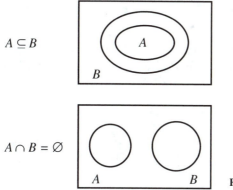

$A \cap B = \emptyset$

Figure 3–2

Drawing Venn diagrams is a good method of suggesting or testing conjectures. For example, because of the way a minus sign works in arithmetic, we might conjecture that $B - (B - A) = A$ for all sets A and B. However, the Venn diagram in the following example suggests that this conjecture is false.

Example 3

Figure 3–3 shows the Venn diagram for $B - (B - A)$.

$B - (B - A)$

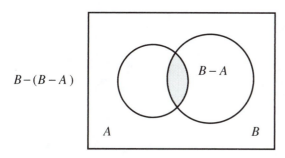

Figure 3–3

Based on this diagram, it is more reasonable to conjecture that $B - (B - A) = A \cap B$, for all sets A and B. In fact, this conjecture is true; however, a formal proof has to await a discussion of pick-a-point proofs in Section 3.3.

The Venn diagram in Example 3 suggests a counterexample to the original conjecture that $B - (B - A) = A$. We need only choose sets A and B so that $A \neq A \cap B$. One counterexample is as follows. Let $A = \{1, 3\}$ and $B = \{1, 2\}$. Then $B - A = \{2\}$ and $B - (B - A) = B - \{2\} = \{1\}$. So $B - (B - A) \neq A$ in this case.

It is important to note that a counterexample to a conjecture about sets is a *specific* list of sets which, when substituted in the conjecture, produces a false statement.

Although a Venn diagram can be useful for suggesting counterexamples, it cannot stand alone as a counterexample.

Each Venn diagram for two unequal intersecting sets has four disjoint regions, as shown in the following practice problem.

Practice Problem 1.   Draw a Venn diagram for two unequal sets A and B with nonempty intersection, showing four disjoint regions. Write each set represented by a disjoint region in terms of A, B, $'$, and $\cap$.

In Section 3.3, it is shown that $(A \cup B) \cup C = A \cup (B \cup C)$ and $(A \cap B) \cap C = A \cap (B \cap C)$ for all sets A, B, and C. Because of these relationships, we omit the parentheses and write $A \cup B \cup C$ and $A \cap B \cap C$.

Practice Problem 2.   Let $A = \{a, b, c, 3, 5\}$, $B = \{a, 3, 4\}$, and $C = \{a, b, 3, 5, 6\}$.
a. Find $A \cup B \cup C$.
b. Find $A \cap B \cap C$.

Let us now consider Venn diagrams for three sets A, B, and D. Figure 3–4 shows that a Venn diagram for three sets has eight disjoint regions.

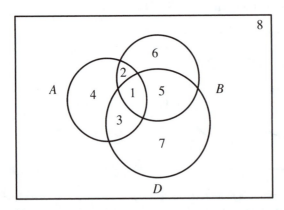

Figure 3–4

The following table describes the Venn diagram for three sets. In this table, "Yes" is denoted by "1" and "No" is denoted by "0." Note that because the eight disjoint regions form a partition of the points in the rectangle (in other words, the elements of U), every possible combination of the three sets A, B, and D can be written as the union of some of the eight sets represented by the eight disjoint regions. Also, it is interesting and instructive to compare the pattern of 1's and 0's in the table with the pattern of T's and F's in an eight-line truth table.

Region	Set represented	Whether an element of the region is also in:		
		A	B	D
1	$A \cap B \cap D$	1	1	1
2	$A \cap B \cap D'$	1	1	0
3	$A \cap B' \cap D$	1	0	1
4	$A \cap B' \cap D'$	1	0	0
5	$A' \cap B \cap D$	0	1	1
6	$A' \cap B \cap D'$	0	1	0
7	$A' \cap B' \cap D$	0	0	1
8	$A' \cap B' \cap D'$	0	0	0

Example 4

a. Draw the Venn diagrams for $D \cup (A \cap B)$ and $D \cap (A \cup B)$.

b. Make a conjecture based on the diagrams.

c. Write both $D \cup (A \cap B)$ and $D \cap (A \cup B)$ as the union of some of the eight sets represented by the disjoint regions in Figure 3–4.

Solution

a.

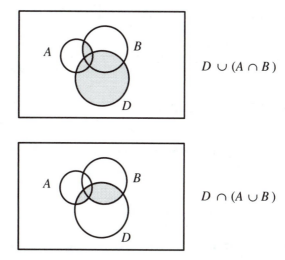

$D \cup (A \cap B)$

$D \cap (A \cup B)$

b. We see that $D \cap (A \cup B) \subseteq D \cup (A \cap B)$.

c. $D \cup (A \cap B) = (A \cap B \cap D) \cup (A \cap B \cap D') \cup (A \cap B' \cap D) \cup (A' \cap B \cap D) \cup (A' \cap B' \cap D)$.
$D \cap (A \cup B) = (A \cap B \cap D) \cup (A \cap B' \cap D) \cup (A' \cap B \cap D)$.

The following example illustrates the relationship between logical connectives and set operations.

Example 5

Let $E = \{x : (x \in A) \land (x \in B \Rightarrow x \in D)\}$, where A, B, and D are arbitrary sets.
a. Draw the Venn diagram for E.
b. Write E in terms of the sets A, B, and D and the operations $\cap$, $\cup$, and $'$.

Solution First, we write $(x \in A) \land (x \in B \Rightarrow x \in D)$ in terms of the connectives $\land$, $\lor$, and $\neg$. By the or-form of an implication, $x \in B \Rightarrow x \in D$ is logically equivalent to $\neg(x \in B) \lor x \in D$. Hence, $(x \in A) \land (x \in B \Rightarrow x \in D)$ is logically equivalent to $(x \in A) \land (x \notin B \lor x \in D)$. So $E = \{x : (x \in A) \land (x \notin B \lor x \in D)\}$.
a. The Venn diagram for E is:

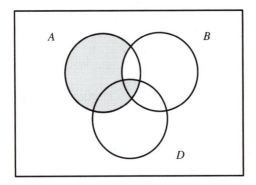

$$E = \{x : (x \in A) \land (x \notin B \lor x \in D)\}$$

b. $E = A \cap (B' \cup D)$

EXERCISE SET 3.2

1. Draw Venn diagrams for the following sets.
  (a) $A \cap (B \cup C)$
  (b) $(A \cap B) \cup C$

2. Draw Venn diagrams for the following sets.
  (a) $A \cup (B \cap C)$
  (b) $(A \cup B) \cap C$

3. Draw Venn diagrams for the following sets.
  (a) $(A \cap B)'$
  (b) $A' \cup B'$

4. Draw Venn diagrams for the following sets.
  (a) $(A \cup B)'$
  (b) $A' \cap B'$

5. Refer to Exercise 1, and decide whether the following conjecture is false (in other words, not always true). If the statement is false, give a counterexample suggested by a Venn diagram. $A \cap (B \cup C) = (A \cap B) \cup C$

6. Refer to Exercise 2, and decide whether the following conjecture is false. If the statement is false, give a counterexample suggested by a Venn diagram. $A \cup (B \cap C) = (A \cup B) \cap C$

7. Refer to Exercise 3, and decide whether the following conjecture is false. If the statement is false, give a counterexample suggested by a Venn diagram. $(A \cap B)' = A' \cup B'$

8. Refer to Exercise 4, and decide whether the following conjecture is false. If the statement is false, give a counterexample suggested by a Venn diagram. $(A \cup B)' = A' \cap B'$.

Decide whether the conjectures in Exercises 9–12 are false. If a statement is false, give a counterexample suggested by a Venn diagram.

9. $A - (B - D) \subseteq (A - B) - D$

10. $(A - D) \cup (B - D) \subseteq (A \cup B) - D$

11. $(A \cup B) - D \subseteq (A - D) \cup (B - D)$

12. $(A - B) \cap D$ and $(A - D) \cap B$ are disjoint.

13. Which of the following statements are true for *all* sets A, B, and C?
    (a) $A \in P(A)$
    (b) $A \cap B \cap C \subseteq B$
    (c) $\emptyset \subseteq B \cap C$

14. Which of the following statements are true for *all* sets E, F, and G?
    (a) $E \cap F \subseteq E \cup G$
    (b) $P(F) = F$
    (c) $\emptyset \in P(E)$

15. Is the following statement true for all sets A, B, C, and D? $(A \cup B) \cap (C \cup D) = (A \cap C) \cup (B \cap D)$

16. Let A, B, and D be arbitrary sets. Write each of the following sets in terms of A, B, and D, and the operations $\cap$, $\cup$, and $'$.
    (a) $E = \{x : x \in D \vee (x \notin A \wedge x \in B)\}$
    (b) $F = \{x : (x \in A \wedge x \in D) \Rightarrow x \in B\}$

17. Let A, B, and D be arbitrary sets. Write each of the following sets in terms of A, B, and D, and the operations $\cap$, $\cup$, and $'$.
    (a) $E = \{x : x \notin D \wedge (x \in A \vee x \in B)\}$
    (b) $F = \{x : (x \in A \Rightarrow x \in D) \vee x \in B\}$

3.3 THE ALGEBRA OF SETS

The Boolean laws for set theory are listed here for reference. We have seen a similar list of rules as the Boolean laws for propositional logic. A similar list of rules will also surface as the axioms for Boolean algebra in Chapter 7. These laws provide a unifying framework for several apparently different structures.

In the laws that follow, U denotes the universal set, $\emptyset$ is the empty set, and A, B, and C are arbitrary subsets of U.

Boolean Laws For Set Theory

Union	Intersection	
S1a. $A \cup B = B \cup A$	S1b. $A \cap B = B \cap A$	[Commutative laws]
S2a. $(A \cup B) \cup C = A \cup (B \cup C)$	S2b. $(A \cap B) \cap C = A \cap (B \cap C)$	[Associative laws]
S3a. $A \cup (B \cap C) = (A \cup B) \cap (A \cup C)$	S3b. $A \cap (B \cup C) = (A \cap B) \cup (A \cap C)$	[Distributive laws]
S4a. $A \cup \emptyset = A$	S4b. $A \cap U = A$	[Identity laws]
S5a. $A \cup A' = U$	S5b. $A \cap A' = \emptyset$	[Complement laws]

The first law states that for all sets A and B, $A \cup B = B \cup A$. It is customary to omit the phrase "for all sets A and B," and a similar omission is assumed for the other Boolean laws. The ten Boolean laws for set theory are listed in parallel columns to exhibit the duality of these laws. Law S1a is the **commutative law** for union, S2a is the **associative** law for union, S3a is the **distributive law** for union over intersection, S4a is the **identity law** for union, and S5a is the **complement law** for union. Law S1b through S5b have identical names, except that "union" is replaced by "intersection."

Practice Problem 1. Let $A = \{a, b, c, d\}$, $B = \{a, d, e, f\}$, and $C = \{d, f, g, k\}$.
a. Find $A \cup B$ and $B \cup C$.
b. Find $(A \cup B) \cup C$ and $A \cup (B \cup C)$, and verify that these two sets are equal.
c. Find $A \cap B$ and $B \cap C$.
d. Verify that $(A \cap B) \cap C$ and $A \cap (B \cap C)$ are equal.

In what follows, we prove the Boolean laws for set theory as well as many other theorems of set theory. The most important technique for a proof of inclusion of one set in another is the **pick-a-point method.** To use this method for proving that $A \subseteq B$, take an arbitrary element u in the universal set U and assume that $u \in A$. Then use any known true statements, including properties of A and B, to prove that $u \in B$.

Since $A \subseteq B$ means $\forall x(x \in A \Rightarrow x \in B)$, it is plain that the pick-a-point method is just a specific type of proof involving universal quantifiers. Following is the definition, in terms of quantifiers, of $A \subseteq B$ and its negation, $A \nsubseteq B$.

Definition	Negation
$A \subseteq B$	$A \nsubseteq B$
$\forall u(u \in A \Rightarrow u \in B)$	$\exists u(u \in A \wedge u \notin B)$

Practice Problem 2. Use one of the rules for negating a quantified statement to derive the negation of $A \subseteq B$.

Example 1

The following pick-a-point proofs illustrate the method.

a. Prove that $A \cap B \subseteq A$.

Proof   Let $u \in U$. (This step is usually not written.)
Assume $u \in A \cap B$.   (Assumption)
Then $u \in A$ and $u \in B$.   (Definition of $A \cap B$)
Therefore, $u \in A$.   ($p \wedge q \Rightarrow p$ is a tautology)
Done!

b. Prove that $\emptyset \subseteq A$.

Proof   (By contradiction)
Suppose $\emptyset \not\subseteq A$.   (Assumption)
Then there is an element u such that
$u \in \emptyset$ and $u \notin A$.   (Definition of $\subseteq$)
But $u \notin \emptyset$.   (Definition of $\emptyset$)
Contradiction!

Practice Problem 3.   Use a pick-a-point proof to show that $A \subseteq A \cup B$.

The next example illustrates a proof of an implication in set theory.

Example 2

Prove: If $A \subseteq B$, then $B' \subseteq A'$. We give two proofs, a direct proof and an indirect proof.
Direct Proof: Assume $A \subseteq B$. Then, by definition of $\subseteq$, $u \in A$ implies $u \in B$. Hence, $u \notin B$ implies $u \notin A$, by the rule of contraposition. Therefore, by the definition of A' and B', $u \in B'$ implies $u \in A'$. Hence, $B' \subseteq A'$.
Indirect Proof: Assume $A \subseteq B$ and suppose $B' \not\subseteq A'$. Then there exists an element u such that $u \in B'$ and $u \notin A'$. By double negation, $u \in A$. Therefore, $u \in B$, since $A \subseteq B$. But $u \notin B$, since $u \in B'$. Hence, we have a contradiction, and the supposition that $B' \not\subseteq A'$ is false. It follows that $B' \subseteq A'$.

The converse of Example 2 is left as an exercise.
The following example shows that any equality in set theory can be proved using a corresponding tautology and the pick-a-point method.

Example 3

a. Prove S4a. In other words, prove that $A \cup \emptyset = A$.

Proof   Let $u \in U$ be arbitrary. We must prove that $u \in A \cup \emptyset \Leftrightarrow u \in A$.
$u \in A \cup \emptyset \Leftrightarrow (u \in A$ or $u \in \emptyset)$ (Definition of $\cup$)
$\qquad\qquad \Leftrightarrow u \in A$ ($p \vee \mathcal{O} \Leftrightarrow p$ is a tautology;
                                                          note that $u \in \emptyset$ is false.)
      Hence, $u \in A \cup \emptyset$ if and only if $u \in A$, and it follows that $A \cup \emptyset = A$. Observe that the tautology $p \vee \mathcal{O} \Leftrightarrow p$ bears a resemblance to $A \cup \emptyset = A$.
b. Prove S3a, the distributive law for union. In other words, prove that $A \cup (B \cap C) = (A \cup B) \cap (A \cup C)$

Proof   Let $u \in U$ be arbitrary. We must prove that

$$u \in A \cup (B \cap C) \Leftrightarrow u \in (A \cup B) \cap (A \cup C)$$

$$u \in A \cup (B \cap C) \Leftrightarrow u \in A \text{ or } u \in B \cap C \qquad \text{(Definition of } \cup \text{)}$$
$$\Leftrightarrow u \in A \text{ or } (u \in B \text{ and } u \in C) \qquad \text{(Definition of } \cap \text{)}$$
$$\Leftrightarrow (u \in A \text{ or } u \in B) \text{ and } (u \in A \text{ or } u \in C) \quad \text{(Tautology)}$$
$$\Leftrightarrow (u \in A \cup B) \text{ and } (u \in A \cup C) \qquad \text{(Definition of } \cup \text{)}$$
$$\Leftrightarrow u \in (A \cup B) \cap (A \cup C) \qquad \text{(Definition of } \cap \text{)}$$

We used the tautology $p \vee (q \wedge r) \Leftrightarrow (p \vee q) \wedge (p \vee r)$, where p is $u \in A$, q is $u \in B$, and r is $u \in C$.

As a universal rule, we can obtain a tautology of propositional logic from an equality of set theory by replacing $\cup$ by $\vee$, $\cap$ by $\wedge$, $\emptyset$ by $\mathcal{O}$, U by $\mathcal{I}$, $=$ by $\Leftrightarrow$, $'$ by $\neg$, and the letters used to denote sets by letters used to denote statements. For example, to prove $(A' \cap B)' = A \cup B'$, we would use the tautology $\neg(\neg p \wedge q) \Leftrightarrow p \vee \neg q$. However, to prove $A - (A \cap B) = A - B$, we would first write it as $A \cap (A \cap B)' = A \cap B'$ and then use the tautology $p \wedge \neg(p \wedge q) \Leftrightarrow p \wedge \neg q$.

Practice Problem 4.

a. State which tautology can be used to prove that $(A \cup B) - B = A \cap B'$.

b. Use the tautology of part a to prove that $(A \cup B) - B = A \cap B'$.

Each of the proofs in Example 3 is a straightforward application of definitions and a known tautology. For many such proofs, it is routine to write the proof as a sequence of $\Leftrightarrow$'s. However, most proofs of set equalities are not so routine.

In most cases, a proof of set equality is easier if done in two parts. This is so because $A = B$ is equivalent to $\forall u (u \in A \Leftrightarrow u \in B)$, which is logically equivalent to $\forall u [(u \in A \Rightarrow u \in B) \wedge (u \in B \Rightarrow u \in A)]$, which in turn is logically equivalent to $\forall u (u \in A \Rightarrow u \in B) \wedge \forall u (u \in B \Rightarrow u \in A)$. This last step is equivalent to $A \subseteq B$ and $B \subseteq A$. So, in order to prove a set equality, we may prove two set inclusions. This gives rise to the **method of double inclusion:** To prove $A = B$, prove $A \subseteq B$ and $B \subseteq A$.

Both steps in the method of double inclusion were carried out in Example 3 even though they are not explicit in the proofs.

Example 4

Prove $A \cup (A \cap B) = A$ by the method of double inclusion.

Proof   First we prove $A \cup (A \cap B) \subseteq A$. Let $x \in A \cup (A \cap B)$. Then $x \in A$ or $x \in A \cap B$. This gives rise to two cases.
Case 1. Assume $x \in A$. We are done.
Case 2. Assume $x \in A \cap B$. Then $x \in A$ and $x \in B$. Hence, $x \in A$.
        The set inclusion $A \subseteq A \cup (A \cap B)$ follows from Practice Problem 3.
        Therefore, by the method of double inclusion, $A \cup (A \cap B) = A$.

Example 5

Prove that if $A \subseteq B$, then $A \cap B = A$.

Proof   Assume $A \subseteq B$. We need to prove that $A \cap B \subseteq A$ and $A \subseteq A \cap B$.
        That $A \cap B \subseteq A$ follows from Example 1a and does not require the hypothesis $A \subseteq B$.

We use the pick-a-point method to prove $A \subseteq A \cap B$. Assume $u \in A$. Then $u \in B$, since $A \subseteq B$. Therefore, $u \in A \cap B$, and we are done.

Practice Problem 5. *Prove:* If $A \cap B = A$, then $A \subseteq B$. (This is the converse of the statement proved in Example 5.)

From Example 5 and Practice Problem 5, we have shown that $A \subseteq B$ and $A \cap B = A$ are logically equivalent. If we now draw a Venn diagram illustrating $A \subseteq B$ and a Venn diagram of $A \cap B$, we can see that this logical equivalence makes sense. Using Venn diagrams, we can also conjecture that $A \subseteq B$ and $A \cup B = B$ are logically equivalent (see the exercises).

Example 6

Prove the transitive property for set inclusion:

$$\text{If } A \subseteq B \text{ and } B \subseteq D, \text{ then } A \subseteq D$$

Proof Assume $A \subseteq B$ and $B \subseteq D$. We use the pick-a-point method to prove that $A \subseteq D$. Assume $u \in A$. Then $u \in B$, since $A \subseteq B$. Now, since $B \subseteq D$, we obtain $u \in D$. Therefore, $A \subseteq D$.

Practice Problem 6. *Prove:* If $A \subseteq B$, then $A \cup (B \cap C) = B \cap (A \cup C)$.

It is possible to prove statements about sets by using the Boolean laws for sets directly, without invoking a pick-a-point proof. Such proofs are called *algebraic proofs*. For this purpose, we note that $A - B = A \cap B'$.

Example 7

Prove that $(A - D) \cup (B - D) = (A \cup B) - D$.

Proof

$$(A - D) \cup (B - D) = (A \cap D') \cup (B \cap D')$$
$$= (D' \cap A) \cup (D' \cap B)$$
$$= D' \cap (A \cup B)$$
$$= (A \cup B) \cap D'$$
$$= (A \cup B) - D$$

Practice Problem 7. Supply a reason for each step of the proof in Example 7.

In Example 8, we prove a statement using the pick-a-point method. The proof uses the fact that if $u \in A \cup B$ and $u \notin A$, then $u \in B$, which follows from the tautology $(p \vee q) \wedge \neg p \Rightarrow q$.

Example 8

Prove that $(A \cup B) \cap (A \cup B') = A$.

Proof We first prove $(A \cup B) \cap (A \cup B') \subseteq A$, using a proof by contradiction. Assume $u \in (A \cup B) \cap (A \cup B')$, but $u \notin A$. Then $u \in (A \cup B)$ and $u \in (A \cup B')$. Since $u \in A \cup B$ and $u \notin A$, we have $u \in B$. Similarly, $u \in (A \cup B')$ and $u \notin A$ implies $u \in B'$, that is, $u \notin B$. But this contradicts $u \in B$. Hence, $u \in A$.

Next, we prove $A \subseteq (A \cup B) \cap (A \cup B')$. Assume $u \in A$. Then, by Practice Problem 3, $u \in A \cup B$. Similarly, $u \in A \cup B'$. Therefore, $u \in (A \cup B) \cap (A \cup B')$.

Practice Problem 8. Prove that $(A \cup B) \cap (A \cup B') = A$ using the algebraic method.

Exercise Set 3.3 has more proofs of theorems of set theory. We list, for ease of reference, some of the most common theorems concerning sets. Recall that the Boolean axioms, S1–S5, were presented earlier.

Theorems of Set Theory

S6a. $A \cup U = U$	S6b. $A \cap \emptyset = \emptyset$	
S7a. $A \cup A = A$	S7b. $A \cap A = A$	[Idempotent laws]
S8a. $U' = \emptyset$	S8b. $\emptyset' = U$	[Universe/Empty Set Complement laws]
S9. $(A')' = A$		[Law of Double Complement]
S10a. $(A \cup B)' = A' \cap B'$	S10b. $(A \cap B)' = A' \cup B'$	[De Morgan's laws]
S11a. $A \cap (A \cup B) = A$	S11b. $A \cup (A \cap B) = A$	[Absorption laws]
S12a. $A \subseteq A \cup B$	S12b. $A \cap B \subseteq B$	
S13. $A \subseteq B \Leftrightarrow B' \subseteq A'$		
S14. $A \subseteq B \Leftrightarrow A' \cup B = U$		
S15. $A \subseteq B \Leftrightarrow A \cap B' = \emptyset$		
S16. $A \subseteq B \Leftrightarrow A \cup B = B$		
S17. $A \subseteq B \Leftrightarrow A \cap B = A$		
S18. $A \subseteq B \wedge B \subseteq D \Rightarrow A \subseteq D$		[Transitive law]

Several of the theorems in the list were proved earlier in this section. For example, S12b was proved in Example 1, S11b was proved in Example 4, S17 was proved in Example 5 and Practice Problem 5, and S18 was proved in Example 6. Many of the remaining theorems in the list are given in Exercise Set 3.3.

EXERCISE SET 3.3

1. Prove each of the following theorems of set theory.
   (a) $B \subseteq A \cup B$
   (b) $(A')' = A$

2. Prove each of the following theorems of set theory.
    (a) Boolean law S1b.
    (b) Boolean law S3b.

3. Prove each of the following theorems of set theory.
    (a) Boolean law S4b
    (b) Boolean law S5b.

4. *Prove:* If $A' \subseteq B'$, then $B \subseteq A$.

5. *Prove:* If $A \subseteq B$, then $D - B \subseteq D - A$.

6. Either prove that if $D - B \subseteq D - A$, then $A \subseteq B$, or find a counterexample.

7. *Prove:* $A \nsubseteq B$ if and only if $A \cap B' \neq \emptyset$.

8.   **(a)** Use a pick-a-point proof to prove that $D - (A \cap B) = (D - A) \cup (D - B)$. (*Hint:* Use the appropriate De Morgan's law of logic.)
    (b) By substituting U for D in the statement in part a, we get the corollary $(A \cap B)' = A' \cup B'$. Use this corollary and the result $(C')' = C$ proved in Exercise 1b to prove that $D - (A \cup B) = (D - A) \cap (D - B)$. This law, in turn, has the corollary $(A \cup B)' = A' \cap B'$, and all four laws are called **De Morgan's laws for sets.**

9. Prove that $A \subseteq B$ and $A \cup B = B$ are logically equivalent.

10. Prove that $(A \cap B) \cup (A - B) = A$
    (a) With the pick-a-point method.
    (b) With an algebraic proof.

11. Prove that $A - (A - B) = A \cap B$
    (a) With the pick-a-point method.
    (b) With an algebraic proof.

12. Prove that $A \cup (B - A) = A \cup B$
    (a) With the pick-a-point method.
    (b) With the pick-a-point method and a corresponding tautology.
    (c) With an algebraic proof.

13. Prove that $(A \cap B) \cup (C - A) \cup (B \cap C) = (A \cap B) \cup (C - A)$
    (a) With the pick-a-point method.
    (b) With the pick-a-point method and a corresponding tautology.
    (c) With an algebraic proof.

14. Prove that the following statements are logically equivalent.

    i. $A \subseteq B$

    ii. $A' \cup B = U$

    iii. $A \cap B' = \emptyset$

(*Hint:* First use De Morgan's laws to prove that ii and iii are logically equivalent. Then it is sufficient to prove that i and ii are logically equivalent. Do this by considering the or-form of an implication.)

15. *Prove:* If $C \cup B = U$ and $A \cap C = \emptyset$, then $A \subseteq B$.

16. *Prove:* If $(C \cup B) \cap (A \cup D) = U$ and $(A \cap C) \cup (B \cap D) = \emptyset$, then $A = B$ and $C = D$.

17. *Prove:* If $B \cap E = \emptyset$, $C \cup B = U$, and $A \cap C = \emptyset$, then $A \cap E = \emptyset$.

18. *Prove:* If $C \cup B = U$, $D \cup F = U$, and $B \cap D = \emptyset$, then $C \cup F = U$.

Exercise Set 3.3   **99**

Finite Unions and Intersections

Most of the laws of set theory that we have examined up to now involve one, two, or three sets. For example, the commutative laws involve two sets, and the associative laws involve three sets.

The associative law for set union is $(A_1 \cup A_2) \cup A_3 = A_1 \cup (A_2 \cup A_3)$. In essence, this says that we may omit parentheses in the expression $A_1 \cup A_2 \cup A_3$. Similarly, the expression $((A_1 \cup A_2) \cup A_3) \cup A_4$ may be written $A_1 \cup A_2 \cup A_3 \cup A_4$. In general, we write $A_1 \cup A_2 \cup \ldots \cup A_n$ without parentheses and then observe that $x \in A_1 \cup A_2 \cup \ldots \cup A_n \Leftrightarrow x \in A_i$ for some i, $i = 1, 2, \ldots, n$.

A similar observation may be made about intersection; that is, $x \in A_1 \cap A_2 \cap \ldots \cap A_n \Leftrightarrow x \in A_i$ for all i, $i = 1, 2, \ldots, n$.

By examining the case when $n = 2$, we see the correctness of the above observations. Note that $x \in A_i$ for at least one $i = 1, 2$ is equivalent to $(x \in A_1 \vee x \in A_2)$; $x \in A_i$ for all $i = 1, 2$ is equivalent to $(x \in A_1 \wedge x \in A_2)$.

Several of the most important laws of set theory, such as De Morgan's laws and the distributive laws, can be generalized to finitely many sets.

Example 1

Prove the distributive laws for finitely many sets. That is, prove:
a. $A \cup (B_1 \cap B_2 \cap \ldots \cap B_n) = (A \cup B_1) \cap (A \cup B_2) \cap \ldots \cap (A \cup B_n)$, where $n = 1, 2, 3, \ldots$.
b. $A \cap (B_1 \cup B_2 \cup \ldots \cup B_n) = (A \cap B_1) \cup (A \cap B_2) \cup \ldots \cup (A \cap B_n)$, where $n = 1, 2, 3, \ldots$.

Proof We give the proof of part a and leave the proof of part b as an exercise. The proof is by mathematical induction.
Basis Step. $(n = 1)$. $A \cup (B_1) = A \cup B_1 = (A \cup B_1)$.
Induction Hypothesis. Assume $A \cup (B_1 \cap B_2 \cap \ldots \cap B_k) = (A \cup B_1) \cap (A \cup B_2) \cap \ldots \cap (A \cup B_k)$. Then $A \cup (B_1 \cap \ldots \cap B_k \cap B_{k+1}) = A \cup [(B_1 \cap \ldots \cap B_k) \cap B_{k+1}] = [A \cup (B_1 \cap \ldots \cap B_k)] \cap [A \cup B_{k+1}]$, by Boolean law S3a. By the induction hypothesis, $[A \cup (B_1 \cap \ldots \cap B_k)] \cap [A \cup B_{k+1}] = [(A \cup B_1) \cap (A \cup B_2) \cap \ldots \cap (A \cup B_k)] \cap [A \cup B_{k+1}]$. Therefore, $A \cup (B_1 \cap \ldots \cap B_k \cap B_{k+1}) = (A \cup B_1) \cap (A \cup B_2) \cap \ldots \cap (A \cup B_k) \cap (A \cup B_{k+1})$.

Union and Intersection of an Arbitrary Collection of Sets

Let $\Upsilon = \{A, B, C, D\}$, where A, B, C, and D are sets. Υ is then a collection of sets. As discussed previously, x is a member of $A \cup B \cup C \cup D$ if and only if x is a member of at least one of A, B, C, and D—in other words, if x is an element of some member of Υ. We then have $x \in A \cup B \cup C \cup D$ if and only if, for some $Y \in \Upsilon$, $x \in Y$. In this example, since Υ is finite, we can write the union of the members of Υ as $A \cup B \cup C \cup D$. However, when Υ is infinite, we encounter a difficulty since we cannot list all the members of Υ. In that case, the notation we use is $\cup \Upsilon$, and we extend this notation to an arbitrary finite collection of sets.

Definition. Let Υ be an arbitrary collection of sets. The **generalized union** of Υ is denoted by $\cup\Upsilon = \{x : \text{for some } Y \in \Upsilon, x \in Y\}$.

Let $\Upsilon = \{A, B, C, D\}$, where A, B, C, and D are sets, and assume x is a member of $A \cap B \cap C \cap D$. Then x is a member of each of A, B, C, and D. Hence, x is an element of every member of Υ; that is, for all $Y \in \Upsilon$, $x \in Y$. In a manner analogous to that for set union, when Υ is an arbitrary collection of sets, we write $\cap\Upsilon$ for the intersection of the members of Υ.

Definition. Let Υ be an arbitrary collection of sets. The **generalized intersection** of Υ is denoted by $\cap\Upsilon = \{x : \text{for all } Y \in \Upsilon, x \in Y\}$.

Example 2

Find $\cup\Upsilon$ and $\cap\Upsilon$ in each case.
a. $\Upsilon = \{\{0,1\}, \{0,1,2\}, \{0,1,3\}\}$
b. $\Upsilon = \{A_n : n = 1, 2, \ldots, 39\}$, where $A_n = \{0, n-1, n\}$

Solution a. $\cup\Upsilon = \{0,1,2,3\}$ and $\cap\Upsilon = \{0,1\}$. For example, $2 \in \cup\Upsilon$, since there is an $A \in \Upsilon$ such that $2 \in A$, namely, $A = \{0,1,2\}$. As another example, $1 \in \cap\Upsilon$, since 1 belongs to every member of Υ.
b. $\cup\Upsilon = \{0,1,2,\ldots,39\}$ and $\cap\Upsilon = \{0\}$.

Since Υ can be a finite collection of sets, $\cup\Upsilon$ and $\cap\Upsilon$ are generalizations of the intersection and union of a finite collection of sets. In other words, if $\Upsilon = \{A_1, A_2, \ldots, A_n\}$, then $\cup\Upsilon = A_1 \cup A_2 \cup \ldots \cup A_n$ and $\cap\Upsilon = A_1 \cap A_2 \cap \ldots \cap A_n$.

Practice Problem 1. Let $\Omega = \{\{a,b,c\}, \{b,c\}, \{b,d,e\}\}$. Find
a. $\cup\Omega$
b. $\cap\Omega$

Practice Problem 2. Let $\Upsilon = \{A, B\}$. Show that
a. $\cap\Upsilon = A \cap B$
b. $\cup\Upsilon = A \cup B$

Example 3

Prove that every element of Ω is a subset of $\cup\Omega$.

Proof We must prove that $A \subseteq \cup\Omega$ for all $A \in \Omega$. Assume that $A \in \Omega$, and let $x \in A$. Then $x \in B$ for some $B \in \Omega$, namely, $B = A$. Hence, $x \in \cup\Omega$.

Practice Problem 3. Prove that $\cap\Omega$ is a subset of every element of Ω.

One of the most important applications of the generalized intersection and union of sets is to sets of real numbers. Before considering several examples, we state a property of $\mathcal{R}$.

Definition.   The ϵ-**property** for $\mathcal{R}$. Given any positive real number ϵ, there exists a positive integer m such that $1/m < \epsilon$.

Recall the standard notation for **intervals** of the real numbers $\mathcal{R}$.

$[a,b] = \{x : x \in \mathcal{R} \text{ and } a \leq x \leq b\}$ $[a,\infty) = \{x : x \in \mathcal{R} \text{ and } a \leq x\}$

$[a,b) = \{x : x \in \mathcal{R} \text{ and } a \leq x < b\}$ $(a,\infty) = \{x : x \in \mathcal{R} \text{ and } a < x\}$

$(a,b] = \{x : x \in \mathcal{R} \text{ and } a < x \leq b\}$ $(-\infty,b] = \{x : x \in \mathcal{R} \text{ and } x \leq b\}$

$(a,b) = \{x : x \in \mathcal{R} \text{ and } a < x < b\}$ $(-\infty,b) = \{x : x \in \mathcal{R} \text{ and } x < b\}$

Example 4

Let $\Omega = \{[0, 1/n) : n = 1, 2, 3, \ldots\}$.
  a. Find $\cup\Omega$.
  b. Find $\cap\Omega$.

Solution   Note that $\Omega = \{[0,1), [0,\frac{1}{2}), [0,\frac{1}{3}), \ldots\}$.

  a. We show that $\cup\Omega = [0,1)$. Let $x \in \cup\Omega$. Then $x \in Y$ for some $Y \in \Omega$. By the definition of Ω, $Y = [0, 1/n)$ for some $n = 1, 2, 3, \ldots$. Now, $[0, 1/n) \subseteq [0,1)$, and hence, $x \in [0,1)$. For the other inclusion, let $x \in [0,1)$. Then since $[0,1) \in \Omega$, $x \in Y$ for some $Y \in \Omega$, namely, $Y = [0,1)$. Hence, $x \in \cup\Omega$.

  b. We show that $\cap\Omega = \{0\}$. First, note that $0 \in \cap\Omega$ since $0 \in [0, 1/n)$ for all $n = 1, 2, 3, \ldots$. So we only need to prove that if $a \neq 0$, then $a \notin \cap\Omega$. Clearly, $a \notin \cap\Omega$ if $a < 0$. So assume $a > 0$. Then, by the ϵ-property, there is a positive integer m such that $1/m < a$. Hence, $a \notin [0, 1/m)$. So a is not an element of every member of Υ. Therefore, $a \notin \cap\Omega$. It follows that 0 is the only element of $\cap\Omega$.

Practice Problem 4.   Let $\Upsilon = \{(0, 1/n) : n = 1, 2, 3, \ldots\}$. Prove $\cap\Upsilon = \emptyset$.

In the following, we will use sets of the form $\Omega = \{D \cup A : A \in \Upsilon\}$. With this usage, $X \in \Omega$ means $X = D \cup A$ for some $A \in \Upsilon$, $x \in \cup\Omega$ means $x \in D \cup A$ for some $A \in \Upsilon$, and $x \in \cap\Omega$ means $x \in D \cup A$ for all $A \in \Upsilon$.

Example 5

Prove the generalized distributive law

$$D \cup \cap\Upsilon = \cap\{D \cup A : A \in \Upsilon\}$$

Proof   As usual, we prove two set inclusions. Let $x \in D \cup \cap\Upsilon$. We have two cases.
Case 1. Assume $x \in D$. Then for every $A \in \Upsilon$, $x \in D \cup A$. Hence, $x \in \cap\{D \cup A : A \in \Upsilon\}$.
Case 2. Assume $x \in \cap\Upsilon$. Then $x \in A$ for every $A \in \Upsilon$. Hence, $x \in D \cup A$ for every $A \in \Upsilon$. Therefore, $x \in \cap\{D \cup A : A \in \Upsilon\}$.
        In each case, we conclude that $x \in \cap\{D \cup A : A \in \Upsilon\}$.
        Now, let $x \in \cap\{D \cup A : A \in \Upsilon\}$. Then $x \in D \cup A$ for all $A \in \Upsilon$. We again have two cases.
Case 1. Assume $x \in D$. Then $x \in D \cup \cap\Upsilon$.
Case 2. Assume $x \notin D$. By assumption, $x \in D \cup A$ for all $A \in \Upsilon$. Therefore, $x \in A$ for all $A \in \Upsilon$. But this shows that $x \in \cap\Upsilon$, and hence, $x \in D \cup \cap\Upsilon$.

Practice Problem 5. Prove the generalized distributive law

$$D \cap \cup \Upsilon = \cup \{D \cap A : A \in \Upsilon\}$$

Example 6

Prove: $\cup(\Upsilon \cup \Omega) = (\cup\Upsilon) \cup (\cup\Omega)$.

Proof First we prove $\cup(\Upsilon \cup \Omega) \subseteq (\cup\Upsilon) \cup (\cup\Omega)$. Let $x \in \cup(\Upsilon \cup \Omega)$. Then $x \in A$ for some $A \in (\Upsilon \cup \Omega)$.
Case 1. Assume $A \in \Upsilon$. Then $x \in \cup\Upsilon$.
Case 2. Assume $A \in \Omega$. Then $x \in \cup\Omega$.
In either case, $x \in (\cup\Upsilon) \cup (\cup\Omega)$.
    Next, we prove $(\cup\Upsilon) \cup (\cup\Omega) \subseteq \cup(\Upsilon \cup \Omega)$. Let $x \in (\cup\Upsilon) \cup (\cup\Omega)$.
Case 1. Assume $x \in \cup\Upsilon$. Then $x \in A$ for some $A \in \Upsilon$.
Case 2. Assume $x \in \cup\Omega$. Then $x \in A$ for some $A \in \Omega$.
In either case, $x \in A$ for some $A \in \Upsilon \cup \Omega$. Hence, $x \in \cup(\Upsilon \cup \Omega)$.

Practice Problem 6. Find sets Υ and Ω such that $\Upsilon \cap \Omega \neq \emptyset$ and

$$\cap(\Upsilon \cap \Omega) \neq (\cap\Upsilon) \cap (\cap\Omega)$$

EXERCISE SET 3.4

1. Let $\Omega = \big\{\{1,2,5\}, \{2,5,6\}, \{3,4,5,6\}\big\}$.
    (a) Find $\cap\Omega$.
    (b) Find $\cup\Omega$.
2. Let $\Omega = \big\{\{a,b,d\}, \{b,c,d,e\}, \{a,b,d,e\}, \{a,b,e\}\big\}$.
    (a) Find $\cap\Omega$.
    (b) Find $\cup\Omega$.
3. Let $\Omega = \{A_n : n = 1,2,3,\ldots\}$, where $A_n = \{0,1,2,\ldots,n\}$.
    (a) Find $\cap\Omega$.
    (b) Find $\cup\Omega$.
4. Let $\Omega = \{A_n : n = 1,2,3,\ldots\}$, where $A_n = \{0,n,n+1,n+2,\ldots\}$.
    (a) Find $\cap\Omega$.
    (b) Find $\cup\Omega$.
5. Let $\Omega = \{A_n : n = 1,2,3,\ldots\}$, where $A_n = (-1+1/n, 1-1/n)$.
    (a) Find $\cap\Omega$.
    (b) Find $\cup\Omega$.
6. Let $\Omega = \{A_n : n = 1,2,3,\ldots\}$, where $A_n = [-1, 1-1/n)$.
    (a) Find $\cap\Omega$.
    (b) Find $\cup\Omega$.
7. Let $\Omega = \{A_n : n \in \mathcal{Z}\}$, where $A_n = (n, 1+n)$.
    (a) Find $\cap\Omega$.
    (b) Find $\mathcal{R} - \cup\Omega$.
8. Let $\Omega = \{A_n : n \in \mathcal{Z}\}$, where $A_n = (n-1, 1+n)$.

(a) Find $\cap\Omega$.

(b) Find $\mathcal{R} - \cup\Omega$.

9. *Prove:* $A\cap(B_1\cup B_2\cup\ldots\cup B_n) = (A\cap B_1)\cup(A\cap B_2)\cup\ldots\cup(A\cap B_n)$, where $n = 1, 2, 3, \ldots$.

10. Use De Morgan's laws for sets and mathematical induction to prove that $(A_1 \cap A_2 \cap \ldots \cap An)' = A_1' \cup A_2' \cup \ldots \cup An'$.

11. Use De Morgan's laws for sets and mathematical induction to prove that $D - (A_1 \cup A_2 \cup \ldots \cup A_n) = (D - A_1)\cap(D - A_2)\cap\ldots\cap(D - A_n)$.

12. Prove the generalization, to an arbitrary collection of sets, of the De Morgan's law $(\cup\Omega)' = \cap\{A' : A \in \Omega\}$.

13. Prove the generalization, to an arbitrary collection of sets, of the De Morgan's law $D - \cap\Omega = \cup\{D - A : A \in \Omega\}$.

14. *Prove:* If $D \subseteq A$ for all $A \in \Omega$, then $D \subseteq \cap\Omega$.

15. *Prove:* If $A \subseteq D$ for all $A \in \Upsilon$, then $\cup\Upsilon \subseteq D$.

16. Recall that in any discussion of sets, we assume that all elements are members of a universal set U. Hence, $\cup\Upsilon = \{x : x \in U \text{ and } (\exists A \in \Upsilon)(x \in A)\}$ and $\cap\Upsilon = \{x : x \in U \text{ and } (\forall A \in \Upsilon)(x \in A)\}$. Assume $\Upsilon = \emptyset$.

(a) Find $\cup\Upsilon$.

(b) Find $\cap\Upsilon$.

17. Assume $\Upsilon \neq \emptyset$. Prove that $\cap\Upsilon \subseteq \cup\Upsilon$.

18. *Prove:* If $\Omega \subseteq \Upsilon$, then $\cup\Omega \subseteq \cup\Upsilon$.

19. *Prove:* If $\Omega \subseteq \Upsilon$, then $\cap\Upsilon \subseteq \cap\Omega$.

The statements in Exercises 20–23 are conjectures. Find either a counterexample or a proof for each of them.

20. If $A \cap B = A \cap C$, then $B = C$.

21. If $A \cup B = A \cap B$, then $A = B$.

22. If $\cup\Omega = \cup\Upsilon$, then $\Omega = \Upsilon$.

23. If $\cap\Omega = \cap\Upsilon$, then $\Omega = \Upsilon$.

KEY CONCEPTS

Sets	Disjoint sets
Elements	Venn diagram
Extensionality	Counterexample
Subset	Pick-a-point method
Power set	Method of double inclusion
Paradoxes	Operations on arbitrary collections of sets
Set Construction	Generalized union,
Set operations	generalized intersection
Complement, difference, intersection, union	

PROOFS TO EVALUATE

See the instructions following the Proofs to Evaluate in Chapter 2.

1. *Conjecture:* If $A \cap B \subseteq D$, then $A \subseteq D$.

Argument: We use a pick-a-point proof. Assume $A \cap B \subseteq D$. Let $u \in A \cap B$. Then $u \in A$. Since $A \cap B \subseteq D$, we have $u \in D$. Hence, $A \subseteq D$.

2. *Conjecture:* If $D - A \subseteq D - B$, then $B \subseteq A$.

Argument: We use a pick-a-point proof. Assume $D - A \subseteq D - B$. Let $u \in D - A$. Then $u \in D - B$, and we have $u \notin B$. Since $u \notin B$, $B \subseteq A$ is vacuously true.

3. *Conjecture:* $A \cap (A \cup B) = A$.

Argument: We use the method of double set inclusion. First, we prove that $A \cap (A \cup B) \subseteq A$. Let $u \in A \cap (A \cup B)$. Then $u \in A$ and $u \in A \cup B$. Hence, $u \in A$. Now, let $u \in A$. Then $u \in A$ or $u \in B$. Hence, $u \in A \cup B$, and we have $u \in A \cap (A \cup B)$.

4. *Conjecture:* If $A \cup B = U$ and $C \cup D = U$, then $B \cap C \neq \emptyset$ or $A \cup D = U$.

Argument: We give an indirect proof. Assume that $A \cup B = U$, $C \cup D = U$, and $B \cap C = \emptyset$. We need to prove that $A \cup D = U$. We use a pick-a-point proof to prove that $U \subseteq A \cup D$. Let $u \in U$. Then $u \in A \cup B$ and $u \in C \cup D$. We use cases to prove $u \in A \cup D$.

Case 1. Assume $u \in A$. Then we are done.

Case 2. Assume $u \notin A$. Then $u \in B$, since $A \cup B = U$. Hence, $u \notin C$, since $B \cap C = \emptyset$. Therefore, $u \in D$, since $C \cup D = U$. In either case, we see that $u \in A \cup D$.

5. *Conjecture:* If $A \cap C = B \cap C$, then $A = B$.

Argument: Assume $A \cap C = B \cap C$. We use an algebraic proof. Cancel C from both sides of $A \cap C = B \cap C$ to obtain $A = B$.

6. *Conjecture:* If $P(X) = P(Y)$, then $X = Y$.

Argument: It is sufficient to prove that $P(X) \subseteq P(Y)$ implies $X \subseteq Y$. Assume that $P(X) \subseteq P(Y)$. Let $x \in X$. Then $\{x\} \subseteq X$, and so $\{x\} \in P(X)$. Hence, $\{x\} \in P(Y)$. Therefore, $\{x\} \subseteq Y$, and so $x \in Y$.

REVIEW EXERCISES

1. Let $S = \{x : x \in \mathcal{Z} \text{ and } x^2 \leq 100\}$. Answer true or false:
   (a) $\{0, 1, 2, 3\} \subseteq S$
   (b) $100 \in S$
   (c) $12 \notin S$
   (d) $\mathcal{Z} \subseteq S$
   (e) $\{x : x \in \mathcal{Z} \text{ and } x < 10\} \subseteq S$

2. List the elements of each set.
   (a) $\{x : x \in \mathcal{Z} \text{ and } 2 < |x| < 9\}$
   (b) $\{3x : x \in \mathcal{N} \text{ and } x^2 = 36\}$
   (c) $\{2x - 5 : x \in \mathcal{Z} \text{ and } -2 \leq x \leq 3\}$

The statements in Exercises 3–11 are conjectures. Find either a counterexample or a proof for each of them.

3. If $A \neq B$ and $B \neq D$, then $A \neq D$.

4. If $A \not\subseteq B$ and $B \subseteq D$, then $A \not\subseteq D$.

5. If $A \in B$ and $B \in D$, then $A \in D$.

6. If $A - B = B - A$, then $A = B$.

7. If $A \subseteq B$, then $B - (B - A) = A$.

8. If $B - (B - A) = A$, then $A \subseteq B$.

9. $(A - D) \cup (B - D) \subseteq (A \cup B) - D$

10. $(A \cup B) - D \subseteq (A - D) \cup (B - D)$

11. $(A - B) \cap D$ and $(A - D) \cap B$ are disjoint.

12. *Prove:* $A \cap D \subseteq D - B$ and $A \cap B \subseteq D$ if and only if A and B are disjoint.

13. *Prove:* $D \cup (A \cap B) = D \cap (A \cup B)$ if and only if $D \subseteq A \cup B$ and $A \cap B \subseteq D$.

For Exercises 14–17, define the **symmetric difference,** $A \triangle B$ of A and B by $A \triangle B = (A - B) \cup (B - A)$.

14. *Prove:* $A \triangle B = (A \cup B) - (A \cap B)$.

15. *Prove:*

    (a) $A \triangle B = B \triangle A$

    (b) $A \triangle \emptyset = A$

    (c) $A \triangle A = \emptyset$

16. *Prove:* $(A \triangle B) \triangle D = A \triangle (B \triangle D)$.

17. *Prove:* $A \cap (B \triangle D) = (A \cap B) \triangle (A \cap D)$.

18. Prove that $\cap(\Upsilon \cup \Omega) = (\cap \Upsilon) \cap (\cap \Omega)$.

19. Supply a proof or a counterexample for the conjecture "If $\cup \Omega = \cap \Omega$, then Ω contains exactly one member."

SOLUTIONS TO PRACTICE PROBLEMS

Section 3.1

1. $F = \{0, 1, 2, 3, \ldots, 18, 19\}$, $G = \{-2, -1, 0, 1, 2, 3, 4, 5\}$.

2. Since the solutions to $x^2 = 100$ are $x = 10$ and $x = -10$, $S = \{-10, 10\} = T$.

3.   **(a)** $\subset$ or $\subseteq$

    (b) $\not\subseteq$

    (c) $\subset$ or $\subseteq$

4. $A' = \{-5, -4, -3, -2, 1, 3, 5\}$

    $B' = \{-5, -4, -3, -2, -1, 0, 5\}$

    $A - B = \{-1, 0\}$

    $B - A = \{1, 3\}$

5. $A \cap \emptyset = \emptyset$ for any set A.

    $B \cup \emptyset = B$ for any set B.

    $A \cap B = \{d, f\}$

    $A \cup B = \{a, b, c, d, e, f, 1, 2\}$

Section 3.2

1. The four disjoint sets, in terms of A, B, $'$, and $\cap$, are $A \cap B$, $A \cap B'$, $A' \cap B$, and $A' \cap B'$, as shown in the following Venn diagram:

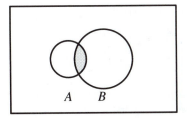

Shaded region is $A \cap B$

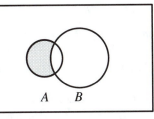

Shaded region is $A \cap B'$

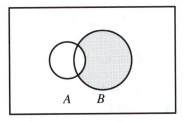

Shaded region is $A' \cap B$

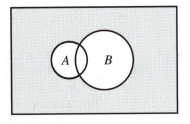

Shaded region is $A' \cap B'$

2. (a) $A \cup B \cup C = \{a, b, c, 3, 4, 5, 6\}$
    (b) $A \cap B \cap C = \{a, 3\}$

Section 3.3

1. (a) $A \cup B = \{a, b, c, d, e, f\}$ $B \cup C = \{a, d, e, f, g, k\}$
    (b) $(A \cup B) \cup C = \{a, b, c, d, e, f, g, k\} = A \cup (B \cup C)$
    (c) $A \cap B = \{a, d\}$ $B \cap C = \{d, f\}$
    (d) $(A \cap B) \cap C = \{d\} = A \cap (B \cap C)$

2. $\neg(A \subseteq B) \Leftrightarrow \neg \forall u(u \in A \Rightarrow u \in B)$

$\Leftrightarrow \exists u(u \in A \wedge u \notin B)$

3. *Proof:* Assume that $u \in A$. Then $u \in A$ or $u \in B$. Hence, $u \in A \cup B$.

4. (a) $(p \vee q) \wedge \neg q \Leftrightarrow p \wedge \neg q$
  (b) *Proof:* Let $u \in U$ be arbitrary. Then:

$$u \in (A \cup B) \cap B' \Leftrightarrow u \in (A \cup B) \text{ and } u \in B'$$

$$\Leftrightarrow (u \in A \text{ or } u \in B) \text{ and } u \notin B$$

$$\Leftrightarrow u \in A \text{ and } u \notin B \qquad [\text{ Use } (p \vee q) \wedge \neg q \Leftrightarrow p \wedge \neg q.]$$

$$\Leftrightarrow u \in A \text{ and } u \in B'$$

$$\Leftrightarrow u \in A \cap B'$$

5. *Proof:* Assume $A \cap B = A$, and prove $A \subseteq B$. Let $u \in U$ be arbitrary, and assume $u \in A$. Then $u \in A \cap B$, since $A \cap B = A$. Hence, $u \in B$. Therefore, $A \subseteq B$.

6. *Proof:* Assume $A \subseteq B$. We first show that $A \cup (B \cap C) \subseteq B \cap (A \cup C)$. Let $x \in A \cup (B \cap C)$. There are two cases.

Case 1. $x \in A$. Then $x \in B$, since $A \subseteq B$. Also, $x \in A \cup C$. Therefore, $x \in B \cap (A \cup C)$.

Case 2. $x \in B \cap C$. Hence, $x \in B$ and $x \in C$. Therefore, $x \in A \cup C$. It follows that $x \in B \cap (A \cup C)$. We now show that $B \cap (A \cup C) \subseteq A \cup (B \cap C)$. Let $x \in B \cap (A \cup C)$. Then $x \in B$ and $x \in A \cup C$. Again, there are two cases.

Case 1. $x \in A$. Then $x \in A \cup (B \cap C)$.

Case 2. $x \in C$. Then $x \in B \cap C$, and hence, $x \in A \cup (B \cap C)$.

7. *Reasons, in order:* definition of relative complement, commutative law, distributive law, commutative law, definition of relative complement.

8. *Proof:* $(A \cup B) \cap (A \cup B') = A \cup (B \cap B') = A \cup \emptyset = A$.

Section 3.4

1. (a) $\cup \Omega = \{a, b, c, d, e\}$
  (b) $\cap \Omega = \{b\}$

2. (a) $x \in \cap \{A, B\} \Leftrightarrow (\forall Y \in \{A, B\})(x \in Y) \Leftrightarrow x \in A \wedge x \in B \Leftrightarrow x \in A \cap B$
  (b) $x \in \cup \{A, B\} \Leftrightarrow (\exists X \in \{A, B\})(x \in X) \Leftrightarrow x \in A \vee x \in B \Leftrightarrow x \in A \cup B$

3. *Proof:* Assume $A \in \Omega$. Let $x \in \cap \Omega$. Then $x \in B$ for all $B \in \Omega$. In particular, $x \in A$.

4. *Proof:* We need to prove that no real number a is an element of $\cap \Upsilon$. Clearly, $a \leq 0$ is not an element of $\cap \Upsilon$, since $a \notin (0, 1/n)$ for any n. The proof that any positive real number is not an element of $\cap \Upsilon$ is identical to the proof using the ϵ-property in Example 4.

5. *Proof:* We use a pick-a-point proof to show that $D \cap \cup \Upsilon \subseteq \cup \{D \cap A : A \in \Upsilon\}$. Let $x \in D \cap \cup \Upsilon$. Then $x \in D$ and $x \in \cup \Upsilon$. Hence, $x \in D$, and there exists $A \in \Upsilon$ such that $x \in A$. So $x \in D \cap A$ for some $A \in \Upsilon$. Therefore, $x \in \cup \{D \cap A : A \in \Upsilon\}$. The other set inclusion is proved by reversing each of the preceding steps.

6. One example is $\Omega = \big\{\{a, b\}, \{b, c\}\big\}$ and $\Upsilon = \big\{\{a, b\}, \{a, c\}\big\}$.

4

Relations and Functions

In this chapter, we consider the concept of a product set and use it to define the concept of a relation, which is used to define the notion of a function. Relations and functions are the most useful mathematical tools discussed in this text.

4.1 PRODUCT SETS

The idea of a function plays a central role in mathematics and many other related fields. Intuitively, a function is a rule that assigns, to each given object in one set, a unique object in another set. For example, consider the rule that assigns, to the real number x, the real number x^2. If the given real number is 3, then the real number assigned to it is 3^2, or 9. Functions of this type are encountered in calculus. In our development of mathematics, with a foundation in set theory, we would like a function to be a set. In particular, a function will be a set with properties that make it act like a rule.

A function must distinguish between two elements, the given one and the assigned one. For example, to define a function that takes 3 and assigns it the value 9, we could try $\{3, 9\}$. However, since $\{3, 9\} = \{9, 3\}$, this does not distinguish between the given, 3, and the assigned, 9. Rather, we need an object that designates 3 as the first number and 9 as the second. Such an object is called an ordered pair.

An **ordered pair** is an object made up of two elements. If the first element is x and the second element is y, then the ordered pair is denoted by (x, y). A rigorous definition is discussed in the exercises. For our purposes, what is important is that ordered pairs satisfy the following property.

Ordered Pair Property $(x, y) = (z, w)$ if and only if $x = z$ and $y = w$.

The **first coordinate** of the ordered pair (x, y) is x and the **second coordinate** is y. Hence, two ordered pairs are identical if and only if their first coordinates are the same and their second coordinates are the same.

Definition.   Given two sets A and B, the **product set** (or **Cartesian product**) of A and B is $A \times B = \{(a, b) : a \in A \text{ and } b \in B\}$.

So $A \times B$ consists of all ordered pairs with first coordinate from A and second coordinate from B. The product set $A \times A$ is frequently written A^2.

In general, $A \times B \neq B \times A$ as illustrated in Example 1.

Example 1

Let $A = \{a, b, c, d\}$ and $B = \{1, 2, 3\}$. Then $A \times B = \{(a, 1), (a, 2), (a, 3), (b, 1), (b, 2), (b, 3), (c, 1), (c, 2), (c, 3), (d, 1), (d, 2), (d, 3)\}$ and $B \times A = \{(1, a), (1, b), (1, c), (1, d), (2, a), (2, b), (2, c), (2, d), (3, a), (3, b), (3, c), (3, d)\}$. There are 4 elements in A and 3 elements in B, while $A \times B$ and $B \times A$ each has 12 elements. Note that $A \times B \neq B \times A$.

Practice Problem 1.   Let $A = \{a, b, c, d, e\}$ and $B = \{3, 4\}$.
a. Find $A \times B$.
b. Find $B \times A$.
c. Observe that $A \times B \neq B \times A$.

If $A = B$, then certainly $A \times B = B \times A$. But what about the converse? Practice Problem 2 shows that $A \times B = B \times A$ does not imply $A = B$.

Practice Problem 2.   Let $A = \emptyset$ and $B = \{1, 2\}$. Clearly $A \neq B$. Verify that $A \times B = \emptyset = B \times A$.

When we pick an element in $A \times B$ we may assume that it is an ordered pair. So instead of writing "Let $v \in A \times B$. Therefore, $v = (x, y)$ with $x \in A$ and $y \in B$," we write "Let $(x, y) \in A \times B$. Therefore, $x \in A$ and $y \in B$."

If one or both of the sets A and B are empty, then $A \times B = \emptyset$. Conversely, if $A \times B = \emptyset$, then at least one of A and B is empty. Following is a proof of these assertions.

Theorem 1 $A \times B = \emptyset$ if and only if $A = \emptyset$ or $B = \emptyset$.

Proof. We prove the equivalent statement that $A \times B \neq \emptyset$ if and only if $A \neq \emptyset$ and $B \neq \emptyset$.
($\Rightarrow$) Assume $A \times B \neq \emptyset$. Then there exists an element $(a, b) \in A \times B$. Hence, $a \in A$ and $b \in B$. Therefore, $A \neq \emptyset$ and $B \neq \emptyset$.
($\Leftarrow$) Assume $A \neq \emptyset$ and $B \neq \emptyset$. Then there exist elements $a \in A$ and $b \in B$. Hence, $(a, b) \in A \times B$. Therefore, $A \times B \neq \emptyset$.   □

Now, assume that $A \neq \emptyset$ and $B \neq \emptyset$. Is the conjecture $A \times B = B \times A$ if and only if $A = B$ true under that condition? In Exercise 9, you are asked to supply a proof or a counterexample.

Let us examine some examples of products of subsets of the real numbers. The set $\mathcal{R} \times \mathcal{R}$ is called the **Cartesian plane** or the **Euclidean plane**. We will usually call the Cartesian plane simply the **plane**. The **x-axis** in $\mathcal{R} \times \mathcal{R}$ is the set of points $\mathcal{R} \times \{0\}$, while $\{0\} \times \mathcal{R}$ is the **y-axis.** Any set of ordered pairs in the plane is a **graph.**

Example 2

Let $A = [1, 3]$ and $B = [2, 5)$. For each of the following sets, give the inequalities specifying the set and draw the graph.

   a. $A \times B$

   b. $B \times A$.

Solution:

   a. $A \times B = \{(x, y) : 1 \le x \le 3 \text{ and } 2 \le y < 5\}$

   b. $B \times A = \{(x, y) : 2 \le x < 5 \text{ and } 1 \le y \le 3\}$. The graphs are shown in Figure 4–1.

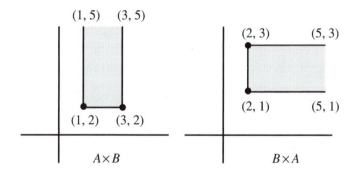

Figure 4–1

Theorem 2   If $A \subseteq D$ and $B \subseteq E$, then $A \times B \subseteq D \times E$.

   Proof. Assume $A \subseteq D$ and $B \subseteq E$. We use the pick-a-point method to prove that $A \times B \subseteq D \times E$. Let $(u, v) \in A \times B$ be an arbitrary element. Then $u \in A$ and $v \in B$. But by the assumption, $u \in D$ and $v \in E$. Therefore, $(u, v) \in D \times E$.   □

Is the converse of Theorem 2 true? Exercise 11 requests a counterexample or a proof for this conjecture. Before trying to prove any conjecture, it is best to test it with several examples. In this case, a counterexample consists of sets A, B, D, and E such that $A \times B \subseteq D \times E$ is true but $A \subseteq D$ and $B \subseteq E$ is false. In Practice Problem 2, we saw that the empty set is sometimes useful for finding counterexamples. Thus, it is possible that in a counterexample, one or more of the sets A, B, D, and E will be the empty set.

Example 3

Conjecture: $(A \cap B) \times C \subseteq (A \times C) \cap (B \times C)$.

Solution   Let us test this conjecture with sample sets in the plane. Since we are trying to find a counterexample, A and B should not be disjoint, because that makes the set inclusion trivially true. Accordingly, consider $A = [2, 5]$, $B = [1, 3] \cup [4, 7]$, and $C = [1, 2]$. Note that $A \cap B = [2, 3] \cup [4, 5]$.

Since the sets A and B consist of the first coordinate elements, we identify the elements of those sets with points on the x-axis. Similarly, we identify elements of C with points on the y-axis. Figure 4–2 shows the set of points.

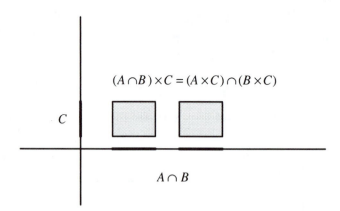

C

$(A \cap B) \times C = (A \times C) \cap (B \times C)$

$A \cap B$

Figure 4–2

This choice of sets makes the inclusion true. Other examples also show that the inclusion is true. In fact, equality holds in all of them. We prove this in Theorem 3.

Theorem 3 $(A \cap B) \times C = (A \times C) \cap (B \times C)$.

Proof.   We use the pick-a-point method to prove that $(A \cap B) \times C \subseteq (A \times C) \cap (B \times C)$. Proof of the other set inclusion is left as Practice Problem 3.

Assume $(x, y) \in (A \cap B) \times C$. Then $x \in A \cap B$ and $y \in C$. Hence, $x \in A$ and $x \in B$ and $y \in C$. So $(x, y) \in A \times C$ and $(x, y) \in B \times C$. Therefore, $(x, y) \in (A \times C) \cap (B \times C)$.   □

Practice Problem 3.   Prove that $(A \times C) \cap (B \times C) \subseteq (A \cap B) \times C$.

After success with one conjecture, we try another. First we introduce a notational convenience.

Notation   For any two sets X and Y, $X \overset{?}{=} Y$ means

1. Prove or disprove the conjecture $X = Y$.
2. If $X \neq Y$, then prove or disprove the conjectures
   $X \subseteq Y$ and $Y \subseteq X$.

Example 4

$(A \cup B) \times (C \cup D) \overset{?}{=} (A \times C) \cup (B \times D).$

Solution First we consider sets in the plane as possible counterexamples to the conjecture $(A \cup B) \times (C \cup D) = (A \times C) \cup (B \times D)$. After a little thought, we decide to pick A and B as disjoint sets and C and D as disjoint sets. We let $A = [1, 2]$, $B = [3, 5]$, $C = [1, 2]$, and $D = [3, 4]$. Figure 4–3 illustrates the sets of points.

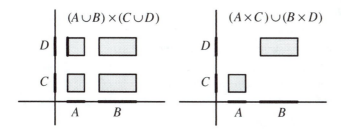

Figure 4–3

Thus, the set equality conjecture is false. The counterexample also shows that, in general, $(A \cup B) \times (C \cup D) \not\subseteq (A \times C) \cup (B \times D)$. However, it is still possible that $(A \times C) \cup (B \times D) \subseteq (A \cup B) \times (C \cup D)$. Exercise 12 requests a counterexample or proof for this conjecture.

Several other theorems concerning products of sets are given in the exercises.

EXERCISE SET 4.1

1. Let $A = [1, 3]$, $B = (2, 5]$, and $C = \{2\}$. Write the inequalities (or equalities) that specify each of the following sets, and draw each graph.
   (a) $(A \cup B) \times C$
   (b) $(A \times C) \cup (B \times C)$

2. Let $A = [1, 3]$, $B = (2, 5]$, and $C = \{2\}$. Write the inequalities (or equalities) that specify each of the following sets, and draw each graph.
   (a) $A \times (B \cap C)$
   (b) $(A \times B) \cap (A \times C)$

3. Let $A = [1, 3]$, $B = \{5\}$, and $C = [4, 6]$. Write the inequalities (or equalities) that specify each of the following sets, and draw each graph.
   (a) $(A \cup B) \times C$
   (b) $(A \times C) \cup (B \times C)$

4. Let $A = [1, 3]$, $B = \{5\}$, and $C = [4, 6]$. Write the inequalities (or equalities) that specify each of the following sets, and draw each graph.
   (a) $A \times (B \cap C)$
   (b) $(A \times B) \cap (A \times C)$

5. The following definition of ordered pair was first given by Kazimierz Kuratowski in 1921: (x, y) is defined to be $\big\{\{x\}, \{x, y\}\big\}$. Prove, using this definition, that $(a, b) = (c, d)$ if and only if $a = c$ and $b = d$.

6. Prove that $(A \cup B) \times C = (A \times C) \cup (B \times C)$.

7. Prove that $A \times (B \cup C) = (A \times B) \cup (A \times C)$.

8. Prove that $A \times (B \cap C) = (A \times B) \cap (A \times C)$.

9. Supply a proof or counterexample for the conjecture: If $A \times B \neq \emptyset$, then $A \times B = B \times A$ if and only if $A = B$.

10. Supply a proof or counterexample for the conjecture: $A \times B$ and $B \times A$ are disjoint if and only if A and B are disjoint.

11. Supply a proof or counterexample for the converse of Theorem 2.

12. Find a proof or counterexample for the conjecture that $(A \times C) \cup (B \times D) \subseteq (A \cup B) \times (C \cup D)$.

13. Find a proof or counterexample for the conjecture that $(A \times C) \cap (B \times D) \subseteq (A \cap B) \times (C \cap D)$.

14. Find a proof or counterexample for the conjecture that $(A \cap B) \times (C \cap D) \subseteq (A \times C) \cap (B \times D)$.

15. $A \times (B - C) \stackrel{?}{=} (A \times B) - (A \times C)$

16. Prove that $A \times \cup \Omega = \cup \{A \times B : B \in \Omega\}$.

17. Prove that $A \times \cap \Omega = \cap \{A \times B : B \in \Omega\}$.

4.2 RELATIONS

In the previous section, we saw that every product set is a set of ordered pairs. We now study sets of ordered pairs that are not necessarily product sets.

Before defining a relation, we give an example.

Example 1

Mr. and Mrs. Smith have four children: John, Sally, Jim, and Jill. We would like to represent the relation "is the sister of" as a set. In such a relation the order in which people are mentioned is important. For example, "Sally is the sister of Jim" is true, but "Jim is the sister of Sally" is false. We can represent the relation "is the sister of" as follows:

$$R = \{(\text{Sally}, \text{John}), (\text{Sally}, \text{Jim}), (\text{Sally}, \text{Jill}), (\text{Jill}, \text{John}), (\text{Jill}, \text{Sally}), (\text{Jill}, \text{Jim})\}$$

Since Sally is the sister of Jim, (Sally, Jim)$\in R$. Since Jim is not the sister of Sally, (Jim, Sally) $\notin R$. If $A = \{\text{Sally}, \text{Jill}\}$ and $B = \{\text{John}, \text{Jim}, \text{Sally}, \text{Jill}\}$, then we say R is a relation from A to B. Note that $R \subseteq A \times B$.

Definition. A **binary relation** is a set of ordered pairs.

Given sets A and B, a **binary relation** R **from** A **to** B is a subset of $A \times B$; that is, $R \subseteq A \times B$. When $A = B$, we say that R is a **binary relation on** A.

In what follows, we will use **relation** interchangeably with binary relation. So a binary relation (or relation) from a set A to a set B is a set of ordered pairs from $A \times B$, and a relation on A is a set of ordered pairs of elements from A.

Practice Problem 1. Refer to Example 1. Let $A = \{\text{Mr. Smith, Mrs. Smith}\}$ and $B = \{\text{Sally, Jill, John, Jim}\}$.

a. Let F be the relation "is the father of." Find F. F is a relation from which set to which set?

b. Let S be the relation "is the spouse of." Find S. S is a relation on which set?

Example 2

Let $A = \{4, 6\}$ and $B = \{2, 8, 12\}$. Then $A \times B = \{(4, 2), (4, 8), (4, 12), (6, 2), (6, 8), (6, 12)\}$. The set $R = \{(4, 8), (6, 12)\}$ is a binary relation from A to B, since $R \subseteq A \times B$. Observe that an ordered pair $(a, b) \in A \times B$ is a member of R if and only if $b = 2a$. Hence, the set of ordered pairs R specifies the relationship $b = 2a$ for $a \in A$ and $b \in B$.

We often write aRb for $(a, b) \in R$. Thus, in Example 2, $4R8$ and $6R12$.

Example 3

Let $A = \{a, b, c, d\}$ and $B = \{x, y, t\}$.

a. The set $R = \{(a, x), (b, t), (d, y), (c, x)\}$ is a binary relation from A to B. Note that aRx and cRx, but it is not the case that dRx.

b. The set $A \times B$ is also a relation from A to B.

c. The set $S = \{(a, b), (a, c), (c, d), (d, b), (d, d)\}$ is a relation on A.

d. The set $T = \{(x, y), (y, t), (t, x)\}$ is a binary relation on B.

Example 4

Let $A = B = \mathcal{R}$. Thus $A \times B$ is the plane $\mathcal{R} \times \mathcal{R}$. The following subsets R, S, and T of $\mathcal{R} \times \mathcal{R}$ are binary relations on $\mathcal{R}$:

a. $R = \{(x, y) : x = y\}$. Note that xRx for all $x \in \mathcal{R}$.

b. $S = \{(x, y) : |x| + |y| \leq 1\}$

c. $T = \{(x, y) : y = 1.5\}$. Note that $xT1.5$ for all $x \in \mathcal{R}$.

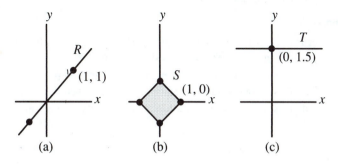

Figure 4–4

The relations R, S, and T of Example 4 are shown as subsets of the plane in Figure 4–4.

Properties of Relations

A binary relation R may satisfy several important properties. The four most important are the reflexive, symmetric, antisymmetric, and transitive properties.

Definition. Let R be a binary relation. The properties listed in the left-hand column of the following table are defined in the right-hand column.

Property of R	Defining Condition
R is **reflexive on** A	R is a relation on A and for all $x \in A$, xRx
R is **symmetric**	for all x, y, if xRy, then yRx
R is **antisymmetric**	for all x, y, if xRy and yRx, then $x = y$
R is **transitive**	for all x, y, z, if xRy and yRz, then xRz

Instead of "R is reflexive on A," we will sometimes write, more simply, "R is **reflexive**" when it is clear to which set A refers. However, it is important to note that whether a relation R is reflexive depends on both R *and* the set A. For example, if $R = \{(1,1),(2,2),(1,2)\}$, then R is reflexive on $\{1,2\}$, but R is not reflexive on $\{1,2,3\}$.

The contrapositive form of the antisymmetric property is frequently useful. That is, R is antisymmetric if and only if for all x, y, if $x \neq y$, then $\neg xRy$ or $\neg yRx$.

It is important to note that each of the four properties is true only when the defining condition is true for *all* elements. Hence, a given property is false if the defining condition fails to hold for even one selection of elements.

Since the symmetric, antisymmetric, and transitive properties are defined in terms of implications, they can be vacuously true. As an extreme example, take the relation $\emptyset$. This relation is not reflexive on $\{a\}$ since it does not contain (a,a). However, $\emptyset$ is symmetric, antisymmetric, and transitive.

Example 5

Let $R = \{(a,a),(c,c),(a,b),(b,a),(a,c)\}$ be a relation on the set $\{a,b,c\}$.
  a. R is not reflexive, since we do not have bRb.
  b. R is not symmetric, since aRc but we do not have cRa.
  c. R is not antisymmetric, since aRb and bRa but $a \neq b$.
  d. R is not transitive, since bRa and aRc but we do not have bRc.

Example 6

Let $R = \{(a,a),(b,b),(c,c),(b,a)\}$ be a relation on the set $\{a,b,c\}$.

a. R is reflexive because aRa, bRb, and cRc.

b. R is not symmetric, since bRa but we do not have aRb.

c. R is antisymmetric because (b, a) is the only pair in R with unequal coordinates.

d. R is transitive because bRa and aRa implies bRa is true, and so is every similar statement.

Definition.   Let A be a set. The **identity relation** on A is

$$1_A = \big\{(x, x) : x \in A\big\}$$

As a consequence of this definition, $x1_Ay$ if and only if $x = y$.

Example 7

Let A be any nonempty set, and let $S = A \times A$.
a. Both 1_A and S are reflexive.

b. Both 1_A and S are symmetric.

c. 1_A is antisymmetric. However, S is not antisymmetric unless A has exactly one element.

d. Both 1_A and S are transitive.

Note that "antisymmetric" does not mean "not symmetric." A binary relation can be both symmetric and antisymmetric, or it can be symmetric without being antisymmetric, or vice versa. Also, a binary relation can be neither symmetric nor antisymmetric. Since this tends to get confusing, it is a good idea to examine several concrete cases for clarification. The following example illustrates some combinations of properties a given binary relation can possess.

Example 8

a. $R = \{(a, a), (b, b), (c, c)\}$ is both symmetric and antisymmetric.

b. $S = \{(a, a), (a, c), (b, b), (c, a)\}$ is symmetric but not antisymmetric. It is not antisymmetric because aSc and cSa but $a \neq c$.

c. $T = \{(a, a), (b, a), (b, b), (c, a), (c, b), (c, c)\}$ is antisymmetric but not symmetric.

d. $U = \{(a, a), (a, c), (b, a), (b, b), (b, c), (c, b)\}$ is neither symmetric nor antisymmetric.

e. The foregoing relations R and T are reflexive on $\{a, b, c\}$, whereas S and U are not reflexive on $\{a, b, c\}$.

Practice Problem 2.   Find relations R, S, T on $\{1, 2, 3, 4\}$ such that:
a. R is reflexive but neither antisymmetric nor symmetric.

b. S is antisymmetric but neither reflexive nor symmetric.

c. T is symmetric and antisymmetric but not reflexive.

When we define a relation R on A we have been writing, for example,

$$\text{Let } R = \{(x, y) : x, y \in A \text{ and } x = y^2\}$$

Instead of this, we shall often write

$$\text{Define } R \text{ on } A \text{ by } xRy \text{ if and only if } x = y^2$$

Of course, x and y are arbitrary elements of A.

Clearly, the first definition can be obtained from the second and vice versa.

Example 9

Define a relation R on $\mathcal{R}$ as follows: xRy if and only if $x - 1 \le y \le x + 1$. Give a proof or a counterexample for each of the following statements:

    a. R is reflexive.

    b. R is symmetric.

    c. R is antisymmetric.

    d. R is transitive.

Solution

a. R is reflexive since $x - 1 \le x \le x + 1$, and hence, xRx.

b. We prove that R is symmetric. Assume xRy. Therefore, $x - 1 \le y \le x + 1$. From $x - 1 \le y$, we obtain $x \le y + 1$ and from $y \le x + 1$, we obtain $y - 1 \le x$. Putting these together, we have $y - 1 \le x \le y + 1$. Hence, yRx.

c. We show that R is not antisymmetric. Observe that $2R1.5$ since $2 - 1 < 1.5 < 2 + 1$. Also, since R is symmetric, $1.5R2$. However, $1.5 \ne 2$.

d. R is not transitive. Since $1 - 1 \le 2 \le 1 + 1$ and $2 - 1 \le 3 \le 2 + 1$, we have $1R2$ and $2R3$. However, we do not have $1R3$, since $1 - 1 \le 3 \le 1 + 1$ is false.

Practice Problem 3. Define a relation R on $\mathcal{Q}$ as follows: xRy if and only if for some $k \in \mathcal{N}$, $x = 3^k y$. Give a proof or a counterexample for each of the following statements.

    a. R is reflexive.

    b. R is symmetric.

    c. R is antisymmetric.

    d. R is transitive.

Operations on Relations

Since relations are sets, we can perform the usual set operations of union, intersection, and set difference on them. However, there are additional operations that can be performed on relations. In what follows, all relations are assumed to be on some set A.

Definition. The **inverse** of the relation R is $R^{-1} = \{(y, x) : xRy\}$.

In other words, R^{-1} is the set of ordered pairs that are the reverse of the ordered pairs in R.

Definition. The **composition of relations** R and S is $R \circ S = \{(x, y) :$ for some z, xSz and $zRy\}$.

Figure 4–5 is one way to envision $x(R \circ S)y$, or $(x, y) \in R \circ S$.

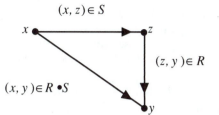

Figure 4–5

Example 10

Let $R = \{(2, 4), (2, 7), (3, 4)\}$ and $S = \{(1, 2), (3, 5), (1, 3)\}$.
Find: a. R^{-1} b. S^{-1} c. $R \circ S$ d. $S \circ R$

Solution
a. $R^{-1} = \{(4, 2), (7, 2), (4, 3)\}$. $[(4, 2) \in R^{-1}$ since $(2, 4) \in R.]$
b. $S^{-1} = \{(2, 1), (5, 3), (3, 1)\}$.
c. $R \circ S = \{(1, 4), (1, 7)\}$. $[(1, 4) \in R \circ S$ since there is an element z such that $1Sz$ and $zR4$, namely, $z = 2$. Note that $z = 3$ would also work.]
d. $S \circ R = \emptyset$. $[(2, 5) \notin S \circ R$ since there is no z such that $2Rz$ and $zS5$.]

It is often useful to diagram relations such as those in Example 10. Figure 4–6 shows such a diagram.

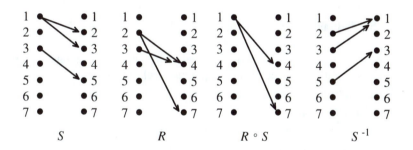

Figure 4–6

Practice Problem 4. Let $R = \{(1, 3), (1, 4), (2, 3)\}$ and $S = \{(3, 1), (2, 2)\}$. Find R^{-1}, S^{-1}, $R \circ S$, $(R \circ S)^{-1}$, $R^{-1} \circ S^{-1}$, and $S^{-1} \circ R^{-1}$. What can be said about $(R \circ S)^{-1}$ and $S^{-1} \circ R^{-1}$?

Notice that $R \circ \emptyset = \emptyset$, for if $(x, y) \in R \circ \emptyset$, then for some z, $(x, z) \in \emptyset$. Similarly, $\emptyset \circ R = \emptyset$.

Example 11

Define two relations on $\mathcal{N}$ as follows:

xRy if and only if $x < y$

xSy if and only if x divides y (see Exercise Set 2.5 for the definition of divides)

Show that $R \circ S = R$.

Solution Assume $x(R \circ S)y$. Then for some z, xSz and zRy; that is, x divides z and $z < y$. Since x divides z, $x \leq z$ (Why?) Hence, $x < y$, and therefore, xRy. Now assume xRy. Then $x < y$. We wish to find a z such that x divides z and $z < y$. We accomplish this by letting $z = x$. Then xSz and zRy. Therefore, $x(R \circ S)y$.

Definition.   The **domain** of a relation R is $\mathbf{dom(R)} = \{x : \text{for some } y, xRy\}$.

Definition.   The **range** of a relation R is $\mathbf{ran(R)} = \{y : \text{for some } x, xRy\}$.

In other words, dom(R) is the set of all first coordinates of the ordered pairs in the relation R, and ran(R) is the set of all second coordinates of the ordered pairs in R.

Example 12

Let $R = \{(1, 2), (1, 3), (2, 3)\}$.
   a. Find dom(R).
   b. Find ran(R).

Solution
a. dom(R) = $\{1, 2\}$. [$1 \in$ dom(R) since $1Ry$ for some y, namely, $y = 2$ (or $y = 3$).]
b. ran(R) = $\{2, 3\}$.

Practice Problem 5.   Let $R = \{(2, 4), (3, 2)\}$.
a. Find dom(R).
b. Find ran(R).

Before proving theorems about relations and the operations on them, it is useful to summarize precisely the result of applying each operation.
   a. $x(R \circ S)y$ if and only if for some z, xSz and zRy.
   b. $xR^{-1}y$ if and only if yRx.
   c. $x \in$ dom(R) if and only if for some y, xRy.
   d. $y \in$ ran(R) if and only if for some x, xRy.

Remember that xRy means $(x, y) \in R$. So we can rewrite, for example, a as follows:

   a. $(x, y) \in R \circ S$ if and only if for some z, $(x, z) \in S$ and $(z, y) \in R$.

Sentences b, c, and d may be rewritten similarly.
For the remainder of this section, R, S, and T are relations.

Theorem 1 $\text{dom}(R \cup S) = \text{dom}(R) \cup \text{dom}(S)$

Proof. Let $x \in \text{dom}(R \cup S)$. Then there is a y such that $x(R \cup S)y$. (That is, $(x, y) \in R \cup S$.) Hence, xRy or xSy.

Case 1. xRy. Then $x \in \text{dom}(R)$, and hence, $x \in \text{dom}(R) \cup \text{dom}(S)$.

Case 2. xSy. Then $x \in \text{dom}(S)$, and hence, $x \in \text{dom}(R) \cup \text{dom}(S)$.

In either case, we obtain $x \in \text{dom}(R) \cup \text{dom}(S)$.

Now let $x \in \text{dom}(R) \cup \text{dom}(S)$.

Case 1. $x \in \text{dom}(R)$. Hence, xRy for some y. Therefore, $x(R \cup S)y$, and it follows that $x \in \text{dom}(R \cup S)$.

Case 2. $x \in \text{dom}(S)$. Then xSy for some y. Therefore, $x(R \cup S)y$, and it follows that $x \in \text{dom}(R \cup S)$.

In either case, we obtain $x \in \text{dom}(R \cup S)$. $\qquad\square$

Practice Problem 6.   Prove that $\text{ran}(R \cup S) = \text{ran}(R) \cup \text{ran}(S)$.

In the following theorem we use $(x, y) \in R$ instead of xRy, since some of the expressions for relations are rather cumbersome and the latter notation is difficult to read.

Theorem 2 $R \circ (S \cap T) \subseteq (R \circ S) \cap (R \circ T)$, and equality does not hold.

Proof. Let $(x, y) \in R \circ (S \cap T)$. Then, $(x, z) \in (S \cap T)$ and $(z, y) \in R$ for some z. Since $(x, z) \in (S \cap T)$, we also have $(x, z) \in S$ and $(x, z) \in T$. From $(x, z) \in S$ and $(z, y) \in R$, we obtain $(x, y) \in (R \circ S)$ and from $(x, z) \in T$ and $(z, y) \in R$, we obtain $(x, y) \in (R \circ T)$. Hence, $(x, y) \in (R \circ S) \cap (R \circ T)$.

The following counterexample shows that equality does not hold. Let $R = \{(2, 1), (3, 1)\}$, $S = \{(1, 2)\}$, and $T = \{(1, 3)\}$. Then $R \circ S = \{(1, 1)\}$ and $R \circ T = \{(1, 1)\}$, and hence, $(R \circ S) \cap (R \circ T) = \{(1, 1)\}$. However, $S \cap T = \emptyset$, and hence, $R \circ (S \cap T) = \emptyset$. $\qquad\square$

Practice Problem 7.   Prove that $R \circ (S \cup T) = (R \circ S) \cup (R \circ T)$.

Theorem 3 R is symmetric if and only if $R^{-1} \subseteq R$.

Proof. Assume R is symmetric and let $xR^{-1}y$. Then yRx. But since R is symmetric, xRy. Therefore, $R^{-1} \subseteq R$. Assume $R^{-1} \subseteq R$ and let xRy. Then $yR^{-1}x$. But since $R^{-1} \subseteq R$, yRx. Hence, R is symmetric. $\qquad\square$

Practice Problem 8.   *Prove:* R is transitive if and only if $R \circ R \subseteq R$.

EXERCISE SET 4.2

1. Let $R = \{(1, 1), (1, 2), (1, 3), (2, 2), (2, 3), (3, 3)\}$ be a relation on $\{1, 2, 3\}$. Determine whether R is reflexive, symmetric, antisymmetric, or transitive.

2. Let $R = \{(1,1), (1,2), (1,3), (2,1), (2,2), (3,1), (3,3)\}$ be a relation on $\{1,2,3\}$. Determine whether R is reflexive, symmetric, antisymmetric, or transitive.

3. Let $A = \{1,2,3\}$ and $B = \{4,9,21,25\}$. Define the binary relation R from A to B by aRb if and only if a divides b for $a \in A$ and $b \in B$.
  (a) List all the members of R.
  (b) Determine whether R is reflexive on $A \cup B$, symmetric, antisymmetric, or transitive.

4. Let $A = \{1,2,3,4\}$. Define the binary relation R on A by xRy if and only if $x \leq y$.
  (a) List all the members of R.
  (b) Determine whether R is reflexive, symmetric, antisymmetric, or transitive.

5. Let S be the binary relation on $\mathcal{R}$ defined by xSy if and only if $y = -2x$.
  (a) Graph the set S in the x–y plane.
  (b) Determine whether S is reflexive, symmetric, antisymmetric, or transitive.

6. Let $A = \{2,3\}$ and $B = \{6,9,10,12\}$.
  (a) Let T be a binary relation from A to B defined by aTb if and only if $a+b$ is an even integer. List all the members of T.
  (b) Let S be a binary relation from A to B defined by aSb if and only if b is a multiple of a. List all the members of S.

7. Let S be a binary relation on the set $\mathcal{R}$ such that $S = \{(x,y) : y < x\}$.
  (a) Graph the set S in the plane.
  (b) Determine whether S is reflexive, symmetric, antisymmetric, or transitive.

8. Let $A = \{1,2,5,10\}$. Define R on A by xRy if and only if y divides x.
  (a) List all the members of R.
  (b) Determine whether R is reflexive, symmetric, antisymmetric, or transitive.

9. Let $A = \{1,2,3,4,5\}$. Define R on A by xRy if and only if $x - y$ is divisible by 2.
  (a) List all the members of R.
  (b) Determine whether R is reflexive, symmetric, antisymmetric, or transitive.

10. Refer to Exercise 8.
  (a) Is $R^{-1} = R$?
  (b) Is $R \circ R = R$?
  (c) Is $R^{-1} \circ R = R$?

11. Refer to Exercise 9.
  (a) Is $R^{-1} = R$?
  (b) Is $R \circ R = R$?
  (c) Is $R^{-1} \circ R = R$?

In Exercises 12–17, determine whether the given relation R on the given set A is reflexive, symmetric, antisymmetric, or transitive.

12. $A = \mathcal{R}$, xRy if and only if $x^2 + y^2 = 1$.

13. $A = \mathcal{Z}$, xRy if and only if $x = ky$ for some $k \in \mathcal{Z}$.

14. $A = P(S)$, where $P(S)$ is the power set of S, XRY if and only if $X \subseteq Y$.

15. $A = \mathcal{R} \times \mathcal{R}$, $(x,y)R(u,v)$ if and only if $x < u$ or $y < v$.

16. $A = \mathcal{R} \times \mathcal{R}$, $(x,y)R(u,v)$ if and only if $x < u$ and $y < v$.

17. $A = \mathcal{R}$, xRy if and only if $x = y^k$ for some $k \in \mathcal{N}$.

18. Let $A = \{1,2,3,4,5,6,7\}$. Define R on A by xRy if and only if $y = 2x$.
  (a) $xR^{-1}y$ if and only if _____.

(b) $x(R \circ R)y$ if and only if _____.

19. Let $A = \{1, 2, 3, 4, 5, 6, 7\}$. Define R on A by xRy if and only if $y = 2 + x$.
    (a) $xR^{-1}y$ if and only if _____.
    (b) $x(R \circ R)y$ if and only if _____.

20. Show that $S \circ R = R$ in Example 11.

21. Assume R is a relation on A. Prove that R is reflexive if and only if $1_A \subseteq R$.

22. Assume R is a relation on A. Prove that $R \circ 1_A = 1_A \circ R = R$.

23. Prove that if $R \subseteq S$, then $R^{-1} \subseteq S^{-1}$.

24. $(R^{-1})^{-1} \stackrel{?}{=} R$

25. $(R \circ T)^{-1} \stackrel{?}{=} R^{-1} \circ T^{-1}$

26. $(R \circ T)^{-1} \stackrel{?}{=} T^{-1} \circ R^{-1}$

27. $\text{dom}(R \cap S) \stackrel{?}{=} \text{dom}(R) \cap \text{dom}(S)$

28. $\text{ran}(R \cap S) \stackrel{?}{=} \text{ran}(R) \cap \text{ran}(S)$

In Exercises 29–31, supply a proof or a counterexample for the given conjecture.

29. R is symmetric if and only if $R^{-1} = R$.

30. If $R \subseteq S$, then $R \circ T \subseteq S \circ T$.

31. R is symmetric and transitive if and only if $R^{-1} \circ R = R$.

4.3 FUNCTIONS

As mentioned in Section 4.1, a function can be considered a rule that takes an element in one set and produces a unique element in another set. A function is thus a special type of binary relation.

Definition. A binary relation f is a **function** if for every $x \in \text{dom}(f)$, there is a unique y such that $(x, y) \in f$.

Another way to say that a relation f is a function is that if $(x, y) \in f$ and $(x, y') \in f$, then $y = y'$.

We say that f is a **function from** D **to** C if f is a function with $\text{dom}(f) = D$, and $\text{ran}(f) \subseteq C$. The set C is called the **codomain** of f. We write

$$f : D \to C \text{ for "} f \text{ is a function from } D \text{ to } C\text{"}$$

Now assume f is a function from D to C, and let $x \in D$. Then, since f is a function, there is a unique $y \in C$ such that $(x, y) \in f$. This unique $y \in C$ is called $f(x)$. It follows that if $(x, w) \in f$, then $w = f(x)$ and $(x, f(x)) \in f$. The function f can be written as a set of ordered pairs, $f = \{(x, f(x)) : x \in D\}$. The element $f(x)$ is read "f of x" and is sometimes called the value of f at x or the *image of x* under f. We refer to $f(x)$ as a *function value* and sometimes say that x is mapped to $f(x)$.

Equality of Functions

Theorem 1   Two functions f and g are equal if and only if $\text{dom}(f) = \text{dom}(g)$ and $f(x) = g(x)$ for all $x \in \text{dom}(f)$.

Proof. ($\Rightarrow$) Assume that $f = g$. Since f and g consist of the same ordered pairs, $\text{dom}(f) = \text{dom}(g)$. Let $x \in \text{dom}(f)$. Then $(x, f(x)) \in f$. Hence, $(x, f(x)) \in g$, since $f = g$. However, $(x, g(x)) \in g$, and by the uniqueness property for g, we have $f(x) = g(x)$.
($\Leftarrow$) Assume that $\text{dom}(f) = \text{dom}(g)$ and $f(x) = g(x)$ for all $x \in \text{dom}(f)$. Pick an arbitrary element $(x, f(x)) \in f$. Now, since $(x, g(x)) \in g$, from the assumption that $f(x) = g(x)$, it follows that $(x, f(x)) = (x, g(x))$. Hence, $(x, f(x)) \in g$. So we have $f \subseteq g$. The inclusion $g \subseteq f$ is proved similarly. □

Corollary.   Two functions $f : D \to C$ and $g : D \to C$ are equal if and only if $f(x) = g(x)$ for all $x \in D$.

Proof. This proof is left as an exercise. □

Functions with finite domains can be pictured in a diagram. We illustrate this in the following example.

Example 1

Let $D = \{a, b, c\}$ and $C = \{1, 2, 3, 4\}$ be the domain and codomain, respectively, of the function f given by $f(a) = 1$, $f(b) = 3$, and $f(c) = 3$. In other words, $f = \{(a, 1), (b, 3), (c, 3)\}$ when expressed as a set of ordered pairs. The function f is pictured as an arrow diagram in Figure 4–7.

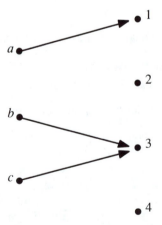

Figure 4–7

The next example illustrates the two main reasons a relation can fail to be a function.

Example 2

a. Let R and S be binary relations from the set $D = \{a, b, c\}$ to the set $C = \{1, 2, 3, 4\}$ given by $R = \{(a, 1), (c, 3)\}$ and $S = \{(a, 1), (b, 3), (c, 3), (c, 4)\}$ (see Figure 4–8). Then neither of R and S is a function from D to C. However, R is a function from $\{a, c\}$ to C.

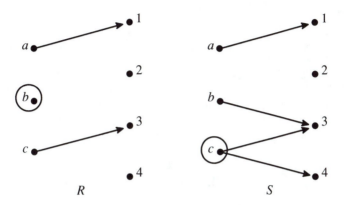

R S

The element $b \in D$, but   The element $c \in D$ does not have
$b \notin$ dom (R).   a unique image.

Figure 4–8

b. Let $R = \{(x, y) : x \in \mathcal{R} - \{0\}$ and $y = 1/x\}$, and $S = \{(x, y) : -1 \le x \le 1, y \in \mathcal{R},$ and $x = \sin y\}$. Then R is not a function from $\mathcal{R}$ to $\mathcal{R}$ since there is no $y \in \mathcal{R}$ such that $(0, y) \in R$. However, R is a function from $\mathcal{R} - \{0\}$ to $\mathcal{R}$. S is not a function from $[-1, 1]$ to $\mathcal{R}$ since $(0, 0) \in S$ and $(0, \pi) \in S$.

Practice Problem 1. Let $D = \{a, b, c, d\}$ and $C = \{1, 2, 3\}$. Let f and g be binary relations defined by $f = \{(a, 1), (b, 3)\}$ and $g = \{(a, 1), (b, 1), (c, 2), (d, 3)\}$.
   a. Represent f as an arrow diagram, and determine whether $f : D \to C$.
   b. Represent g as an arrow diagram, and determine whether $g : D \to C$.
Give reasons for your answers.

Practice Problem 2. Let $R = \{(x, y) : x, y \in \mathcal{R}$ and $y^2 = x\}$ and $S = \{(x, y) : x \in [0, \infty), y \in \mathcal{R}$ and $\sqrt{x} = y\}$.
   a. Determine whether $R : \mathcal{R} \to \mathcal{R}$.
   b. Determine whether $S : \mathcal{R} \to \mathcal{R}$.
   c. Determine whether $R : [0, \infty) \to \mathcal{R}$.
   d. Determine whether $S : [0, \infty) \to \mathcal{R}$.

Many useful functions have domains with an infinite number of elements. To define such functions, we usually do not give f as a set. For example, instead of $f = \{(x, x^2) : x \in \mathcal{N}\}$, we write

$$\text{Let } f : \mathcal{N} \to \mathcal{R} \text{ be defined by } f(x) = x^2$$

Of course, x is an arbitrary element in $\mathcal{N}$.

From a definition of this type, we can easily define f as a set.

Example 3

a. Let $D = \mathcal{N} = \{0, 1, 2, \ldots\} = C$. Define $f : D \to C$ by $f(n) = 2^n$. Then $f(0) = 1, f(1) = 2, f(2) = 4, \ldots$, and in this case, $f = \{(n, m) : n \in \mathcal{N} \text{ and } m = 2^n\}$.

b. Let $D = \mathcal{R}^+$ and $C = \mathcal{R}$. Define $g : D \to C$ by $g(x) = \log_2 x$, where $\log_2$ is the logarithm base 2 function. Then $g(1) = 0, g(2) = 1, g(4) = 2, \ldots$. Also, $g(\frac{1}{2}) = -1$.

c. Let A be a set. The identity relation 1_A is a function from A to A. It can also be defined by $1_A(x) = x$. 1_A is also called the **identity function.**

The function in Example 3b is a real-valued function, according to the following definition.

Definition. A **real-valued function** f is a function $f : D \to \mathcal{R}$.

We restrict ourselves to real-valued functions f such that $\text{dom}(f) \subseteq \mathcal{R}$. In particular, if f is a real-valued function, then $f \subseteq D \times \mathcal{R} \subseteq \mathcal{R} \times \mathcal{R}$. So every real-valued function has a graph (see Figure 4–9).

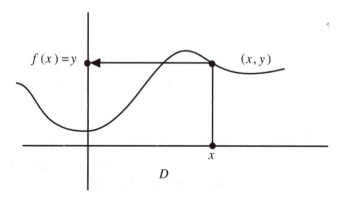

Figure 4–9

To use the graph in the figure to evaluate the function f at $x \in D$, we draw a vertical line from $(x, 0)$. This line will cross the graph at (x, y) where $y = f(x)$. If we draw a horizontal line from (x, y) to the y-axis, it will cross the y-axis at $(0, y)$, which corresponds to an element y of the codomain. So for real-valued functions, we can consider the domain as a subset of the x-axis and the codomain as a subset of the y-axis.

The preceding ideas lead to the following test of whether a relation on $\mathcal{R}$ is a real-valued function.

Vertical line test. Assume $D \subseteq \mathcal{R}$ and $f \subseteq D \times \mathcal{R}$. In the plane, consider D as a subset of the x-axis, and draw the graph of f. Then f is a real-valued function with $\text{dom}(f) = D$ if and only if for every point $x \in D$, the vertical line drawn through $(x, 0)$ intersects the graph exactly once.

This test is a way of visualizing the fact that a relation f is a function if for each $x \in \text{dom}(f)$, there is exactly one y such that $(x, y) \in f$.

Example 4

Figure 4–10 shows a relation that fails the vertical line test. Let $a, b, c \in D$. Then the vertical lines through the points a and b intersect the graph of f in more than one point, while the vertical line through the point c does not intersect the graph of f anywhere.

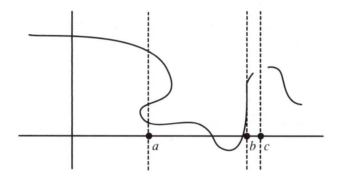

Figure 4–10

Practice Problem 3. Let $D = [0, \infty)$, and let f be a relation from D to $\mathcal{R}$ defined by $f = \{(x, y) : x = y^2 + 1\}$.

a. Apply the vertical line test to f.

b. Is f a real-valued function with domain D?

Functions Defined as Rules

In algebra or calculus texts, real-valued functions are frequently specified by giving a rule of association. For example, a definition such as "let f be the function defined by $f(x) = \sqrt{x - 2}$" is very common. As we have seen, however, this sentence alone is not enough to specify the function f. What is missing is any mention of the domain. The domain is assumed to be the subset of the real numbers that contains precisely those values of x for which the stated rule is meaningful. In this case, $\text{dom}(f) = \{x : x \geq 2\}$. When a real-valued function g is specified by a rule, the set of all real numbers for which the rule is meaningful is called the **domain of definition** of g.

Definition by cases is useful for defining certain functions.

Example 5

Suppose we want to define a function s from the nonnegative integers to the integers which assigns to each even integer half of it and assigns to each odd integer the negative of half of its successor. Then we can give a two-case definition for s as follows:

$$s(n) = \begin{cases} n/2 & \text{if } n \text{ is even} \\ -(n+1)/2 & \text{if } n \text{ is odd} \end{cases}$$

Note that the cases do not overlap; that is, n cannot be both even and odd. If the cases do overlap, then it is important that the definitions agree on the overlap.

Example 6

Define a function f from the real numbers to the real numbers as follows:

$$f(x) = \begin{cases} x^2 & \text{if } x \leq 0 \\ x & \text{if } x \geq 0 \end{cases}$$

In this example we have overlapping cases, since the value 0 is included in both cases. However, the definitions agree on the overlap, that is, when $x = 0$.

Sequences

The function in Example 5 is a sequence.

Definition. A **sequence** is a function s with $\text{dom}(s) = \{m, m+1, m+2, \ldots\} \subseteq \mathcal{Z}$.

Usually, the domain of a sequence is the nonnegative integers or the positive integers. When $n \in \text{dom}(s)$, we write s_n for the image $s(n)$ of n. To distinguish the sequence (a function) from the sequence values (the images), we write $(s_m, s_{m+1}, \ldots)$ for the sequence.

The sequence in Example 5 can be written $(s_0, s_1, s_2, \ldots) = (0, -1, 1, -2, 2, -3, 3, \ldots)$.

Practice Problem 4. Let $(s_0, s_1, s_2, \ldots)$ be a sequence defined by $s_0 = \pi$ and $s_j = j \cdot s_{j-1}$ for $j = 1, 2, 3, \ldots$.
   a. Find s_1, s_2, s_3, and s_4.
   b. Use mathematical induction to prove $s_n = \pi \cdot n!$.

Operations

Definition. A function $h : A \to A$ is sometimes called a **unary operation** on A.

For example, the function $h : \mathcal{Z} \to \mathcal{Z}$ defined by $h(x) = -x$ is a unary operation on $\mathcal{Z}$. In this case, we say that the operation $-$ (minus) is a unary operation on $\mathcal{Z}$. Note that if h is a unary operation on A, then $h(x)$ is defined for *every* $x \in A$ and $h(x)$ is a unique element of A.

Definition. A function $g : A \times A \to A$ is called a **binary operation** on A.

Suppose $x, y \in A$ and g is a binary operation on A. Then $(x, y) \in A \times A$, and the image of (x, y) should be written as $g((x, y))$. However, we write $g(x, y)$ instead. It is also common to write xgy for $g(x, y)$. Note that if g is a binary operation on A, then xgy is defined for *every* $x, y \in A$ and xgy is a unique element of A.

As a consequence of the preceding definition, $g : \mathcal{R} \times \mathcal{R} \to \mathcal{R}$ defined by $g(x, y) = x + y$ is a binary operation on $\mathcal{R}$. In this case, we say that $+$ (plus) is a binary operation on $\mathcal{R}$.

Definition. Let h be a unary operation on A and $B \subseteq A$. Then the set B is **closed under** h if for all $x \in B$, $h(x) \in B$. Let g be a binary operation on A and $C \subseteq A$. Then the set C is **closed under** g if for all $x, y \in C$, $g(x, y) \in C$.

In other words, set B is closed under a unary operation h if and only if the image under h of any element of B is in B, and set C is closed under a binary operation g if and only if the image of any pair of elements in C is in C.

Example 7

a. Define $h : \mathcal{Z} \to \mathcal{Z}$ by $h(x) = -x$, and let B be the set of even integers. Show that B is closed under h, but $\mathcal{N}$ is not closed under h.

Solution Let $x \in B$. Then $x = 2y$ for some $y \in \mathcal{Z}$. Now, $h(x) = h(2y) = -(2y) = 2(-y)$. Since $2(-y)$ is even, $h(x) \in B$. $\mathcal{N}$ is not closed under h, since $1 \in \mathcal{N}$ but $h(1) = -1 \notin \mathcal{N}$.

b. Define $g : \mathcal{Z} \times \mathcal{Z} \to \mathcal{Z}$ by $g(x, y) = x - y$, and let $C = \{3k : k \in \mathcal{Z}\}$. Show that C is closed under g, but $\mathcal{N}$ is not closed under g.

Solution Let $x, y \in C$. Then $x = 3k$ and $y = 3m$ for some $k, m \in \mathcal{Z}$. Now, $g(x, y) = x - y = 3k - 3m = 3(k - m)$, and hence, $g(x, y) \in C$. $\mathcal{N}$ is not closed under g, since $2 \in \mathcal{N}$ and $3 \in \mathcal{N}$ but $g(2, 3) = 2 - 3 = -1 \notin \mathcal{N}$.

Practice Problem 5.

a. Define a unary operation h on $\mathcal{Z}$ by $h(x) = x^2 - 2x$. Is $\mathcal{N}$ closed under h?
b. Define a binary operation g on $\mathcal{Z}$ by $g(x, y) = x^2 - xy + y^2$. Is $\mathcal{N}$ closed under g?

EXERCISE SET 4.3

1. Let $f = \{(a, b), (b, b), (c, b), (d, c)\}$ be a binary relation from the set $A = \{a, b, c, d\}$ to the set $B = \{a, b, c\}$. Determine whether f is a function from A to B.

2. Let $g = \{(1, a), (2, c), (3, d), (4, c)\}$ be a binary relation from the set $D = \{1, 2, 3, 4, 5\}$ to the set $C = \{a, b, c, d\}$. Determine whether g is a function from D to C.

3. Assume f is a real-valued function defined by a rule. Find the domain of definition of f if:
   (a) $f(x) = 1/(x + 3)$
   (b) $f(x) = \sqrt{x + 5}$

4. Assume f is a real-valued function defined by a rule. Find the domain of definition of f if:
   (a) $f(x) = 1/(x^2 - 9)$
   (b) $f(z) = \sqrt{z - 2}$

5. Define a sequence (s_n) by $(s_0, s_1, s_2, \ldots) = (0, 1/1, 2, 1/3, 4, 1/5, \ldots)$. Give a two-case definition of s_n.

6. Define a sequence (t_n) by $(t_0, t_1, t_2, \ldots) = (0, 1, 1/2, 9, 4, 1/5, 36, 7, 1/8, \ldots)$. Give a three-case definition of t_n.

7. Define a sequence $(t_1, t_2, \ldots)$ by $t_n =$ the sum of the first n positive integers.
   (a) Find t_1
   (b) Find t_5
   (c) Test the conjecture $t_m = (m + 1)m/2$ for several values of m.

8. Define a sequence $(t_1, t_2, \ldots)$ by $t_1 = 1$ and $t_j = j + t_{j-1}$, for $j = 2, 3, 4, \ldots$.
   (a) Find t_2, t_3, and t_5.
   (b) Test the conjecture $t_m = (m + 1)m/2$ for several values of m.
   (c) Either prove the conjecture in part b, or find a counterexample.

9. Define a sequence $(s_1, s_2, \ldots)$ by $s_1 = 1$ and $s_j = (2j - 1) + s_{j-1}$, for $j = 2, 3, 4, \ldots$.
   (a) Find s_2, s_3, and s_5.
   (b) Discover a simple formula for s_n, and prove your conjecture.

10. Let f be a real-valued function with domain $\mathcal{R}$ defined by $f(x) = x^2$. Define a sequence $(s_1, s_2, \ldots)$ by $s_1 = b$ and $s_j = f(s_{j-1})$, for $j = 2, 3, 4, \ldots$.
    (a) Assume that $b = 1$, and find a formula for s_n.
    (b) Assume that $b = 3$, and find a formula for s_n.
    (c) Prove the conjecture in part b.

11. Define the **floor function** or **greatest integer function** $G(x) = \lfloor x \rfloor$ by $G : \mathcal{R} \to \mathcal{Z}$, where $G(x) = n$ if $n \le x < n + 1$.
    (a) Find $G(2.3) = \lfloor 2.3 \rfloor$.
    (b) Find $\lfloor 3.0 \rfloor$.
    (c) Find $\lfloor -1.3 \rfloor$.

12. Define the **ceiling function** $H(x) = \lceil x \rceil$ by $H : \mathcal{R} \to \mathcal{Z}$, where $H(x) = n$ if $n - 1 < x \le n$.
    (a) Find $H(3.7) = \lceil 3.7 \rceil$.
    (b) Find $\lceil 5.0 \rceil$.
    (c) Find $\lceil -6.3 \rceil$.

13. (a) Define a unary operation h on $\mathcal{Z}$ by $h(x) = x^2 - 2x + 2$. Is $\mathcal{N}$ closed under h?
    (b) Define a binary operation g on $\mathcal{Z}$ by $g(x, y) = x^2 - xy$. Is $\mathcal{N}$ closed under g?

14. (a) Define a unary operation h on $\mathcal{R}$ by $h(x) = x^2 - 6x + 5$. Is $\mathcal{R}^+$ closed under h?
    (b) Define a binary operation g on $\mathcal{R}$ by $g(x, y) = x/(y^2 + 1)$. Is $\mathcal{Q}$ closed under g?

15. Prove the corollary to Theorem 1.

Assume that f and g are functions in Exercises 16–18.

16. Prove that $f \cap g$ is a function.

17. Prove that $f \cup g$ is a function if and only if $f(x) = g(x)$, for all $x \in \text{dom}(f) \cap \text{dom}(g)$.

18. Prove that $f \subseteq g$ if and only if $\text{dom}(f) \subseteq \text{dom}(g)$ and $f(x) = g(x)$, for all $x \in \text{dom}(f)$.

4.4 COMPOSITIONS, BIJECTIONS, AND INVERSE FUNCTIONS

In this section we consider functions with special and useful properties, as well as functions composed of other functions.

One-to-One Functions

Definition.   A function f is **one-to-one** (or **1–1** or **injective**) if for all $x, z \in \text{dom}(f)$, $f(x) = f(z)$ implies $x = z$.

The following example gives an insight into the concept of a 1–1 function.

Example 1

a. Let $D = \{a, b, c\}$ and $C = \{1, 2, 3, 4\}$, with $f : D \rightarrow C$ defined by $f(a) = 1$, $f(b) = 3$, $f(c) = 3$. f is not 1–1 since $f(b) = f(c)$ and $b \neq c$ (see Figure 4–11).

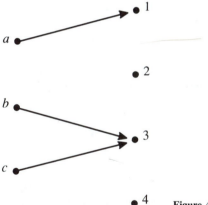

Figure 4–11

b. The functions $f(n) = 2^n$ and $g(x) = \log_2 x$ of Example 3 in Section 4.3 are 1–1. To verify that $f : \mathcal{R} \rightarrow \mathcal{R}$ is 1–1 let $f(x) = f(z)$. Then $2^x = 2^z$. By taking the logarithm base 2 of both sides of the equation $2^x = 2^z$, we obtain $x = \log_2(2^x) = \log_2(2^z) = z$. Therefore, $x = z$.

c. Define $f : \mathcal{R} \rightarrow \mathcal{R}$ by $f(x) = x^2 - x$. A counterexample, namely, $f(1) = 0 = f(0)$, shows that f is not 1–1.

Practice Problem 1. Prove that f is 1–1, or provide a counterexample showing that it is not:

(a) $f : \mathcal{N} \to \mathcal{N}$ is given by $f(n) = n^2 - 4n + 5$.

(b) $f : \mathcal{R} - \{1\} \to \mathcal{R}$ is given by $f(x) = 1/(x - 1)$.

The following test will determine whether any real-valued function is 1–1.

Horizontal-line test. Draw the graph for a real-valued function f in the plane. Then f is a 1–1 function if and only if for every y in the codomain of f, the horizontal line drawn through y on the y-axis intersects the graph no more than once.

The horizontal line test is valid because f is a 1–1 function if for $x, z \in \text{dom}(f), x \neq z$ implies $f(x) \neq f(z)$.

Example 2

Figure 4–12 shows a function that fails the horizontal-line test. The horizontal line through the point a intersects the graph in more than one point. Thus, for this graph, we have $f(b) = a = f(c)$, but $b \neq c$.

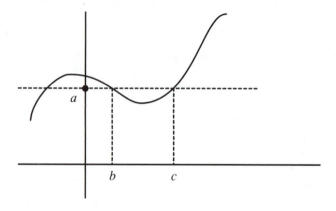

Figure 4–12

Onto Functions

Definition. The function $f : D \to C$ is **onto** (or **surjective**) if for every $y \in C$, there is at least one $x \in D$ such that $f(x) = y$.

Recall that when $f(x) = y$, we call y the image of x; x is called the **preimage** of y. Hence, a function f is onto if and only if every element in the codomain of f has a preimage.

The following example illuminates the concept of an onto function.

Example 3

a. Let $D = \{a, b, c\}$ and $C = \{1, 2\}$, with $f : D \to C$ defined by $f(a) = 1$, $f(b) = 1$, $f(c) = 2$. Then f is onto.

b. The function of Example 1a is not onto since, for example, $2 \in C$ and there is no $x \in D$ such that $f(x) = 2$. Figure 4–11 provides some insight into what makes a function fail to be onto.

c. The function $g : \mathcal{R}^+ \to \mathcal{R}$ defined by $g(x) = \log_2 x$ is onto. To prove that g is onto, let $y \in \mathcal{R}$. We seek an $x \in \mathcal{R}^+$ such that $g(x) = y$, that is, $\log_2 x = y$. This last equation is equivalent to $2^y = x$. So let $x = 2^y \in \mathcal{R}^+$. Then $g(x) = g(2^y) = \log_2(2^y) = y$.

d. Define $f : \mathcal{N} \to \mathcal{R}$ by $f(x) = x^3$. We show that f is not onto by finding a $y \in \mathcal{R}$ such that for no $x \in \mathcal{N}$ do we have $f(x) = y$. Let $y = 3$. Then there is no $x \in \mathcal{N}$ such that $f(x) = 3$; that is, there is no $x \in \mathcal{N}$ such that $x^3 = 3$, since $\sqrt[3]{3} \notin \mathcal{N}$.

Practice Problem 2. Show that f is onto, or provide a counterexample that shows that f is not onto:

   a. $f : \mathcal{N} \to \mathcal{N}$ is given by $f(n) = n + 1$.

   b. $f : \mathcal{R} \to [0, \infty)$ is given by $f(x) = x^2$.

Let $f : D \to C$. Then, since f is a relation, $\mathrm{ran}(f)$ is defined; in fact, $\mathrm{ran}(f) \subseteq C$. Furthermore, by Practice Problem 3 following, if f is onto, then $\mathrm{ran}(f) = C$.

Practice Problem 3. Prove that $f : D \to C$ is onto if and only if $\mathrm{ran}(f) = C$.

Bijective Functions

Definition. A function $f : D \to C$ is **bijective** if f is both 1–1 and onto.

A bijective function is also called a **bijection**.

Example 4

a. Show that the function $f : \mathcal{R} \to \mathcal{R}^+$ defined by $f(x) = 2^x$ is bijective.

Solution Consider the function of Example 1b. Even though the codomain of that function is not the same as the codomain of f, the same proof shows that f is 1–1. To show that f is onto, let $y \in \mathcal{R}^+$. We seek an $x \in \mathcal{R}$ such that $f(x) = y$, that is, $2^x = y$. Solving $2^x = y$ for x, we obtain $x = \log_2 y$. Hence, $f(x) = f(\log_2 y) = 2^{\log_2 y} = y$.

b. Show that the function $g : \mathcal{R}^+ \to \mathcal{R}$ defined by $g(x) = \log_2 x$ is bijective.

Solution Example 3c shows that g is onto. To verify that g is 1–1, let $g(x) = g(y)$. Then $\log_2 x = \log_2 y$. Hence, $2^{\log_2 x} = 2^{\log_2 y}$ and it follows that $x = y$.

Example 5

Let A be any set. The identity function $1_A : A \to A$ is bijective.

Practice Problem 4. Define $f : (2, \infty) \to (1, \infty)$ by $f(x) = x/(x - 2)$. Show that f is bijective.

Permutation Functions

A useful class of functions is the class of bijective functions from a finite set to the same finite set.

Definition. A bijective function $f : A \to A$, where A is finite, is called a **permutation function on** A (or simply, **permutation of** A).

Example 6

Let $A = \{1, 2, 3\}$.
    a. 1_A is a permutation of A.
    b. $f : A \to A$ defined by the table

x	1	2	3
$f(x)$	2	3	1

is also a permutation of A.

Practice Problem 5. There are four other permutations of $\{1, 2, 3\}$. Find them, and write out the table for each.

Writing a permutation in its tabular form, as in Example 6, suggests a useful notational convention for permutations. Let $A = \{1, 2, \ldots, n\}$, and assume $f : A \to A$ is a permutation. Then f may be written as a two-dimensional array,

$$\begin{pmatrix} 1 & 2 & 3 & \ldots & n \\ f(1) & f(2) & f(3) & \ldots & f(n) \end{pmatrix}$$

Thus, the array form for f in Example 6b is $f = \begin{pmatrix} 1 & 2 & 3 \\ 2 & 3 & 1 \end{pmatrix}$.

Compositions of Functions

Definition. Let $f : D \to A$ and $g : B \to C$, where $A \subseteq B$. Define the **composition** $g \circ f : D \to C$ by $(g \circ f)(x) = g(f(x))$ for all $x \in D$.

Because the codomain A of f is a subset of the domain B of g, we may conclude that $g(f(x))$ is a well-defined element of C for all possible values $f(x)$. Hence, the composition $g \circ f$ is a function with domain D and codomain C.

Since a function is a relation, we may think of $g \circ f$ as a composition of relations (see Section 4.2). As a composition of relations do we obtain the function in the above definition? You are asked to verify that the answer is yes in Exercise 24.

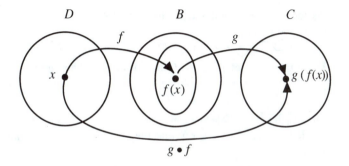

Figure 4–13

The diagram in Figure 4–13 gives a pictorial representation of $g \circ f$.

Example 7

Let $f : \mathcal{R} \rightarrow [1, \infty)$ be defined by $f(x) = x^2 + 1$ and $g : \mathcal{R}^+ \rightarrow \mathcal{R}$ be defined by $g(x) = \log_2 x$. Then $(g \circ f)(x) = g(f(x)) = g(x^2 + 1) = \log_2(x^2 + 1)$.

Example 8

Let

$$f = \begin{pmatrix} 1 & 2 & 3 \\ 2 & 3 & 1 \end{pmatrix}$$

and

$$g = \begin{pmatrix} 1 & 2 & 3 \\ 2 & 1 & 3 \end{pmatrix}$$

Then

$$g \circ f = \begin{pmatrix} 1 & 2 & 3 \\ 1 & 3 & 2 \end{pmatrix}$$

since

$(g \circ f)(1) = g(f(1)) = g(2) = 1$
$(g \circ f)(2) = g(f(2)) = g(3) = 3$
$(g \circ f)(3) = g(f(3)) = g(1) = 2$

Practice Problem 6. Verify that

$$f \circ g = \begin{pmatrix} 1 & 2 & 3 \\ 3 & 2 & 1 \end{pmatrix}$$

where f and g are as in Example 8.

Example 9

Let $f : \mathcal{R} \rightarrow \mathcal{R}^+$ be defined by $f(x) = 2^x$ and $g : \mathcal{R}^+ \rightarrow \mathcal{R}$ be defined by $g(x) = \log_2 x$. Then $g \circ f : \mathcal{R} \rightarrow \mathcal{R}$ and $f \circ g : \mathcal{R}^+ \rightarrow \mathcal{R}^+$ are well-defined compositions. Furthermore, $(g \circ f)(x) = g(f(x)) = g(2^x) = \log_2(2^x) = x$ and $(f \circ g)(x) = f(\log_2 x) = 2^{\log_2 x} = x$. Hence, $g \circ f = 1_{\mathcal{R}}$ and $f \circ g = 1_{\mathcal{R}^+}$.

As the next definition states, functions like f and g of Example 9 are said to be inverses of one another.

Definition. Let $f : D \rightarrow C$ be a function. Then $g : C \rightarrow D$ is the **inverse** of f if $g \circ f = 1_D$ and $f \circ g = 1_C$. When g is the inverse of f, we write $g = f^{-1}$.

Since a function f is also a relation, we may think of f^{-1} as the inverse of a relation. In Exercise 22, we show that the two notions are the same.

Practice Problem 7. Let $f : \mathcal{R} \rightarrow \mathcal{R}$ be a function defined by $f(x) = 8x + 1$ for each $x \in \mathcal{R}$. Let $g : \mathcal{R} \rightarrow \mathcal{R}$ be a function defined by $g(x) = (x - 1)/8$ for each $x \in \mathcal{R}$. Show that f and g are inverses of each other.

We will prove shortly that any function having an inverse must be bijective. In addition, you will be asked to prove in Exercise Set 4.4 that for a bijective function f, the inverse f^{-1} exists, is unique, and is also bijective.

Finding Inverses of Bijective Functions

Let $f : D \rightarrow C$ be a bijective function, and suppose $v = f(u)$. Then $f^{-1}(v) = f^{-1}(f(u)) = (f^{-1} \circ f)(u) = 1_D(u) = u$. Thus, the inverse f^{-1} maps the image of u under f back to u (see Figure 4–14).

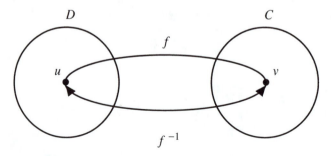

Figure 4–14

The idea just illustrated can be exploited to find the inverse of some bijective functions.

Example 10

Define $f : \mathcal{R} \rightarrow \mathcal{R}$ by $f(x) = 3x + 5$. Find f^{-1}.

Solution Any inverse function f^{-1} must have the property $f^{-1}(v) = u$ if $f(u) = v$. To find f^{-1}, we set $v = f(u) = 3u + 5$. Solve for u to get $u = (v - 5)/3$. But $u = f^{-1}(v)$, so $f^{-1}(v) = (v - 5)/3$. Hence, $f^{-1}(x) = (x - 5)/3$. It is easy to verify that $(f \circ f^{-1})(x) = x = (f^{-1} \circ f)(x)$.

Relations and Functions   Chap. 4

Note that the solution of the equation $v = f(u)$ for u depends on the fact that f is bijective. In particular, the existence of u for each v is equivalent to f being onto, and the uniqueness of the solution u is equivalent to f being 1–1.

Practice Problem 8. Define $f : (1, \infty) \to (2, \infty)$ by $f(x) = 2x/(x-1)$. Find f^{-1}, and verify that $(f \circ f^{-1})(x) = x = (f^{-1} \circ f)(x)$.

Since any function that has an inverse must be bijective, successful completion of Practice Problem 8 enables us to conclude that the function f is bijective.

Example 11

a. Let
$$f = \begin{pmatrix} 1 & 2 & 3 \\ 3 & 1 & 2 \end{pmatrix}$$

Then
$$f^{-1} = \begin{pmatrix} 1 & 2 & 3 \\ 2 & 3 & 1 \end{pmatrix}$$

b. Let
$$g = \begin{pmatrix} 1 & 2 & 3 \\ 2 & 1 & 3 \end{pmatrix}$$

Then
$$g^{-1} = \begin{pmatrix} 1 & 2 & 3 \\ 2 & 1 & 3 \end{pmatrix}$$

Observe that permutation functions always have inverses and the inverse can be found by exchanging the top row and the bottom row in the array form of the function and then rearranging the columns so that the top row is in the correct order. Thus, for
$$f = \begin{pmatrix} 1 & 2 & 3 \\ 3 & 1 & 2 \end{pmatrix}$$

we exchange rows to obtain
$$\begin{pmatrix} 3 & 1 & 2 \\ 1 & 2 & 3 \end{pmatrix}$$

and then rearrange columns so that the top row is in order, to obtain the inverse
$$f^{-1} = \begin{pmatrix} 1 & 2 & 3 \\ 2 & 3 & 1 \end{pmatrix}$$

Practice Problem 9. For the function f of Example 11, verify that
$$f \circ f^{-1} = \begin{pmatrix} 1 & 2 & 3 \\ 1 & 2 & 3 \end{pmatrix} = f^{-1} \circ f$$

Theorem 1   If $f : D \to C$ has an inverse f^{-1}, then f is bijective.

Proof. Assume $f : D \to C$ has an inverse $f^{-1} : C \to D$. Then $f \circ f^{-1} = 1_C$ and $f^{-1} \circ f = 1_D$. To prove that f is 1–1, we assume that $x, z \in D$ and $f(x) = f(z)$. Then

$f^{-1}(f(x)) = f^{-1}(f(z))$, so by the definition of composition, $(f^{-1} \circ f)(x) = (f^{-1} \circ f)(z)$. Now, since $f^{-1} \circ f = 1_D, x = 1_D(x) = 1_D(z) = z$. Therefore, f is 1–1.

To show that f is onto, we let $y \in C$. Then $(f \circ f^{-1})(y) = y$ since $f \circ f^{-1} = 1_C$ and $1_C(y) = y$. Now, by the definition of composition, $f(f^{-1}(y)) = y$. However, $f^{-1}(y) \in D$, since $f^{-1} : C \to D$. Hence, $y = f(f^{-1}(y))$ and $f^{-1}(y) \in D$, and it follows that f is onto. □

Theorem 2 If $g : D \to A$ is onto and $f : A \to C$ is onto, then $f \circ g : D \to C$ is onto.

Proof. Let $y \in C$. Then, since f is onto, $y = f(z)$ for some $z \in A$. Also, since g is onto, $z = g(x)$ for some $x \in D$. Therefore, $y = f(g(x)) = (f \circ g)(x)$ for some $x \in D$, and hence, $f \circ g$ is onto. □

Theorem 3 If $g : D \to A$ is 1–1 and $f : A \to C$ is 1–1, then $f \circ g : D \to C$ is 1–1.

Proof. Assume $(f \circ g)(x) = (f \circ g)(z)$. Then $f(g(x)) = f(g(z))$. Since f is 1–1, $g(x) = g(z)$; and since g is 1–1, $x = z$. Therefore, $f \circ g$ is 1–1. □

It is natural to ask whether the converses of Theorem 2 and Theorem 3 hold. These questions are answered in Exercise Set 4.4.

EXERCISE SET 4.4

1. Let $f = \{(a, b), (b, b), (c, b), (d, c)\}$ be a binary relation from the set $A = \{a, b, c, d\}$ to the set $B = \{a, b, c\}$. Determine whether f is a function. If it is a function, determine whether it is 1–1.

2. Let $g = \{(1, a), (2, c), (3, d), (4, b)\}$ be a binary relation from the set $D = \{1, 2, 3, 4\}$ to the set $C = \{a, b, c, d\}$. Determine whether g is a function. If it is a function, determine whether it is 1–1.

3.

$$f = \begin{pmatrix} 1 & 2 & 3 & 4 \\ 2 & 1 & 4 & 3 \end{pmatrix}$$

and

$$g = \begin{pmatrix} 1 & 2 & 3 & 4 \\ 1 & 3 & 4 & 2 \end{pmatrix}$$

are two permutation functions on $\{1, 2, 3, 4\}$.
   (a) Find $f \circ g$ and $g \circ f$.
   (b) Find f^{-1}, and verify that

$$f \circ f^{-1} = \begin{pmatrix} 1 & 2 & 3 & 4 \\ 1 & 2 & 3 & 4 \end{pmatrix} = f^{-1} \circ f$$

(c) Find g^{-1}, and verify that

$$g \circ g^{-1} = \begin{pmatrix} 1 & 2 & 3 & 4 \\ 1 & 2 & 3 & 4 \end{pmatrix} = g^{-1} \circ g$$

(d) Verify that $(f \circ g)^{-1} = g^{-1} \circ f^{-1}$.

4.

$$f = \begin{pmatrix} 1 & 2 & 3 & 4 \\ 2 & 1 & 3 & 4 \end{pmatrix}$$

and

$$g = \begin{pmatrix} 1 & 2 & 3 & 4 \\ 3 & 1 & 4 & 2 \end{pmatrix}$$

are permutation functions on $\{1, 2, 3, 4\}$.

(a) Find $f \circ g$ and $g \circ f$.

(b) Find f^{-1}, and verify that

$$f \circ f^{-1} = \begin{pmatrix} 1 & 2 & 3 & 4 \\ 1 & 2 & 3 & 4 \end{pmatrix} = f^{-1} \circ f$$

(c) Find g^{-1}, and verify that

$$g \circ g^{-1} = \begin{pmatrix} 1 & 2 & 3 & 4 \\ 1 & 2 & 3 & 4 \end{pmatrix} = g^{-1} \circ g$$

(d) Verify that $(f \circ g)^{-1} = g^{-1} \circ f^{-1}$.

5. Let $f : \mathcal{R} \to \mathcal{R}$ be defined by $f(x) = 2 - 5x$. Assume that f is bijective, and use the method of Example 10 to find f^{-1}. Verify that $f \circ f^{-1} = 1_{\mathcal{R}} = f^{-1} \circ f$.

6. Let $f : \mathcal{R} \to \mathcal{R}$ be defined by $f(x) = 1 + x^3$. Assume that f is bijective, and use the method of Example 10 to find f^{-1}. Verify that $f \circ f^{-1} = 1_{\mathcal{R}} = f^{-1} \circ f$.

For each of the pairs of functions f and g in Exercises 7–12, check whether $f \circ g$ is well defined; that is, whether the range of g is a subset of the domain of f. Write $(f \circ g)(x)$ explicitly if it is well defined. Also, check whether $g \circ f$ is well defined. Write $(g \circ f)(x)$ explicitly if it is well defined.

7. $f : \mathcal{Z} \to \mathcal{Z}$ is defined by $f(x) = 3x - 2$.
$g : \mathcal{N} \to \mathcal{Z}$ is defined by $g(n) = 2 - n$.

8. $f : \mathcal{N} \to \mathcal{R}$ is defined by $f(n) = 2 - \sqrt{n}$.
$g : \mathcal{Z} \to \mathcal{Z}$ is defined by $g(x) = x + 7$.

9. $f : \mathcal{N} \to \mathcal{R}$ is defined by $f(n) = \log_2(n + 1)$.
$g : \mathcal{N} \to \mathcal{N}$ is defined by $g(n) = 2^n - 1$.

10. $f : \mathcal{Z} \to \mathcal{Z}$ is defined by $f(x) = 3x^2 + 2$.
$g : \mathcal{N} \to \mathcal{Z}$ is defined by $g(n) = 2 - n$.

11. $f : \mathcal{N} \to \mathcal{R}$ is defined by $f(n) = 2 - \sqrt{n}$.
$g : \mathcal{Z} \to \mathcal{Z}$ is defined by $g(x) = x^2 + 7$.

12. $f : \mathcal{N} \to \mathcal{R}$ is defined by $f(n) = \log_3(n + 1)$.
$g : \mathcal{N} \to \mathcal{N}$ is defined by $g(n) = 3^n - 1$.

In Exercises 13–18; (a) Show that the given function is 1–1, or give a counterexample; (b) Show that the given function is onto, or give a counterexample.

13. $f : \mathcal{R} \to \mathcal{R}$ is defined by $f(x) = x^2 + 2x + 1$.

14. $f : [-1, \infty) \to [0, \infty)$ is defined by $f(x) = x^2 + 2x + 1$.

15. Let $A = \{X : X \subseteq \mathcal{Z}$ and X is finite and $X \neq \emptyset\}$. Define $f : A \to \mathcal{R}$ by $f(X) = $ the average of the members of X.

16. Let $A = \{X : X \subseteq \mathcal{N}$ and X is finite and $X \neq \emptyset\}$. Define $f : A \to \mathcal{Q} \cap [0, \infty)$ by $f(X) = $ the average of the members of X.

17. $f : \mathcal{N} - \{0\} \to A$ is defined by $f(n) = \{k : k$ divides n and $k \in \mathcal{N}\}$, where A is the set in Exercise 16.

18. $f : P(\mathcal{N}) \to P(\mathcal{N})$ is defined by $f(X) = \mathcal{N} - X$.

19. Let $f : D \to C$ be a function. Use quantifiers to write each of the following definitions.
    (a) f is 1–1.
    (b) f is onto.

20. Let $f : D \to C$ be a function.
    (a) Prove that $f \circ 1_D = f$.
    (b) Prove that $1_C \circ f = f$.

In Exercises 21–23, the relationship between a bijective function and its inverse is investigated.

21. Let $f : D \to C$ be a bijective function. Prove that f has an inverse.
    (*Hint:* Think of f as a relation. Then f^{-1} is a relation. Prove that f^{-1} is a function from C to D and that $f \circ f^{-1} = 1_C$ and $f^{-1} \circ f = 1_D$.)

22. Let $f : D \to C$ be a bijective function. Show that the inverse of f is unique. That is, prove that if $f \circ g = 1_C$ and $g \circ f = 1_D$, then $g = f^{-1}$.

23. Prove: If $f : D \to C$ is a bijective function, then $f^{-1} : C \to D$ is also bijective.

24. Let $f : D \to A$ and $g : B \to C$, where $A \subseteq B$. Think of f and g as relations and $g \circ f$ as the composition of relations. Show that $g \circ f : D \to C$.

25. This exercise concerns the converse of Theorem 3. Assume that $g : D \to A$ and $f : A \to C$ are functions. Prove, or provide a counterexample to, each of the following conjectures.
    (a) If $f \circ g : D \to C$ is 1–1, then $g : D \to A$ is 1–1.
    (b) If $f \circ g : D \to C$ is 1–1, then $f : A \to C$ is 1–1.

26. This exercise concerns the converse of Theorem 2. Assume that $g : D \to A$ and $f : A \to C$ are functions. Prove, or provide a counterexample to, each of the following conjectures.
    (a) If $f \circ g : D \to C$ is onto, then $g : D \to A$ is onto.
    (b) If $f \circ g : D \to C$ is onto, then $f : A \to C$ is onto.

4.5 IMAGES AND INVERSE IMAGES OF SETS

For a function $f : D \to C$ and a subset A of D, it is sometimes useful to consider the set of images under f of all the elements in A. If B is a subset of C, it is likewise useful to consider the set of all elements of D that are preimages of elements in B.

Definition.   Let $f : D \to C$ and let $A \subseteq D$. Then $f[A] = \{f(x) : x \in A\}$ is the **image** of A under f.

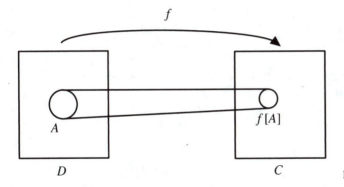

Figure 4–15

Figure 4–15 depicts the image of A under f.

Definition. Let $f : D \to C$ and let $B \subseteq C$. Then $f^{-1}[B] = \{x : x \in D$ and $f(x) \in B\}$ is the **inverse image** of B under f.

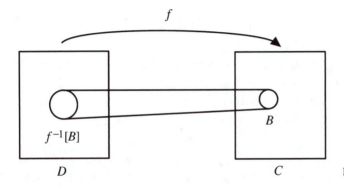

Figure 4–16

Figure 4–16 shows the inverse image of B under f.

Example 1

Define $f : D \to C$, where $D = \{1, 2, 3, 4\}$, $C = \{a, b, c\}$, and f is defined by the following table:

x	1	2	3	4
$f(x)$	a	c	c	a

Let $A = \{1, 2, 3\}$ and $B = \{a, b\}$. Find $f[A]$ and $f^{-1}[B]$.

Solution $f[A] = \{f(x) : x \in A\} = \{f(1), f(2), f(3)\} = \{a, c, c\} = \{a, c\}$. $f^{-1}[B] = \{x : x \in D$ and $f(x) \in B\} = \{x : f(x) = a$ or $f(x) = b\} = \{1, 4\}$.

Example 2

Define $g : \mathcal{R} \to [0, \infty)$ by $g(x) = x^2$, $A = (5, \infty)$, and $B = (2, 7)$. Find $g[A]$ and $g^{-1}[B]$.

Solution $g[A] = \{f(x) : x \in (5, \infty)\} = \{x^2 : x > 5\} = \{t : t > 25\} = (25, \infty)$. $g^{-1}[B] = \{x : x \in \mathcal{R} \text{ and } f(x) \in B\} = \{x : x \in \mathcal{R} \text{ and } 2 < x^2 < 7\} = \{(x : x \in \mathcal{R} \text{ and } (\sqrt{2} < x < \sqrt{7}) \text{ or } (-\sqrt{7} < x < -\sqrt{2})\} = (-\sqrt{7}, -\sqrt{2}) \cup (\sqrt{2}, \sqrt{7})$.

Practice Problem 1. a. Define $f : D \to C$, where $D = \{1, 2, 3, 4\}$, $C = \{a, b, c, d\}$, and f is defined by the following table:

x	1	2	3	4
$f(x)$	a	b	c	a

Let $A = \{1, 3, 4\}$ and $B = \{a, c, d\}$. Find $f[A]$ and $f^{-1}[B]$.

b. Define $f : \mathcal{N} \to \mathcal{R}$ by $f(n) = \sqrt{n}$. Let $A = \{n : n \in \mathcal{N} \text{ and } 10 < n < 26\}$ and $B = [2, 37)$. Find $f[A]$ and $f^{-1}[B]$.

Example 3

Let $D = \{0, 1, 2, 3, 4\}$ and $C = \{a, b, c, d\}$. Define $f : D \to C$ by the following table:

x	0	1	2	3	4
$f(x)$	b	a	b	c	a

Let $A = \{0, 1\}$, $B = \{1, 2\}$, $E = \{a, b\}$, and $F = \{b, d\}$.
   a. Find $f[A \cup B]$ and compare it to $f[A] \cup f[B]$.
   b. Find $f[A \cap B]$ and compare it to $f[A] \cap f[B]$.
   c. Find $f^{-1}[E \cup F]$ and compare it to $f^{-1}[E] \cup f^{-1}[F]$.
   d. Find $f^{-1}[E \cap F]$ and compare it to $f^{-1}[E] \cap f^{-1}[F]$.

Solution a. $f[A \cup B] = f\big[\{0, 1, 2\}\big] = \{f(0), f(1), f(2)\} = \{b, a\}$; $f[A] \cup f[B] = \{f(0), f(1)\} \cup \{f(1), f(2)\} = \{b, a\}$. Hence, $f[A \cup B] = f[A] \cup f[B]$.
b. $f[A \cap B] = f[\{1\}] = \{f(1)\} = \{a\}$; $f[A] \cap f[B] = \{b, a\} \cap \{a, b\} = \{a, b\}$. Hence, $f[A \cap B] \subseteq f[A] \cap f[B]$.
c. $f^{-1}[E \cup F] = f^{-1}[\{a, b, d\}] = \{0, 1, 2, 4\}$; $f^{-1}[E] \cup f^{-1}[F] = \{0, 1, 2, 4\} \cup \{0, 2\} = \{0, 1, 2, 4\}$. Hence, $f^{-1}[E \cup F] = f^{-1}[E] \cup f^{-1}[F]$.
d. $f^{-1}[E \cap F] = f^{-1}[\{b\}] = \{0, 2\}$; $f^{-1}[E] \cap f^{-1}[F] = \{0, 1, 2, 4\} \cap \{0, 2\} = \{0, 2\}$. Hence, $f^{-1}[E \cap F] = f^{-1}[E] \cap f^{-1}[F]$.

In what follows, we shall see that all of the statements in Example 3 involving images and preimages of sets are, in fact, theorems.

In the remaining theorems and practice problems, we will need the following:

$y \in f[A]$ if and only if $y = f(t)$ for some $t \in A$.
$x \in f^{-1}[B]$ if and only if $f(x) \in B$.

From these remarks, we see that $f[\emptyset] = \emptyset$ and $f^{-1}[\emptyset] = \emptyset$.

The theorems and practice problems that follow not only are useful, but they provide an excellent source for developing skills in reading, understanding, and constructing proofs. In all of them, let $f : D \to C$, $A \subseteq D$, $B \subseteq D$, $E \subseteq C$, and $F \subseteq C$.

Theorem 1   If $A \subseteq B$, then $f[A] \subseteq f[B]$.

Proof. Assume $A \subseteq B$. Let $y \in f[A]$. Then $y = f(t)$ for some $t \in A$. But since $A \subseteq B$, $t \in B$, and hence, $y = f(t) \in f[B]$. □

Theorem 2   If $E \subseteq F$, then $f^{-1}[E] \subseteq f^{-1}[F]$.

Proof. Assume $E \subseteq F$. Let $x \in f^{-1}[E]$. Then $f(x) \in E$. But since $E \subseteq F$, $f(x) \in F$. It follows that $f^{-1}[E] \subseteq f^{-1}[F]$. □

Theorem 3 $f[A \cap B] \subseteq f[A] \cap f[B]$.

Proof. Let $y \in f[A \cap B]$. Then $y = f(x)$ for some $x \in A \cap B$. Hence, $x \in A$ and $x \in B$. Now, since $x \in A$, $y = f(x) \in f[A]$. Similarly, $y \in f[B]$. Therefore, $y \in f[A] \cap f[B]$. □

Practice Problem 2.   Find an example that shows that equality does not necessarily hold in Theorem 3. (*Hint:* Use finite sets.) Note that in Exercise 5, you are asked to prove that $f[A \cap B] = f[A] \cap f[B]$ if f is 1–1.

Practice Problem 3.   Prove that $f^{-1}[E \cap F] = f^{-1}[E] \cap f^{-1}[F]$.

Theorem 4 $f[A \cup B] = f[A] \cup f[B]$.

Proof. Let $x \in f[A \cup B]$. Then $x = f(t)$ for some $t \in A \cup B$. There are two cases.
Case 1. $t \in A$. Then $f(t) \in f[A]$, and hence, $x = f(t) \in f[A] \cup f[B]$.
Case 2. $t \in B$. This case is similar to case 1.
Now let $x \in f[A] \cup f[B]$. Again, there are two cases.
Case 1. $x \in f[A]$. Then $x = f(t)$ for some $t \in A$. But $t \in A \cup B$, so it follows that $x = f(t) \in f[A \cup B]$.
Case 2. $x \in f[B]$. This case is similar to case 1. □

Practice Problem 4.   Prove that $f^{-1}[E \cup F] = f^{-1}[E] \cup f^{-1}[F]$.

Theorem 5 $f[f^{-1}[E]] \subseteq E$.

Proof. Let $x \in f[f^{-1}[E]]$. Then $x = f(t)$ for some $t \in f^{-1}[E]$. Hence, $f(t) \in E$. It follows that $x \in E$. □

Theorem 6   If f is onto, then $f[f^{-1}[E]] = E$.

Proof. Let f be onto. Then by Theorem 5, $f[f^{-1}[E]] \subseteq E$. To prove that $E \subseteq f[f^{-1}[E]]$, let $y \in E$. Since $E \subseteq C$, $y \in C$, and since f is onto, $y = f(t)$ for some $t \in D$. Therefore, $f(t) \in E$, and it follows that $t \in f^{-1}[E]$. Hence, $y = f(t) \in f[f^{-1}[E]]$. □

Theorem 7 $A \subseteq f^{-1}\big[f[A]\big]$.

Proof. Let $t \in A$. Then $f(t) \in f[A]$. Hence, $t \in f^{-1}\big[f[A]\big]$. $\qquad\qquad\square$

Practice Problem 5. Prove: If f is 1–1, then $A = f^{-1}\big[f[A]\big]$.

EXERCISE SET 4.5

In Exercises 1–4, find $f[A]$ and $f^{-1}[B]$.

1. $f : \mathcal{R} \to \mathcal{R}$ is defined by $f(x) = 2x + 1$, $A = (0, 1)$, $B = [-1, 1]$.
2. $f : \mathcal{Z} \to \mathcal{N}$ is defined by $f(x) = |x|$, $A = \{-2, -1, 0, 1, 2\}$, $B = \{0, 1, 2\}$.
3. $f : \mathcal{R} \to \mathcal{R}$ is defined by $f(x) = \sin x$, $A = [0, \pi]$, $B = [0, 2]$.
4. $f : \mathcal{R} \to \mathcal{Z}$ is defined by $f(x) = \lfloor x \rfloor$, where $\lfloor x \rfloor$ is the floor function defined in Exercise 11 of Section 4.3, $A = (0, 4]$, $B = \{0, 1, 2\}$.

In Exercises 5–8, let $f : D \to C$, $A \subseteq D$, $B \subseteq D$, $E \subseteq C$, and $F \subseteq C$.

5. *Prove:* If f is 1–1, then $f[A \cap B] = f[A] \cap f[B]$.
6. $f[A - B] \overset{?}{=} f[A] - f[B]$.
7. *Prove:* If f is 1–1, then $f[A - B] = f[A] - f[B]$.
8. $f^{-1}[E - F] \overset{?}{=} f^{-1}[E] - f^{-1}[F]$.

In Exercises 9–13, let $f : D \to C$, $\Omega \subseteq P(D)$, and $\Upsilon \subseteq P(C)$.

9. $f[\cup\Omega] \overset{?}{=} \cup \{f[A] : A \in \Omega\}$.
10. Assume that $\Omega \neq \emptyset$. $f[\cap\Omega] \overset{?}{=} \cap \{f[A] : A \in \Omega\}$.
11. $f^{-1}[\cup\Upsilon] \overset{?}{=} \cup \{f^{-1}[A] : A \in \Upsilon\}$.
12. Assume that $\Upsilon \neq \emptyset$. $f^{-1}[\cap\Upsilon] \overset{?}{=} \cap \{f^{-1}[A] : A \in \Upsilon\}$.
13. Assume that $\Omega \neq \emptyset$. Prove: If f is 1–1, then $f[\cap\Omega] = \cap\{f[A] : A \in \Omega\}$.

The notion of the image of a set under a function can be generalized to relations in the following way. Let R be a relation, and let A be a set. Then $R[A] = \{y : \text{for some } x \in A, xRy\}$.

14. Let $f : D \to C$ and $A \subseteq D$. Think of f as a relation, and use the preceding definition to show that $f[A] = \{f(x) : x \in A\}$.
15. Let $f : D \to C$ and $B \subseteq C$. Think of f as a relation, and use the preceding definition to show that $f^{-1}[B] = \{x : f(x) \in B\}$.

In Exercises 16–22, let R be a relation from D to C, $A \subseteq D$, $B \subseteq D$, and $\Omega \subseteq P(D)$.

16. $R[A \cup B] \overset{?}{=} R[A] \cup R[B]$.
17. $R[A \cap B] \overset{?}{=} R[A] \cap R[B]$.
18. $R[A - B] \overset{?}{=} R[A] - R[B]$.
19. Prove, or find a counterexample to, the following conjecture: If $R^{-1} \circ R \subseteq 1_D$, then $R[A \cap B] = R[A] \cap R[B]$.

20. Prove, or find a counterexample to, the following conjecture: If $R^{-1} \circ R \subseteq 1_D$, then $R[A - B] = R[A] - R[B]$.

21. $R[\cup\Omega] \stackrel{?}{=} \cup\{R[A] : A \in \Omega\}$.

22. Assume that $\Omega \neq \emptyset$. $R[\cap\Omega] \stackrel{?}{=} \cap\{R[A] : A \in \Omega\}$.

KEY CONCEPTS

Ordered pair	Onto function
Product of two sets	Bijective function, bijection
Relation	Composition of functions
Properties of relations	Inverse of a bijective function
Operations on relations	Image of a set under a function
Function	Inverse image of a set
One-to-one function	under a function

PROOFS TO EVALUATE

See the instructions following the Proofs to Evaluate in Chapter 2.

1. *Conjecture:* $(A \times C) \cup (B \times D) = (A \cup B) \times (C \cup D)$.

Argument: We prove $(A \times C) \cup (B \times D) \subseteq (A \cup B) \times (C \cup D)$. The other set inclusion follows by reversing the steps of this proof. Let $(x, y) \in (A \times C) \cup (B \times D)$. Then $(x, y) \in (A \times C)$ or $(x, y) \in (B \times D)$.

Case 1. Assume $(x, y) \in (A \times C)$. Then $x \in A$ and $y \in C$. Hence, $x \in A \cup B$ and $y \in C \cup D$. Therefore, $(x, y) \in (A \cup B) \times (C \cup D)$.

Case 2. Assume $(x, y) \in (B \times D)$. The proof of this case is similar to that of Case 1.

2. *Conjecture:* If R is transitive, then $R \subseteq R \circ R$.

Argument: Assume R is transitive. We use a pick-a-point proof to prove that $R \subseteq R \circ R$. Let $(x, y) \in R$. Then $(x, z) \in R$ and $(z, y) \in R$ for some z, since R is transitive. Hence, $(x, y) \in R \circ R$, by the definition of the composition of relations.

3. *Conjecture:* Let R be a relation on A. If R is transitive and symmetric, then R is reflexive on A.

Argument: Assume R is a transitive and symmetric relation on A. Let $(x, y) \in R$. Since R is symmetric, $(y, x) \in R$. Also, since R is transitive, $(x, y) \in R$ and $(y, x) \in R$ implies $(x, x) \in R$. Therefore, R is reflexive on A.

4. *Conjecture:* If g is the real-valued function defined by $g(x) = -x + b$, where b is arbitrary, then $g^{-1} = g$.

Argument: We need only show that $g \circ g(x) = x$, for all x. But it is easy to verify that $g \circ g(x) = g(g(x)) = g(-x + b) = -(-x + b) + b = x$.

5. *Conjecture:* Let f and g be functions. Then $f \cup g$ is a function if and only if $f(x) = g(x)$ for all $x \in \text{dom}(f \cap g)$.

Argument: ($\Rightarrow$) Assume $f \cup g$ is a function. Let $x \in \text{dom}(f \cap g)$. Then $(x, w) \in f$ and $(x, z) \in g$. Hence, $(x, w) \in f \cup g$ and $(x, z) \in f \cup g$. But, since $f \cup g$ is a function, $f(x) = w = z = g(x)$.

($\Leftarrow$) Assume $f(x) = g(x)$ for all $x \in \text{dom}(f \cap g)$. Let $x \in \text{dom}(f \cup g)$. We must show that for $(x, w) \in f \cup g$ and $(x, z) \in f \cup g, w = z$. There are three cases.

Case 1. Assume $x \in \text{dom}(f \cap g)$. Then by assumption, $w = z$.

Case 2. Assume $x \in \text{dom}(f \cup g) - \text{dom}(g)$. Then $x \in \text{dom}(f)$, and since f is a function, we have $w = z$.

Case 3. Assume $x \in \text{dom}(f \cup g) - \text{dom}(f)$. The proof of this case is similar to the proof of Case 2.

6. *Conjecture:* $f[A] \cap f[B] \subseteq f[A \cap B]$.

Argument: Assume $w \in f[A] \cap f[B]$. Then $w \in f[A]$ and $w \in f[B]$. Hence, $w = f(x)$ for some x such that $x \in A$ and $x \in B$. It follows that $x \in A \cap B$, and thus, $w \in f[A \cap B]$.

REVIEW EXERCISES

1. Let $A = \{2, 3\}$ and $B = \{6, 9, 10, 13\}$, and let R be a binary relation from A to B defined by aRb if and only if $b = 3a$.
   (a) List all the members of R.
   (b) Determine whether R is reflexive on $A \cup B$, symmetric, antisymmetric, or transitive.

2. Let $A = \{2, 3\}$ and $B = \{6, 9, 10, 13\}$, and let S be a binary relation from A to B defined by aSb if and only if $a < b - 4$.
   (a) List all the members of S.
   (b) Determine whether S is reflexive on $A \cup B$, symmetric, antisymmetric, or transitive.

In Exercises 3–6 determine whether the given relation R on the given set A is reflexive, symmetric, antisymmetric, or transitive:

3. $A = \mathcal{R}$, xRy if and only if $x = y$ or $y = 0$.

4. $A = \mathcal{Z}$, xRy if and only if $x = y^{2^k}$ for some $k \in \mathcal{Z}$.

5. $A = \mathcal{R}$, xRy if and only if $x^2 + y^2 = 1$ or $x = y$.

6. $A = \mathcal{R}$, xRy if and only if $x^2 + y^2 = 1$ or $|x| = |y|$.

In Exercises 7–10, prove, or find a counterexample to, the given conjectures.

7. *Conjecture:* If R and S are symmetric, then $R \cup S$ is symmetric.

8. *Conjecture:* If R and S are symmetric, then $R \cap S$ is symmetric.

9. *Conjecture:* If R and S are transitive, then $R \cup S$ is transitive.

10. *Conjecture:* If R and S are transitive, then $R \cap S$ is transitive.

11. Let R be a relation from A to B. Prove that the following conditions are equivalent:
    (i) $R \circ R^{-1} \subseteq 1_B$.
    (ii) For all $x \in A$ and for all $y, z \in B$, xRy and xRz implies $y = z$.

12. $\text{dom}(R \circ T) \stackrel{?}{=} \text{dom}(T)$.

13. $(R \circ S) \circ T \overset{?}{=} R \circ (S \circ T)$.

14. Define the **predecessor function** $p : \mathcal{Z} \to \mathcal{Z}$ by $p(x) = x - 1$ and the **successor function** $s : \mathcal{Z} \to \mathcal{Z}$ by $s(x) = x + 1$. Verify that p and s are inverses of one another.

15. Let C and D be subsets of $\mathcal{N}$. Define the predecessor function $p : D \to \mathcal{N}$ by $p(x) = x - 1$ and the successor function $s : \mathcal{N} \to C$ by $s(x) = x + 1$. Find the sets D and C such that p and s are onto, and verify that p and s are inverses of one another.

In Exercises 16–18, (a) show that the given function is 1–1, or give a counterexample, and (b) show that the given function is onto, or give a counterexample.

16. Let $A = \{X : X \subseteq \mathcal{Z} \text{ and } X \text{ is finite and } X \neq \emptyset\}$. Define $f : A \to \mathcal{Q}$ by $f(X) =$ the average of the members of X.

17. Define $f : P(\mathcal{N}) - \{\emptyset\} \to \mathcal{N}$ by $f(X) =$ the smallest member of X.

18. Define $f : \mathcal{R} \to P(\mathcal{R})$ by $f(x) = (-\infty, x)$.

SUPPLEMENTARY EXERCISES

For the following exercises, assume h is a unary operation on a set A and $B \subseteq A$. Define the subsets, B_* and B^* of A as follows:
$B_* = \cup\{C_n : n \in \mathcal{N}\}$, where the sets C_n are defined recursively by $C_0 = B$ and $C_{n+1} = C_n \cup h[C_n]$.
$B^* = \cap\{X : B \subseteq X \subseteq A \text{ and } X \text{ is closed under } h\}$.

1. Let $A = \mathcal{Z}, B = \{2\}$, and $h(x) = x + 2$.
   (a) Find B_*.
   (b) Find B^*.

2. Let $A = \mathcal{R}$, $B = [\frac{1}{3}, 1]$, and $h(x) = x^2$.
   (a) Find B_*.
   (b) Find B^*.

3. Prove that B_* is closed under h.

4. Prove that B^* is closed under h.

5. Prove that $B^* \subseteq B_*$. (*Hint:* Use Exercise 3 above and Practice Problem 3 of Section 3.4.)

6. Prove that $B_* \subseteq B^*$ and conclude, by Exercise 5, that $B^* = B_*$. (*Hint:* Show by induction that $C_n \subseteq B^*$ for all $n \in \mathcal{N}$. Then use Exercise 15 of Section 3.4.)

SOLUTIONS TO PRACTICE PROBLEMS

Section 4.1

1. (a) $A \times B = \{(a, 3), (a, 4), (b, 3), (b, 4), (c, 3), (c, 4), (d, 3), (d, 4), (e, 3), (e, 4)\}$
   (b) $B \times A = \{(3, a), (3, b), (3, c), (3, d), (3, e), (4, a), (4, b), (4, c), (4, d), (4, e)\}$
   (c) $A \times B \neq B \times A$, and in addition, $A \times B$ and $B \times A$ are disjoint.

2. Let $A = \emptyset$ and $B = \{1, 2\}$. We see that $A \times B = \emptyset$ because any pair in $A \times B$ must have a first coordinate from A, but $A = \emptyset$. Therefore, there are no ordered pairs in $A \times B$. A similar argument shows that $B \times A = \emptyset$.

3. *Proof:* Assume $(x, y) \in (A \times C) \cap (B \times C)$. Then $(x, y) \in A \times C$ and $(x, y) \in B \times C$. Hence, $x \in A$ and $x \in B$ and $y \in C$. So $x \in A \cap B$ and $y \in C$. Therefore, $(x, y) \in (A \cap B) \times C$.

Section 4.2

1. (a) $F = \{$(Mr. Smith, John), (Mr. Smith, Sally), (Mr. Smith, Jim), (Mr. Smith, Jill)$\}$ is a relation from A to B.

   (b) $S = \{$(Mr. Smith, Mrs. Smith), (Mrs. Smith, Mr. Smith)$\}$ is a relation on A.

2. (a) Let $R = \{(1, 1), (2, 2), (3, 3), (4, 4), (1, 2), (2, 1), (1, 3)\}$. Then R is reflexive, but neither antisymmetric nor symmetric.

   (b) Let $S = \{(1, 2)\}$. Then S is antisymmetric (vacuously), but neither reflexive nor symmetric.

   (c) Let $T = \{(1, 1)\}$. Then T is symmetric and antisymmetric but is not reflexive on A.

3. (a) xRx since $x = 3^0 x$

   (b) R is not symmetric since $3R1$ but it is not the case that $1R3$.

   (c) R is antisymmetric. Assume xRy and yRx. Then $x = 3^k y$ and $y = 3^j x$, where k and j are nonnegative integers. Hence, $x = 3^{k+j} x$ and so $k + j = 0$. But since k and j are nonnegative, $k = j = 0$. Therefore, $3^k = 1 = 3^j$. So $x = y$.

   (d) R is transitive. Assume xRy and yRz. Then $x = 3^k y$ and $y = 3^j z$, where k and j are nonnegative integers. Hence, $x = 3^{k+j} z$. Therefore, xRz.

4. $R^{-1} = \{(3, 1), (4, 1), (3, 2)\}, S^{-1} = \{(1, 3), (2, 2)\}, R \circ S = \{(3, 3), (3, 4), (2, 3)\}, (R \circ S)^{-1} = \{(3, 3), (4, 3), (3, 2)\}, R^{-1} \circ S^{-1} = \{(1, 2), (1, 1)\}$, and $S^{-1} \circ R^{-1} = \{(3, 3), (4, 3), (3, 2)\}$. So $(R \circ S)^{-1} = S^{-1} \circ R^{-1}$.

5. (a) $\text{dom}(R) = \{2, 3\}$.

   (b) $\text{ran}(R) = \{2, 4\}$.

6. *Proof:* Assume $y \in \text{ran}(R \cup S)$. Then $x(R \cup S)y$ for some x.

   Case 1. xRy. This implies that $y \in \text{ran}(R)$.

   Case 2. xSy. This implies that $y \in \text{ran}(S)$. In either case, we obtain $x \in \text{ran}(R) \cup \text{ran}(S)$.
   Now let $y \in \text{ran}(R) \cup \text{ran}(S)$.

   Case 1. $y \in \text{ran}(R)$. Hence xRy for some x. Therefore, $x(R \cup S)y$, and hence $y \in \text{ran}(R \cup S)$.

   Case 2. $y \in \text{ran}(S)$. The proof that $y \in \text{ran}(R \cup S)$ is similar to that of case 1.

7. *Proof:* Assume $(x, y) \in R \circ (S \cup T)$. Then $(x, z) \in S \cup T$ and $(z, y) \in R$ for some z.

   Case 1. $(x, z) \in S$ and $(z, y) \in R$. Then $(x, y) \in R \circ S$.

   Case 2. $(x, z) \in T$ and $(z, y) \in R$. Then $(x, y) \in R \circ T$.
   Now assume $(x, y) \in (R \circ S) \cup (R \circ T)$.

   Case 1. $(x, y) \in R \circ S$. Then $(x, z) \in S$ and $(z, y) \in R$ for some z. Hence, $(x, z) \in S \cup T$ and $(z, y) \in R$. Therefore, $(x, y) \in R \circ (S \cup T)$.

   Case 2. $(x, y) \in R \circ T$. The proof is similar to that of case 1.

8. *Proof:* Assume R is transitive. Let $(x, y) \in R \circ R$. Then $(x, z) \in R$ and $(z, y) \in R$ for some z. But since R is transitive, $(x, y) \in R$. Therefore, $R \circ R \subseteq R$.
   Now assume $R \circ R \subseteq R$. Suppose $(x, z) \in R$ and $(z, y) \in R$. Then $(x, y) \in R \circ R$. Hence, $(x, y) \in R$. Therefore, R is transitive.

Section 4.3

1.

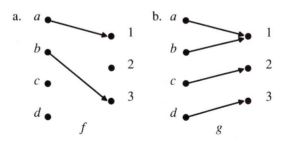

a.

b.

f

g

f is not a function from D to C; g is a function from D to C.

2. (a) No. (b) No. (c) No. (d) Yes.

3. (a) f does not pass the vertical line test. For example, $0 \in D$, but there is no y such that $0fy$. As another example, $2 \in D$, and there are two values y, namely, $y = 1$ and $y = -1$, such that $2fy$.

(b) No.

4. (a) $s_1 = \pi$, $s_2 = 2\pi$, $s_3 = 6\pi$, $s_4 = 24\pi$.

(b) The basis step is true since $0! = 1$. The induction hypothesis is $s_k = \pi \cdot k!$. By definition, $s_{k+1} = (k+1) \cdot s_k$. Hence, $s_{k+1} = (k+1) \cdot \pi \cdot k! = \pi \cdot (k+1)!$.

5. (a) No, since $h(1) = -1 \notin \mathcal{N}$.

(b) Yes, since $x^2 - xy + y^2 = (x-y)^2 + xy \geq 0$ for $x, y \in \mathcal{N}$.

Section 4.4

1. (a) f is not 1–1. $f(n) = (n-2)^2 + 1$, and hence $f(1) = 2 = f(3)$.

(b) f is 1–1. Assume $f(a) = f(b)$. Then $1/(a-1) = 1/(b-1)$. Hence, $a - 1 = b - 1$, and therefore, $a = b$.

2. (a) f is not onto because there is no element $k \in \mathcal{N}$ such that $f(k) = 0$.

(b) f is onto. Let $a \in [0, \infty)$. Then $\sqrt{a} \in \mathcal{R}$ and $f(\sqrt{a}) = a$.

3. Recall that $\text{ran}(f) = \{y : (x, y) \in f \text{ for some } x \in D\}$. Hence, $y \in \text{ran}(f)$ if and only if $y = f(x)$ for some $x \in D$. Since $\text{ran}(f) \subseteq C$, $\text{ran}(f) = C$ is equivalent to $C \subseteq \text{ran}(f)$. However, $C \subseteq \text{ran}(f)$ is equivalent to for all $y \in C$, there is an $x \in D$ such that $y = f(x)$. But the right side of this last equivalence is the definition of onto. So $C \subseteq \text{ran}(f)$ is equivalent to f is onto.

4. *Proof:* 1. f is 1–1. Assume $f(a) = f(b)$. Then $a/(a-2) = b/(b-2)$. Hence $a(b-2) = b(a-2)$. With a little algebra, we conclude that $a = b$.

2. f is onto. Assume $w > 1$. We find an x such that $x/(x-2) = w$. Solving for x, we obtain $x = 2w/(w-1) = 2 + 2/(w-1)$. Hence, $x > 2$. It is then easily verified that $f(x) = w$.

5.

x	1	2	3
$f(x)$	1	3	2

x	1	2	3
$f(x)$	2	1	3

x	1	2	3
$f(x)$	3	2	1

x	1	2	3
$f(x)$	3	1	2

6. Since $(f \circ g)(1) = f\big(g(1)\big) = f(2) = 3$, $(f \circ g)(2) = f\big(g(2)\big) = f(1) = 2$, and $(f \circ g)(3) = f\big(g(3)\big) = f(3) = 1$, we have

$$f \circ g = \begin{pmatrix} 1 & 2 & 3 \\ 3 & 2 & 1 \end{pmatrix}$$

7. We show that $g \circ f = 1_{\mathcal{R}}$ and $f \circ g = 1_{\mathcal{R}}$. For each $x \in \mathcal{R}$, $(g \circ f)(x) = g\big(f(x)\big) = g(8x+1) = \big[(8x+1)-1\big]/8 = x = 1_{\mathcal{R}}(x)$, and $(f \circ g)(x) = f\big(g(x)\big) = 8 \cdot (x-1)/8 + 1 = x = 1_{\mathcal{R}}(x)$. Hence, g is the inverse of f, and f is the inverse of g.

8. Let $v = 2u/(u-1)$. Solving for u we obtain $u = v/(v-2)$. So $f^{-1}(x) = x/(x-2)$, and it follows that

$$(f \circ f^{-1})(x) = f\big(f^{-1}(x)\big) = f\left(\frac{x}{x-2}\right) = \frac{2\left(\frac{x}{x-2}\right)}{\left(\frac{x}{x-2}\right)-1} = x$$

and

$$(f^{-1} \circ f)(x) = f^{-1}\big(f(x)\big) = f^{-1}\left(\frac{2x}{x-1}\right) = \frac{\left(\frac{2x}{x-1}\right)}{\left(\frac{2x}{x-1}\right)-2} = x$$

9. $(f \circ f^{-1})(1) = f\big(f^{-1}(1)\big) = f(3) = 1$, $(f \circ f^{-1})(2) = f(f^{-1}(2)) = f(1) = 2$, and $(f \circ f^{-1})(3) = f(f^{-1}(3)) = f(2) = 3$. Hence,

$$f \circ f^{-1} = \begin{pmatrix} 1 & 2 & 3 \\ 1 & 2 & 3 \end{pmatrix}$$

and similarly,

$$f^{-1} \circ f = \begin{pmatrix} 1 & 2 & 3 \\ 1 & 2 & 3 \end{pmatrix}$$

Section 4.5

1. **(a)** $f[A] = \{f(x) : x \in A\} = \{f(1), f(3), f(4)\} = \{a, c, a\} = \{a, c\}$. $f^{-1}[B] = \{x : f(x) \in B\} = \{1, 3, 4\}$.

(b) $f[A] = \{f(n) : n \in A\} = \{f(n) : 10 < n < 26\} = \{\sqrt{n} : 10 < n < 26\} = \{x : \sqrt{10} < x < \sqrt{26}\} = (\sqrt{10}, \sqrt{26})$. $f^{-1}[B] = \{n : f(n) \in B\} = \{n : 2 \le \sqrt{n} < 37\} = \{n : 4 < n \le 1,369\} = \{4, 5, 6, \ldots, 1,368\}$.

2. Consider $f : \{0, 1\} \to \{0\}$ defined by $f(0) = 0$ and $f(1) = 0$. Let $A = \{0\}$ and $B = \{1\}$. Then $f[A \cap B] = f[\{0\} \cap \{1\}] = f[\emptyset] = \emptyset$. However, $f[A] \cap f[B] = f[\{0\}] \cap f[\{1\}] = \{0\} \cap \{0\} = \{0\}$.

3. *Proof:* Let $x \in f^{-1}[E \cap F]$. Then $f(x) \in E \cap F$, and hence, $f(x) \in E$ and $f(x) \in F$. It follows that $x \in f^{-1}[E]$ and $x \in f^{-1}[F]$. Therefore, $x \in f^{-1}[E] \cap f^{-1}[F]$. To obtain the other inclusion, reverse the preceding steps.

4. *Proof:* Let $x \in f^{-1}[E \cup F]$. Then $f(x) \in E \cup F$. There are two cases. We consider the case $f(\dot{x}) \in E$. (The other case is similar.) Then $x \in f^{-1}[E]$, and so $x \in f^{-1}[E] \cup f^{-1}[F]$. Now let $x \in f^{-1}[E] \cup f^{-1}[F]$. There are again two cases. We consider the case $x \in f^{-1}[E]$. Then $f(x) \in E$, and hence, $f(x) \in E \cup F$. Therefore, $x \in f^{-1}[E \cup F]$.

5. *Proof:* Assume f is 1–1. By Theorem 7, $A \subseteq f^{-1}[f[A]]$. So let $t \in f^{-1}[f[A]]$. Then $f(t) \in f[A]$. Hence, $f(t) = f(s)$ for some $s \in A$. But since f is 1–1, we have $t = s$, and hence, $t \in A$.

5

Equivalence Relations and Partial Orders

As we have seen, a function is a relation satisfying certain conditions that give it useful properties. Two other special types of relations—equivalence relations and partial orders—are essential to the development of mathematics. Each of these special types of relations satisfies conditions designed to give it desirable properties.

5.1 EQUIVALENCE RELATIONS AND PARTITIONS

Often in mathematics a situation is encountered where the objects under consideration have some affinity for each other. For example, those people with the same last name are similar to one another. As another example, the even integers have an affinity for one another, and so do the odd integers. A special kind of relation, called an equivalence relation, is useful for describing the similarities between members of a given set.

Definition. A binary relation R on a set A is an **equivalence relation** on A if R is reflexive on A, symmetric, and transitive.

Example 1

Let $A = \{$Robert Kent, Joyce Davis, Sam Rogers, Tom Rogers, Mary Rogers, James Davis, Steve Kent$\}$, and let R be a binary relation on A such that $(x, y) \in R$ if and only if x and y have the same last name. It is easily verified that R is an equivalence relation on A.

Practice Problem 1. Let R be defined as follows: xRy if and only if $x-y \in \mathbb{Z}$, for $x, y \in \mathbb{R}$. Verify that R is an equivalence relation on $\mathbb{R}$.

As mentioned, equivalence relations describe similarities between elements of sets. It is also very useful to collect those similar elements in specified subsets of the given set.

Definition. Let R be an equivalence relation on A and $b \in A$. Then the **equivalence class** of b under R is $[b]_R = \{x : x \in A \text{ and } bRx\}$.

For an equivalence relation R on a set A, the equivalence class of b under R contains all the elements in A to which b is related by R. When there is no danger of confusion, we will drop the subscript and write $[b]$ for $[b]_R$.

The equivalence relation R in Example 1 determines three equivalence classes: {Robert Kent, Steve Kent}, {James Davis, Joyce Davis}, and {Sam Rogers, Tom Rogers, Mary Rogers}. The elements in each equivalence class are related to each other by the relation R. Also, the equivalence class [Robert Kent] is identical with the equivalence class [Steve Kent]; in other words, [Steve Kent] = {Robert Kent, Steve Kent} = [Robert Kent]. In general, as we shall see later, the equivalence classes $[x]$ and $[y]$ are identical whenever xRy.

For any set A, it is routine to verify that the identity relation 1_A is an equivalence relation. Moreover, each equivalence class is a singleton set $\{a\}$, where a is an element of A (see Exercise 5).

For the equivalence relation in Practice Problem 1, $[1] = \{x : 1 - x \in \mathbb{Z}\} = \mathbb{Z}$ and $[\pi] = \{\pi, \pi \pm 1, \pi \pm 2, \ldots\}$.

Practice Problem 2. From Practice Problem 1, find
    a. $[0]$ b. $[1/2]$ c. $[\sqrt{2}]$ d. $[3]$

Example 2

Let $A = \{1, 2, 3, 4, 5\}$. Define R as follows: mRn if and only if $m - n = 2k$ for some $k \in \mathbb{Z}$ for $m, n \in A$. That is, mRn if and only if $m - n$ is divisible by 2. Show that R is an equivalence relation on A, and find all the equivalence classes.

Solution:
a. The relation R is reflexive, since $m - m = 0 = 2 \cdot 0$ and $0 \in \mathbb{Z}$.
b. To show that R is symmetric, assume mRn. Then $m - n = 2k$ for some $k \in \mathbb{Z}$. Hence, $n - m = 2(-k)$, and since $-k \in \mathbb{Z}$, we have nRm.
c. To show that R is transitive, we first assume mRn and nRp. We then need to prove mRp. By assumption, $m - n = 2k$ and $n - p = 2j$ for some $k, j \in \mathbb{Z}$. But $m - p = (n-p) + (m-n) = 2j + 2k = 2(j + k)$. Since $j + k \in \mathbb{Z}$, mRp. Therefore, R is an equivalence relation on A.

    The equivalence classes in this example are $[1] = \{1, 3, 5\}$ and $[2] = \{2, 4\}$, that is, the odd integers and the even integers, respectively, from the set A.

Practice Problem 3. Let $A = \{1, 2, 3, 4, 5, 6, 7, 8, 9, 10\}$. Define R as follows: aRb if and only if $a - b = 4k$ for some $k \in \mathcal{Z}$ for $a, b \in A$. Verify that R is an equivalence relation on A, and specify all the equivalence classes.

The relations in Example 2 and Practice Problem 3 are defined on a finite set of integers. Such relations are special cases of a very useful relation on $\mathcal{Z}$.

Definition. Assume d is a positive integer. Define the relation $\equiv_d$ on $\mathcal{Z}$ as follows: $x \equiv_d y$ if and only if $x - y$ is divisible by d.

Theorem 1 The relation $\equiv_d$ is an equivalence relation on $\mathcal{Z}$.

Proof. Assume that d is a positive integer. The relation $\equiv_d$ is reflexive because $x - x = 0 \cdot d$, and hence, $x \equiv_d x$.

To prove that $\equiv_d$ is symmetric, we assume $x \equiv_d y$. Then there is an integer k such that $x - y = kd$. Hence, $y - x = (-k)d$. Therefore, $y - x$ is divisible by d, and it follows that $y \equiv_d x$.

To prove that $\equiv_d$ is transitive, we assume $x \equiv_d y$ and $y \equiv_d z$. Then $x - y = jd$ and $y - z = hd$ for some integers j and h. We need to prove that $x \equiv_d z$, in other words, that $x - z$ is a multiple of d. Another look at our assumptions leads us to consider the sum $(x - y) + (y - z)$. We obtain $x - z = (x - y) + (y - z) = (j + h)d$. Therefore, $x - z$ is divisible by d, and $x \equiv_d z$. $\qquad\square$

Practice Problem 4. Specify the equivalence classes determined by
a. The relation $\equiv_2$ on $\mathcal{Z}$.
b. The relation $\equiv_3$ on $\mathcal{Z}$.

We now turn our attention to some properties of equivalence classes.

Theorem 2 Let R be an equivalence relation on A, and let $a, b \in A$. Then $[a] = [b]$ if and only if aRb.

Proof. ($\Rightarrow$) Assume $[a] = [b]$. Now, $a \in [a]$ since aRa. Hence, $a \in [b]$. So bRa, and it follows that aRb.

($\Leftarrow$) Assume aRb. Then bRa. We prove that $[a] \subseteq [b]$. Let $x \in [a]$. Then aRx. So bRa and aRx. Hence, bRx, since R is transitive. Therefore, $x \in [b]$. The proof that $[b] \subseteq [a]$ is similar. Therefore, $[a] = [b]$. $\qquad\square$

Since aRb is equivalent to $b \in [a]$, Theorem 2 may be restated as follows.

Theorem 2 (Restated) Let R be an equivalence relation on A, and let $a, b \in A$. Then $[a] = [b]$ if and only if $b \in [a]$.

Theorem 3 Let R be an equivalence relation on A. If $[a] \neq [b]$, then $[a] \cap [b] = \emptyset$.

Proof. We prove the contrapositive. Assume $[a] \cap [b] \neq \emptyset$. Then there is an $x \in [a] \cap [b]$. By Theorem 2, $[x] = [a]$, since $x \in [a]$. Also, $x \in [b]$, and hence, $[x] = [b]$. Therefore, $[a] = [b]$. ☐

Definition.   Let R be an equivalence relation on A. Then $\mathbf{A/R} = \{[a] : a \in A\}$.

The set A/R is the collection of equivalence classes under R. We read A/R as A **modulo** R, or, more simply, A **mod** R.

Example 3

From Example 2, we have $[1] = [3] = [5]$ and $[2] = [4]$. So $A = [1] \cup [2]$, and $[1] \cap [2] = \emptyset$. Another way to put this is that the members of the collection $\{[1], [2]\}$ are disjoint, and $\cup \{[1], [2]\} = A$. In this example, $A/R = \{[1], [2]\}$ (see Figure 5–1).

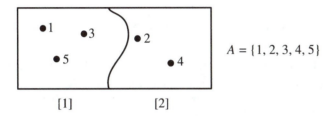

$A = \{1, 2, 3, 4, 5\}$

[1]   [2]

Figure 5–1

In Example 3, any two members of A/R are disjoint, and $\cup(A/R) = A$. This leads to the following definition.

Definition. Π is a **partition** of a set A if Π is a collection of subsets of A such that

1. $\emptyset \notin \Pi$ (No member of Π is empty.)
2. If $X, Y \in \Pi$ and $X \neq Y$, then $X \cap Y = \emptyset$. (The members of Π are **pairwise disjoint.**)
3. $\cup \Pi = A$ (The union of the members of Π is A.)

It is frequently useful to state condition (2) in the definition of a partition in its contrapositive form, if $X \cap Y \neq \emptyset$, then $X = Y$. This yields a very useful alternative statement of (2):

If there exists an $x \in X \cap Y$, then $X = Y$.

Since $X \subseteq A$ for all $X \in \Pi$, the union of the members of Π is a subset of A ($\cup \Pi \subseteq A$). Hence, to prove condition (3), we need only prove that $A \subseteq \cup \Pi$. When condition (3) holds, we say that Π **covers** A.

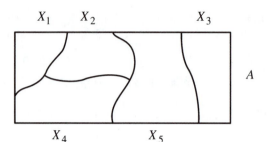

Figure 5–2 $\Pi = \{X_1, X_2, X_3, X_4, X_5\}$ is a partition of A.

Figure 5–2 shows a partition Π of a set A consisting of five subsets of A.

For the rest of this section, we assume that A is some given set and R is an equivalence relation on A. The collection of equivalence classes, A/R, in Example 3 is a partition of A. This fact is true of all equivalence relations.

Theorem 4   The collection of equivalence classes A/R is a partition of A.

Proof.

1. No member of A/R is empty. That is, $[z] \neq \emptyset$ for all $z \in A$, since $z \in [z]$.
2. The members of A/R are pairwise disjoint. Since the members of A/R are equivalence classes, we let $[a], [b] \in A/R$, where $a, b \in A$. It then follows from Theorem 3 that if $[a] \neq [b]$, then $[a] \cap [b] = \emptyset$.
3. The union of the members of A/R is A. We need to prove that A is a subset of the union of the equivalence classes. Let $x \in A$. Since $x \in [x]$ and $[x] \in A/R$, we have $x \in \cup(A/R)$. Hence, $A \subseteq \cup(A/R)$. $\square$

Theorem 4 says that any equivalence relation on a given set determines a partition of the set. It is also true that any partition of a set determines an equivalence relation on that set. Before proving this fact, we need a definition.

Definition.   Let Π be a partition of A. The **relation A/Π on A induced by** Π is defined by $x(A/\Pi)y$ if and only if for some $X \in \Pi$, $x \in X$ and $y \in X$.

In other words, two elements are related by A/Π whenever those elements are in the same set of the partition Π. As with A mod R, we can read A/Π as A **mod** Π.

Example 4

Let $\Pi = \{(x, x+1] : x \in \mathcal{Z}\}$.
   a. Show that Π is a partition of $\mathcal{R}$.
   b. Give a definition for $\mathcal{R}/\Pi$ without referring to Π.
   c. Show that $\mathcal{R}/\Pi$ is an equivalence relation on $\mathcal{R}$.

Solution
      a. Assume $m, n \in \mathcal{Z}$. Now, the set $(m, m+1]$ is not empty, since $m < m+\frac{1}{2} < m+1$. To see that the members of Π are pairwise disjoint, assume $t \in (m, m+1] \cap (n, n+1]$.

Suppose $m \neq n$; for specificity, let $m < n$. Then $m + 1 \leq n$, since $m \in \mathcal{Z}$. But then, $t \leq m + 1 \leq n < t$, and hence, $t < t$, a contradiction. It follows that $m = n$ and $(m, m + 1] = (n, n + 1]$.

To show that $\mathcal{R} \subseteq \cup\Pi$, let $t \in \mathcal{R}$. Then for some $m \in \mathcal{Z}, m < t$. Choose the largest such m. Then $m < t \leq m + 1$. Therefore, $t \in (m, m + 1]$, and so $t \in \cup\Pi$.

b. By the definition of $\mathcal{R}/\Pi$, $x(\mathcal{R}/\Pi)y$ if and only if for some $m \in \mathcal{Z}, m < x \leq m+1$ and $m < y \leq m + 1$. We can write this as $m < x, y \leq m + 1$.

c. To show that $\mathcal{R}/\Pi$ is reflexive, let $x \in \mathcal{R}$. Choose m to be the greatest integer such that $m < x$. Then $m < x$ and $x \leq m + 1$. Hence, $x(\mathcal{R}/\Pi)x$.

It is clear that $\mathcal{R}/\Pi$ is symmetric.

To show that $\mathcal{R}/\Pi$ is transitive, let $x(\mathcal{R}/\Pi)y$ and $y(\mathcal{R}/\Pi)z$; that is, $m < x, y \leq m+1$ and $n < y, z \leq n + 1$ for some n and m in z. Then $m < y \leq m + 1$ and $n < y \leq n + 1$, and the argument of part a shows that $m = n$. Hence, $m < x, z \leq m + 1$, and $x(\mathcal{R}/\Pi)z$.

In Example 4, we showed that A/Π is an equivalence relation on A, where $A = \mathcal{R}$. The following theorem states that A/Π is an equivalence relation on A for any set A and any partition Π of A.

Theorem 5   Let Π be a partition of A. Then the relation A/Π is an equivalence relation on A.

Proof. The proof is left as an exercise.

In Theorem 4, we proved that if R is an equivalence relation on A, the collection of equivalence classes $A/R = \{[z] : z \in A\}$ is a partition of A. By Theorem 5, the partition A/R induces an equivalence relation $A/(A/R)$ on A. So an equivalence relation R induces a partition A/R which in turn induces an equivalence relation $A/(A/R)$. Let us examine how this works with a specific equivalence relation.

Example 5

Let $A = \{1, 2, 3, 4, 5, 6, 7\}$. Define R on A by aRb if and only if $b - a$ is divisible by 3. It is easily shown that R is an equivalence relation. This equivalence relation R determines a partition A/R of the set A into three disjoint subsets: $[1] = \{1, 4, 7\}$, $[2] = \{2, 5\}$, and $[3] = \{3, 6\}$. The partition A/R then induces an equivalence relation $A/(A/R)$ defined by $x(A/(A/R))y$ if and only if x and y are in the same set of the partition. For example, $x(A/(A/R))2$ if and only if $x \in [2]$ since $2 \in [2]$. Hence, $x(A/(A/R))2$ if and only if $x - 2$ is a multiple of 3. In general, $x(A/(A/R))y$ if and only if $x - y$ is divisible by 3; that is, if and only if xRy. Therefore, $A/(A/R) = R$.

The next theorem states that the result of Example 5 is true in general.

Theorem 6 $A/(A/R) = R$.

Proof. The proof is left as an exercise.

A schematic synopsis of Theorems 4, 5, and 6 might look like Figure 5–3.

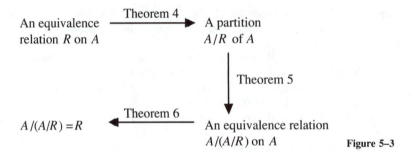

$$A/(A/R) = R \xleftarrow{\text{Theorem 6}} \text{An equivalence relation } A/(A/R) \text{ on } A$$

Figure 5–3

On the other hand, a partition Π of A induces an equivalence relation A/Π on A. But every equivalence relation induces a partition—in this case, the partition $A/(A/\Pi)$.

Theorem 7 $A/(A/\Pi) = \Pi$.

Proof. The proof is left as an exercise.

Figure 5–4 summarizes Theorems 4, 5, and 7.

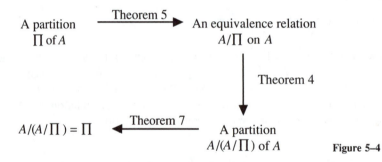

Figure 5–4

Equivalence Relations and Functions

For the remainder of this section, we use the following definition.

> **Definition.**   Let $f : D \to C$. Define a relation $\approx$ on D by $x \approx y$ if and only if $f(x) = f(y)$.

Thus, two elements of D are related by $\approx$ if they have equal images under f.

Example 6

Consider the function $f : \mathcal{Z} \to \{0, 1\}$ defined by $f(m) = 0$ if m is even and $f(m) = 1$ if m is odd. It is easy to verify that $\approx$ is an equivalence relation. Specify the equivalence classes under $\approx$.

Solution: There are two equivalence classes, [0] and [1]. Moreover, $[0] = \{m : m$ is even$\} = f^{-1}[\{0\}]$, and $[1] = \{m : m$ is odd$\} = f^{-1}[\{1\}]$. So we have $Z/\approx = \{[0], [1]\}$.

Note that in Example 6 the collection of equivalence classes $Z/\approx = \{[0], [1]\}$ "looks like" the codomain of $f : Z \to \{0, 1\}$. When the function f is onto, as it is in that example, there is a bijective function from the collection of equivalence classes under $\approx$ onto the codomain of f. We prove this result in the following theorem.

Theorem 8 Let $f : D \to C$. Then
a. $\approx$ is an equivalence relation on D.
b. The function $g : D \to D/\approx$ defined by $g(x) = [x]$ is onto.
c. If, in addition, f is onto, then there is a unique function $h : D/\approx \to C$ such that h is bijective and $h \circ g = f$.

Proof.
a. See Exercise 17a.
b. The function g is onto since, if $[x] \in D/\approx$ for some $x \in D$, then clearly, $g(x) = [x]$.
c. Assume $f : D \to C$ is onto. We first show the uniqueness of any function $h : D/\approx \to C$ such that $h \circ g = f$. Suppose there are two functions $h : D/\approx \to C$ and $h' : D/\approx \to C$ such that $h \circ g = f$ and $h' \circ g = f$. To verify that $h = h'$, we show that $h([x]) = h'([x])$ for all $x \in D$ (that is, for all members $[x]$ of $D/\approx$). We have $h([x]) = h(g(x)) = h \circ g(x) = f(x) = h' \circ g(x) = h'(g(x)) = h'([x])$.

To show existence, define $h : D/\approx \to C$ by $h([x]) = f(x)$. To prove that h is a function, we must settle a difficulty that arises here. What guarantees that $h([x]) = f(x) = f(y) = h([y])$ if $[x] = [y]$? We verify that this is, in fact, the case. Assume $[x] = [y]$. Then, by Theorem 2, $x \approx y$, and hence, $f(x) = f(y)$. Hence, $[x] = [y]$ implies $h([x]) = h([y])$. We then say that the function h is well defined.

We now verify that $h \circ g = f$. Let $x \in D$. Then $h \circ g(x) = h(g(x)) = h([x]) = f(x)$. That h is 1–1 and onto is left as an exercise. (See Exercise 17b.) $\square$

The functions in Theorem 8 are diagrammed in Figure 5–5. We say the **diagram** in the figure **commutes** if $h \circ g = f$.

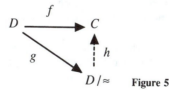

Figure 5–5

Example 7

Let $D = \{0, 1, 2, 3, 4, 5\}$ and $C = \{a, b, c\}$, and let f be defined by the following table:

x	0	1	2	3	4	5
$f(x)$	a	b	c	a	b	c

Define $\approx$ as above. Then $D/\approx = \{[0], [1], [2]\} = \{\{0,3\}, \{1,4\}, \{2,5\}\}$. Note that $[0] = \{0,3\} = f^{-1}[\{a\}], [1] = \{1,4\} = f^{-1}[\{b\}]$, and $[2] = \{2,5\} = f^{-1}[\{c\}]$. The function $h : D/\approx \to C$ is then defined by the following table:

$[x]$	$[0]$	$[1]$	$[2]$
$h([x])$	a	b	c

Notice that h is 1–1 and onto.

Example 8

Let $D = \mathcal{R}$ and $C = \mathcal{Z}$, and let $f : D \to C$ be defined by $f(x) = \lceil x \rceil$. The function f is the ceiling function previously defined in Exercise Set 4.3. Find $D/\approx$.

Solution: As an example, note that $[2] = f^{-1}[\{2\}] = \{x : 1 < x \leq 2\} = (1, 2]$. We see that $D/\approx = \{(m, m+1] : m \in \mathcal{Z}\}$. Notice from Example 4 that $D/\approx = \Pi$, where Π is the partition of $\mathcal{R}$ given in that example.

Practice Problem 5. Let $f : [0, 4) \to \{0, 1, 2, 3\}$ be defined by $f(x) = \lfloor x \rfloor$. Then f is the floor function defined in Exercise Set 4.3. Find $D/\approx$, and make a table for $h : [0, 4)/\approx \to \{0, 1, 2, 3\}$ as defined in Theorem 8. Observe that h is 1–1 and onto.

EXERCISE SET 5.1

1. Define a binary relation $\equiv_4$ on $\mathcal{Z}$ by $m \equiv_4 n$ if $m - n$ is divisible by 4. In Theorem 1, we proved that $\equiv_4$ is an equivalence relation on $\mathcal{Z}$.
    (a) Determine all the equivalence classes under $\equiv_4$. How many are there?
    (b) Verify that the equivalence classes form a partition of $\mathcal{Z}$.

2. Let d be a fixed positive integer. Define a binary relation $\equiv_d$ on $\mathcal{Z}$ by $m \equiv_d n$ whenever $m - n$ is divisible by d. In Theorem 1, we proved that $\equiv_d$ is an equivalence relation on $\mathcal{Z}$.
    (a) Determine all the equivalence classes under $\equiv_d$. How many are there?
    (b) Verify that the equivalence classes form a partition of $\mathcal{Z}$.

3. Let

$$M = \begin{pmatrix} a & b & c & d \\ e & f & g & h \\ i & j & k & l \\ m & n & o & p \end{pmatrix}$$

be a 4×4 matrix. Define R on the set $A = \{a, b, c, d, e, f, g, h, i, j, k, l, m, n, o, p\}$ by xRy if and only if x and y are on the same row of the matrix M.
   (a) Prove that R is an equivalence relation on A.
   (b) What are the equivalence classes under R?

4. Let A be the set of all rabbits alive on Catalina Island. Define a relation T on A by xTy if and only if x and y have the same biological parents.
   (a) Prove that T is an equivalence relation on A.
   (b) Characterize the equivalence classes under T.

5. Let A be any nonempty set.
   (a) Verify that 1_A is an equivalence relation.
   (b) Show that $[a] = \{a\}$ for all $a \in A$.

6. Let A be any nonempty set. Define S on A by $S = A \times A$.
   (a) Verify that S is an equivalence relation.
   (b) Show that $[a] = A$ for all $a \in A$.

7. Let D be the set of digits $\{0, 1, 2, \ldots, 9\}$. Define a relation R on the power set $P(D)$ by ARB if and only if A and B have the same number of elements.
   (a) Prove that R is an equivalence relation on D.
   (b) List the elements in the equivalence class $[\emptyset]$ and in $[D]$.
   (c) List the elements in the equivalence class $[\{4\}]$.
   (d) What is in the equivalence class $[\{4, 7\}]$?
   (e) How many equivalence classes are there under R?

8. Define a relation S on $\mathcal{R}$ by xSy if and only if $\sin x = \sin y$.
   (a) Prove that S is an equivalence relation.
   (b) List the elements in the equivalence classes $[0]$, $[\pi/2]$, $[\pi]$, and $[\pi/3]$.
   (c) Let $a \in \mathcal{R}$ be given. Specify the elements in $[a]$ without reference to the sine function.

9. Define a relation T on $\mathcal{R}$ by xTy if and only if $\sin^2 x + \cos^2 y = 1$.
   (a) Prove that T is an equivalence relation.
   (b) List the elements in the equivalence classes $[0]$, $[\pi/2]$, and $[\pi]$.
   (c) Let $a \in \mathcal{R}$ be given. Specify the elements in $[a]$ without reference to the sine function or the cosine function.

10. Let C be the set of all differentiable real-valued functions. Define a relation R on C by fRg if and only if $f' = g'$.
   (a) Prove that R is an equivalence relation.
   (b) Define the real-valued function g by $g(x) = x^2$. Use a theorem of the calculus to characterize the elements in $[g]$.
   (c) Let $f \in C$ be given. Specify the elements in $[f]$ without reference to a derivative.

11. Define a relation R on $\mathcal{Z} \times (\mathcal{N} - \{0\})$ by $(u, v)R(z, w)$ if and only if $w \cdot u = z \cdot v$.
   (a) Prove that R is an equivalence relation.
   (b) Specify the elements in $[(3, 6)]$, $[(6, 3)]$, $[(0, 1)]$, and $[(-7, 2)]$.

12. Suppose S is an equivalence relation on $\mathcal{R}$. Given $a \in \mathcal{R}$, it is possible to obtain a representation of $[a]$ (a subset of $\mathcal{R}$) as a subset $[a] \times \{0\}$ of the x-axis in the plane as follows. Draw the graph of S. The horizontal line $y = a$ (the set $\mathcal{R} \times \{a\}$) must intersect the graph of S. Draw vertical lines from the points of intersection to the x-axis. These points of intersection on the x-axis form the set $[a] \times \{0\}$.
   Justify each step of the preceding construction.

13. Let R be a symmetric and transitive binary relation on a set A. Assume that $\text{dom}(R) = A$.

Prove that R is an equivalence relation on A.

14. Prove Theorem 5.

15. Prove Theorem 6.

16. Prove Theorem 7.

17. In Theorem 8:

   (a) Show that $\approx$ is an equivalence relation on D.
   (b) Show that h is 1–1 and onto.
   (c) Show that $[a] = f^{-1}[\{f(a)\}]$.

   Exercises 18–20 deal with Theorem 8.

18. Define $f : \mathcal{Z} \to \{0, 1\}$ by

$$f(x) = \begin{cases} 0 & \text{if } x \text{ is divisible by 2} \\ 1 & \text{if } x - 1 \text{ is divisible by 2} \end{cases}$$

   (a) Find $\mathcal{Z}/\approx$.
   (b) Specify the function h.
   (c) Prove that $\approx$ is the same equivalence relation as $\equiv_2$.

19. Define $f : \mathcal{Z} \to \{0, 1, 2\}$ by

$$f(x) = \begin{cases} 0 & \text{if } x \text{ is divisible by 3} \\ 1 & \text{if } x - 1 \text{ is divisible by 3} \\ 2 & \text{if } x - 2 \text{ is divisible by 3} \end{cases}$$

   (a) Find $\mathcal{Z}/\approx$.
   (b) Specify the function h.
   (c) Prove that $\approx$ is the same equivalence relation as $\equiv_3$.

20. Let $f : [0, 1] \times [0, 1] \to [0, 1]$ be defined by $f(x, y) = x$.

   (a) Draw a graph of $[(1/2, 2/3)]_\approx$.
   (b) Describe $([0, 1] \times [0, 1])/\approx$.
   (c) Specify the function h.

21. Prove, or find a counterexample to, the following conjecture: If R and S are equivalence relations on A, then $R \circ S$ is an equivalence relation on A.

22. Prove, or find a counterexample to, the following conjecture: If R and S are equivalence relations on A, then $R \cup S$ is an equivalence relation on A.

23. Let R and S be equivalence relations on A. Prove:

   (a) $R \cap S$ is an equivalence relation on A.
   (b) $[x]_{R \cap S} = [x]_R \cap [x]_S$.

5.2 PARTIALLY ORDERED SETS

The standard ordering $\leq$ for any subset A of the real numbers has three important properties: (1) $x \leq x$ for all $x \in A$; (2) if $x \leq y$ and $y \leq x$, then $x = y$; (3) if $x \leq y$ and $y \leq z$, then $x \leq z$. In other words, the relation $\leq$ is reflexive, antisymmetric, and transitive.

Definition. A binary relation R on a set A that is reflexive on A, antisymmetric, and transitive is called a **partial ordering** on A (or **partial order** of A).

The set A is **partially ordered** by R if R is a partial order of A. The ordered pair (A, R) is called a partially ordered set, or **poset.** The word poset stands for **partially ordered set.**

Example 1

Let $A = \{1, 2, 3, 4\}$. Then the binary relation $R = \{(a, b) : a \leq b\}$ on A is a partial ordering on A. Thus, A is a partially ordered set under R. Note that $R = \{(1, 1), (1, 2), (1, 3), (1, 4), (2, 2), (2, 3), (2, 4), (3, 3), (3, 4), (4, 4)\}$.

Example 2

The binary relation $R = \{(m, n) : m \leq n\}$ on $\mathcal{Z}$ is a partial ordering on $\mathcal{Z}$. Therefore, $\mathcal{Z}$ is a partially ordered set under R. The partial order R is the **standard partial order** of $\mathcal{Z}$.

Definition. Let A be a partially ordered set under R. Elements $x, y \in A$ are **comparable** if either xRy or yRx.

Definition. R is a **linear ordering** on A if R is a partial ordering on A and every pair of elements in A is comparable.

The partial orderings defined in Examples 1 and 2 are also linear orderings. The next example gives a very important partial ordering that is not a linear ordering.

Example 3

The **set-inclusion partial order.** Let $P(X)$ be the set of all subsets of X, that is, the power set of X. Define a binary relation R on $P(X)$ as follows: For A, B in $P(X)$, ARB if and only if $A \subseteq B$. Verify that $P(X)$ is a partially ordered set (poset) under R.

Solution: 1. $A \subseteq A$. 2. $A \subseteq B$ and $B \subseteq A$ implies $A = B$. 3. $A \subseteq B$ and $B \subseteq C$ implies $A \subseteq C$. We also see that, in general, the set-inclusion partial order is not a linear order since, if $X = \mathcal{Z}$, then the sets $\{1, 2\}$ and $\{0, 1\}$ are not comparable.

Practice Problem 1. In Example 3:
a. Find two more sets in $P(\mathcal{Z})$ that are not comparable.
b. Find a set S in $P(\mathcal{Z})$ that is comparable to every other set in $P(\mathcal{Z})$.
How many such sets are there in $P(\mathcal{Z})$?

We have used R to stand for a generic partial ordering and have given several examples with $\leq$ used to denote a specific partial ordering. From now on, we will use $\preceq$ to denote a generic partial ordering.

A partial order of a finite set can be represented by a diagram.

Example 4

Let $A = \{1, 2, 3, 4, 8\}$. Define a partial order by $x \preceq y$ if and only if x divides y. Hence, $a \preceq a$ and $1 \preceq a$ for all $a \in A$. Also, $2 \preceq 4$, $2 \preceq 8$, and $4 \preceq 8$. A diagram of this poset is shown in Figure 5–6.

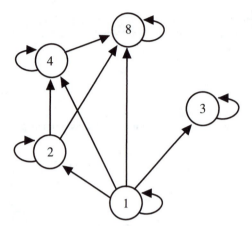

Figure 5–6

Each arrow from a labeled node to a labeled node in the diagram corresponds to an ordered pair in the partial-order relation. For example, the arrow from the node labeled "2" to the node labeled "8" corresponds to the pair $(2, 8)$ in the relation. Similarly, there is an arrow from the node labeled "1" to the node labeled "3" because $1 \preceq 3$. An arrow from a node labeled "x" to a node labeled "y" is called an **edge** (x, y).

We can simplify diagrams of partial orders considerably by eliminating edges that are in the diagram because of the reflexive property or the transitive property. In other words, we eliminate all edges (x, x), and any edge (x, z) whenever (x, y) and (y, z) are edges. We also draw the nodes as points and eliminate arrowheads from the edges by always drawing the directed edges so that they point upwards. With these conventions, a diagram of a partial order is called a **Hasse diagram.**

The Hasse diagram in Figure 5–7 represents the partial order pictured in Figure 5–6.

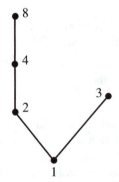

Figure 5–7

Let $U = \{a, b\}$, and recall that set inclusion $\subseteq$ is a partial order for the power set $P(U)$. Figure 5–8 shows the Hasse diagram for the set inclusion partial order on U.

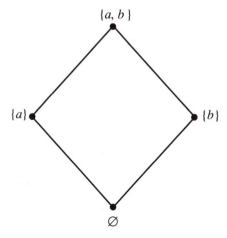

Figure 5–8

For $U = \{a, b, c\}$ and the set-inclusion partial order, we get the Hasse diagram shown in Figure 5–9.

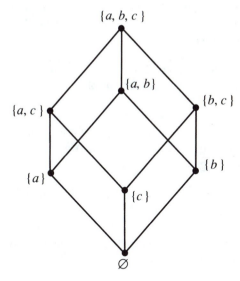

Figure 5–9

For a finite set A, the Hasse diagram is enough to specify the partial ordering on A. Two substantially different partial orders of the same set are given in Example 5.

Example 5

Let $A = \{0, 1, 2, 3, 4\}$.

a. Figure 5–10 shows the Hasse diagram for the standard linear order of A.

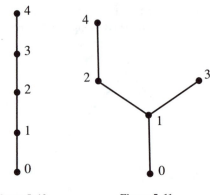

Figure 5-10   **Figure 5-11**

b. Figure 5-11 shows the Hasse diagram for the partial order of A defined by $x \preceq y$ if and only if $x = 0$ or x divides y.

Extremal Elements in Posets

Certain elements in posets possess properties that make them especially useful. In what follows, we will discuss least and greatest elements, minimal and maximal elements, lower bounds and upper bounds, and least upper bounds and greatest lower bounds. These are all called **extremal elements.**

Definition.   Let $(A, \preceq)$ be a poset. Then v is a **least element** of A if $v \in A$ and $v \preceq x$ for all $x \in A$; also, u is a **greatest element** of A if $u \in A$ and $x \preceq u$ for all $x \in A$.

In Example 5a, the least element is 0 and the greatest element is 4. In Example 5b, the least element is 0 but there is no greatest element. A given partial order may not possess either a least element or a greatest element.

Practice Problem 2.   Draw the Hasse diagram for a poset in which there is neither a least element nor a greatest element.

If a least element does exist for a poset, it must be unique, and similarly for a greatest element. The following theorem asserts this fact.

Theorem 1   There is at most one least element and at most one greatest element for a given poset.

Proof. Let $(A, \preceq)$ be a poset. We prove that there is at most one least element; the proof that there is at most one greatest element is left as an exercise. Assume that v and v' are both least elements of $(A, \preceq)$. Then $v \preceq v'$, since v is a least element and $v' \in A$. Similarly, $v' \preceq v$. Hence, $v = v'$, since $\preceq$ is antisymmetric. $\square$

Any subset A of the real numbers can be considered a poset $(A, \leq)$, where $\leq$ is the standard order for real numbers in A. For example, $[1, 5)$ is partially ordered by $\leq$. The least element of $[1, 5)$ is 1, but $[1, 5)$ has no greatest element.

Before discussing minimal and maximal elements, it is convenient to introduce "strictly less than" and "strictly greater than." For the standard order of $\mathcal{R}$, we read $x < y$ as "x is strictly less than y" or "y is strictly greater than x."

Definition   We write $a \prec b$ when $a \preceq b$ and $a \neq b$. When $a \prec b$, we say that a is **strictly less than** b, or, equivalently, b is **strictly greater than** a.

Definition   Let $(A, \preceq)$ be a poset. Then m is a **minimal** element of A if there is no $x \in A$ such that $x \prec m$; also, q is a **maximal** element of A if there is no $x \in A$ such that $q \prec x$.

In other words, a minimal element is an element such that no element is strictly less than it. Similarly, a maximal element is an element with no element strictly greater than it.

It is instructive to see where minimal and maximal elements must appear in Hasse diagrams. Unlike least and greatest elements, minimal and maximal elements are not necessarily unique. We illustrate this in the next example.

Example 6

Let $A = \{a, b, c, d, e, f, g\}$, and let the partial order $\preceq$ be defined by the Hasse diagram pictured in Figure 5–12.

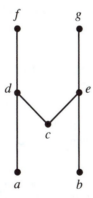

Figure 5–12

The elements a, b, and c are all minimal elements; f and g are maximal elements. Clearly, minimal and maximal elements need not be unique.

It is possible that a poset not have a minimal or maximal element, but finite nonempty posets always contain maximal and minimal elements.

Theorem 2   If A is a finite nonempty set and $(A, \preceq)$ is a poset, then A has at least one minimal element and at least one maximal element.

Proof. We prove that there is at least one minimal element in A. The proof that there is at least one maximal element is similar.

Pick any element $a_1 \in A$. If a_1 is minimal, we are done. If not, we can find a_2 such that $a_2 \prec a_1$. If a_2 is minimal, we are done. If not, we can find a_3 such that $a_3 \prec a_2 \prec a_1$. Similarly, we can find $a_4, \ldots$. But this process must terminate because A is finite. So the last a_k we pick will leave us with a linearly ordered set $a_k \prec \ldots \prec a_3 \prec a_2 \prec a_1$. Clearly, a_k must be a minimal element. $\qquad\square$

Practice Problem 3.   Find a poset that has neither a minimal element nor a maximal element. (*Hint:* See Theorem 2.)

Definition.   Let $(A, \preceq)$ be a poset, and let $B \subseteq A$. Then v is a **lower bound** of B if $v \preceq b$ for all $b \in B$; also, u is an **upper bound** of B if $b \preceq u$ for all $b \in B$.

The interval $[1, 5) \subseteq \mathcal{R}$, where $\mathcal{R}$ is partially ordered by $\leq$, has an infinite number of upper bounds and lower bounds. For example, 1, $\frac{1}{2}$, 0, and $-\pi$ are all lower bounds of $[1, 5)$.

For the poset $(A, \preceq)$ defined in Example 6, the set $\{a, c\}$ has no lower bounds, but it has upper bounds d and f. The elements d and f are both upper bounds of $\{a, c\}$ because they are above both a and c in the diagram. There are no lower bounds because no elements are below both a and c in the diagram. For a given poset described by a Hasse diagram, it is easy, with a little practice, to find lower and upper bounds of subsets by examining the Hasse diagram.

Practice Problem 4.   Let $A = \{a, b, c, d, e, f, g\}$, and let the partial order $\preceq$ be defined as in Example 6.
  a. Find all lower and upper bounds of $B = \{d, f\}$.
  b. Find all lower and upper bounds of $B = \{c, d, e\}$.
  c. Find all lower and upper bounds of $B = \{a, b\}$.

Definition.   Let $(A, \preceq)$ be a poset, and let $B \subseteq A$. Then g is a **greatest lower bound** of B, written $g = \mathrm{glb}(B)$, if g is a lower bound of B and $v \preceq g$ for all lower bounds v of B; also, h is a **least upper bound** of B, written $h = \mathrm{lub}(B)$, if h is an upper bound of B and $h \preceq u$ for all upper bounds u of B.

The interval $[1, 5) \subseteq \mathcal{R}$, where $\mathcal{R}$ is partially ordered by $\leq$, has 1 as a greatest lower bound and 5 as a least upper bound.

For the poset $(A, \preceq)$ defined in Example 6, the set $\{a, c\}$ has no greatest lower bound since it has no lower bounds. The element d is the least upper bound of $\{a, c\}$; that is, $\text{lub}(\{a, c\}) = d$.

Practice Problem 5. Let $A = \{a, b, c, d, e, f, g\}$, and let the partial order $\preceq$ be defined as in Example 6.
  a. Find all greatest lower bounds and least upper bounds of $B = \{d, f\}$.
  b. Find all greatest lower bounds and least upper bounds of $B = \{c, d, e\}$.
  c. Find all greatest lower bounds and least upper bounds of $B = \{a, b\}$.

For a poset $(A, \preceq)$ and a subset $B \subseteq A$, there may be no $\text{lub}(B)$ or $\text{glb}(B)$. However, when a least upper bound or a greatest lower bound does exist, then it is unique. Theorem 3 states this fact.

Theorem 3   For any poset $(A, \preceq)$ and subset $B \subseteq A$, there is at most one $\text{lub}(B)$ and at most one $\text{glb}(B)$.

Proof. We prove that there is at most one $\text{lub}(B)$. Suppose $h = \text{lub}(B)$ and $h' = \text{lub}(B)$. Then both h and h' are upper bounds of B. Since h is a least upper bound of B, we have $h \preceq h'$. Similarly, $h' \preceq h$. Therefore, $h = h'$. The proof that there is at most one greatest lower bound is similar and is left as an exercise.   □

Practice Problem 6. Let $B \subseteq \mathcal{R}$, where $\mathcal{R}$ is partially ordered by $\preceq$ and $B = \{\pi - 1/n : n \text{ is a positive integer}\}$.
  a. Find $\text{glb}(B)$.
  b. Find $\text{lub}(B)$.

For a given poset $(A, \preceq)$ and a subset $B \subseteq A$, it is not necessarily the case that the $\text{lub}(B)$ or the $\text{glb}(B)$, if it exists, belongs to the set B. For instance, in Example 6, the $\text{lub}(\{b, c\}) = e$.

Least upper bounds and greatest lower bounds are especially important in the real numbers $\mathcal{R}$ with the standard order $\leq$.

Example 7

Consider $\mathcal{R}$ with the standard order $\leq$. Let $B = \{r : r \text{ is a positive rational number and } r^2 > 2\}$. Then $\text{glb}(B) = \sqrt{2}$. By Supplementary Exercise 4 in Chapter 2, $\sqrt{2}$ is not a rational number. Note that B is both a subset of the rational numbers $\mathcal{Q}$ and a subset of the real numbers $\mathcal{R}$.

For the poset $(\mathcal{R}, \leq)$, with B a subset of $\mathcal{R}$, the $\text{glb}(B)$ exists, since $\text{glb}(B) = \sqrt{2}$, but $\text{glb}(B)$ is not an element of B.

Now let us restrict our attention to the poset $(Q, \leq)$. When B is considered a subset of Q, there does not exist a glb(B) since glb(B) = $\sqrt{2}$, which is not a rational number.

Example 8

By Example 3, $(P(X), \subseteq)$ is a poset. Let $\Omega \subseteq P(X)$; that is, Ω is a collection of subsets of X. Show that lub(Ω) = $\cup\Omega$.

Solution   We first show that $\cup\Omega$ is an upper bound of Ω. Let $A \in \Omega$. By Example 3 of Section 3.4, we have $A \subseteq \cup\Omega$, and hence, $\cup\Omega$ is an upper bound. To show that $\cup\Omega$ is the least upper bound of Ω, let B be an upper bound of Ω. Then $A \subseteq B$ for all $A \in \Omega$. By Exercise 15 of Section 3.4, we have $\cup\Omega \subseteq B$. Hence, $\cup\Omega =$ lub(Ω).

Example 8 thus shows that the least upper bound of a collection of sets, under the set-inclusion partial order, is the union of the collection. In Exercise 22, you are asked to prove that the greatest lower bound of a collection of sets is the intersection of the collection.

EXERCISE SET 5.2

1. Let $D_{35} = \{1, 5, 7, 35\}$, and define a partial order by $x \preceq y$ if and only if x divides y. Draw the Hasse diagram for the poset $(D_{35}, \preceq)$. Compare this Hasse diagram with that of Figure 5–8.

2. Let $D_{30} = \{1, 2, 3, 5, 6, 10, 15, 30\}$, and define a partial order by $x \preceq y$ if and only if x divides y. Draw the Hasse diagram for the poset $(D_{30}, \preceq)$. Compare this Hasse diagram with that of Figure 5–9.

3. A partial order $\preceq$ is defined on $A = \{a, b, c, d, e\}$ by the Hasse diagram in Figure 5–13.
   (a) Find any maximal or minimal elements.
   (b) Find any greatest or least elements.
   (c) Find all lower bounds and upper bounds of $\{c, e\}$.
   (d) Find the glb($\{c, e\}$), if it exists.
   (e) Find the lub($\{c, e\}$), if it exists.

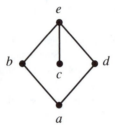

a   **Figure 5–13**

4. A partial order $\preceq$ is defined on $A = \{a, b, c, d, e, f, g\}$ by the Hasse diagram in Figure 5–14.
   (a) Find any maximal or minimal elements.
   (b) Find any greatest or least elements.
   (c) Find all lower bounds and upper bounds of $\{b, e, f\}$.
   (d) Find the glb($\{b, e, f\}$), if it exists.
   (e) Find the lub($\{b, f, g\}$), if it exists.

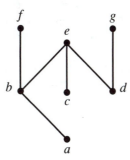

Figure 5–14

5. Let $A = \{2, 3, 6, 12\}$, and define a partial order by $x \preceq y$ if and only if x divides y.
   (a) Draw the Hasse diagram for the poset $(A, \preceq)$.
   (b) Find any maximal or minimal elements.
   (c) Find any greatest or least elements.
   (d) Find all lower bounds and upper bounds of $\{2, 3\}$.
   (e) Find the glb($\{2, 3\}$), if it exists.
   (f) Find the lub($\{3, 6\}$), if it exists.

6. Let $A = \{1, 2, 4, 5, 10\}$, and define a partial order by $x \preceq y$ if x divides y.
   (a) Draw the Hasse diagram for the poset $(A, \preceq)$.
   (b) Find any maximal or minimal elements.
   (c) Find any greatest or least elements.
   (d) Find all lower bounds and upper bounds of $\{2, 4, 5\}$.
   (e) Find the glb($\{2, 4\}$), if it exists.
   (f) Find the lub($\{2, 4, 5\}$), if it exists.

7. Let $(A, \preceq)$ be a poset, and assume $B \subseteq A$. Prove that there is at most one glb(B).

8. Prove that there is at most one greatest element for a given poset.

9. Let $B = \{r : r \in \mathcal{Q} \text{ and } r^2 \leq 2\}$.
   (a) For the poset $(\mathcal{R}, \leq)$, find glb(B) and lub(B), if they exist.
   (b) For the poset $(\mathcal{Q}, \leq)$, find glb(B) and lub(B), if they exist.

10. Let $B = \{r : r \in \mathcal{Q} \text{ and } r^2 > 3\}$.
   (a) For the poset $(\mathcal{R}, \leq)$, find glb(B) and lub(B), if they exist.
   (b) For the poset $(\mathcal{Q}, \leq)$, find glb(B) and lub(B), if they exist.

11. Consider the poset $(\mathcal{R}, \leq)$. Let $B = \{2 + 1/x : x \in \mathcal{R}^+\}$. Find glb($B$) and lub($B$), if they exist.

12. Consider the poset $(\mathcal{R}, \leq)$. Let $B = \{(1 - 3n)/(1 + n) : n \in \mathcal{N}\}$. Find glb($B$) and lub($B$), if they exist.

13. Let $(A, \preceq)$ be a poset. Define a relation $\propto$ on A^2 by $(a, b) \propto (c, d)$ if and only if $a \preceq c$ and $b \preceq d$.

(a) Is $(A^2, \varpropto)$ a poset? Justify your answer.

(b) Is $\varpropto$ a linear order on A^2? Justify your answer.

14. Let $\preceq$ be a linear order on A. Define a relation $\varpropto$ on A^2 by $(a, b) \varpropto (c, d)$ if and only if $a \prec c$ or $(a = c$ and $b \preceq d)$.

(a) Prove that $(A^2, \varpropto)$ is a poset. The partial order $\varpropto$ is called the **lexicographic order** or **dictionary order.**

(b) Is $\varpropto$ a linear order on A^2? Justify your answer.

15. Let R be a relation on A. Then R is **irreflexive** on A if $\neg x R x$ for all $x \in A$. Let R be a transitive and irreflexive relation on A, and define a relation $\preceq$ on A as follows: $x \preceq y$ if and only if $x R y$ or $x = y$. Show that $\preceq$ is a partial ordering on A.

16. This exercise shows that every poset "looks like" a subset of a poset of the type given in Example 3. Let $(A, \preceq)$ be a poset. Define a function $f : A \to P(A)$ by $f(a) = \{x : x \in A$ and $x \preceq a\}$.

(a) Prove that f is 1–1.

(b) Prove that $a \preceq b$ if and only if $f(a) \subseteq f(b)$. (We say that f is an **order-preserving** function.)

17. Let $(A, \preceq)$ be a poset, and let R be an equivalence relation on A such that if $x R x'$ and $y R y'$, then $x \preceq y$ if and only if $x' \preceq y'$. Define a relation $\varpropto$ on A/R as follows: $[x] \varpropto [y]$ if and only if $x \preceq y$.

(a) Show that $\varpropto$ is well defined; that is, if $[x] = [x']$ and $[y] = [y']$, then $x \preceq y$ if and only if $x' \preceq y'$.

(b) Show that $(A/R, \varpropto)$ is a poset.

18. Let R be a transitive and reflexive relation on A. Define a relation $\approx$ on A by $x \approx y$ if and only if $x R y$ and $y R x$.

(a) Show that $\approx$ is an equivalence relation on A.

(b) Define a relation $\preceq$ on $A/\approx$ by $[x] \preceq [y]$ if and only if $x R y$. Show that $\preceq$ is well defined (see Exercise 17a).

(c) Show that $(A/\approx, \preceq)$ is a poset.

19. Prove, or find a counterexample to, the following conjecture: If R and S are partial orderings on A, then $R \circ S$ is a partial ordering on A.

20. Prove, or find a counterexample to, the following conjecture: If R and S are partial orderings on A, then $R \cup S$ is a partial ordering on A.

21. Prove, or find a counterexample to, the following conjecture: If R and S are partial orderings on A, then $R \cap S$ is a partial ordering on A.

22. For the poset $(P(A), \subseteq)$, prove that $\text{glb}(\Omega) = \cap\Omega$, where $\Omega \subseteq P(A)$.

5.3 ALGEBRAS, HOMOMORPHISMS, AND ISOMORPHISMS

Algebras

Definition. An **algebra** is an ordered pair $(A, *)$, where A is a nonempty set and $*$ is a binary operation on A.

Note that A is closed under $*$ in this definition, since $*$ is a binary operation on A. In fact, to show that $(A, *)$ is an algebra, all that is required is to prove that A is closed under $*$.

Example 1

    a. $(\mathcal{Z}, +)$, where $+$ is the usual addition on $\mathcal{Z}$, is an algebra since $\mathcal{Z}$ is closed under $+$.

    b. $(\mathcal{N}, \cdot)$, where $\cdot$ is the usual multiplication on $\mathcal{N}$, is an algebra since $\mathcal{N}$ is closed under $\cdot$.

    c. $(\mathcal{R}, -)$, where $-$ is the usual subtraction on $\mathcal{R}$, is an algebra since $\mathcal{R}$ is closed under $-$.

    d. $(\mathcal{N}, -)$ is not an algebra since $\mathcal{N}$ is not closed under subtraction.

Example 2

Let $B = \{p, q, r\}$, and consider the Hasse diagram in Figure 5–15. Define a binary operation $\wedge$ on B by $x \wedge y = \text{glb}(\{x, y\})$. The operation $\wedge$ is called the **meet** operation, and the element $x \wedge y$ is called the meet of x and y (see Table 5–1)

Figure 5–15

TABLE 5–1

$\wedge$	p	q	r
p	p	r	r
q	r	q	r
r	r	r	r

The table yields, for example, $p \wedge q = r$ and $r \wedge q = r$. We can tell at a glance that B is closed under $\wedge$, since the only entries in the table are p's, q's, and r's. Hence, $(B, \wedge)$ is an algebra.

Example 3

Let $A = \{a, b, c, d, e\}$. Consider the Hasse diagram in Figure 5–16. An algebra $(A, \vee)$ can be formed from the poset in the figure by defining a binary operation on A by $x \vee y = \text{lub}(\{x, y\})$. The operation $\vee$ is called the **join** operation, and the element $x \vee y$ is called the join of x and y. Table 5–2 is the table for the operation $\vee$.

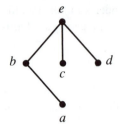

b c d

e

a

Figure 5–16

TABLE 5–2

∨	a	b	c	d	e
a	a	b	e	e	e
b	b	b	e	e	e
c	e	e	c	e	e
d	e	e	e	d	e
e	e	e	e	e	e

Example 4

In Section 5.1, we defined an equivalence relation $\equiv_3$ on $\mathcal{Z}$. Let Z_3 designate the set of equivalence classes $\{[0], [1], [2]\}$. We define "addition" $+_3$ on Z_3 as follows: $[x] +_3 [y] = [x+y]$. We need to show that $+_3$ is a binary operation; that is, that $+_3$ is independent of the choice of elements in the equivalence classes $[x]$ and $[y]$. Let x' be any element in $[x]$ and y' be any element in $[y]$. We must show that $[x+y] = [x'+y']$. By Theorem 2 of Section 5.1, this can be restated as follows: If $[x] = [x']$ and $[y] = [y']$, then $[x+y] = [x'+y']$.

To prove this last statement, assume that $[x] = [x']$ and $[y] = [y']$. Then $x \equiv_3 x'$ and $y \equiv_3 y'$. By the definition of $\equiv_3$, $x - x' = 3k$ for some integer k, and $y - y' = 3j$ for some integer j. Adding these equations yields $(x + y) - (x' + y') = 3k + 3j = 3(k + j)$. Hence, $x + y \equiv_3 x' + y'$. Therefore, $[x + y] = [x' + y']$. But this shows that $+_3$ is an operation on Z_3, and hence, $(Z_3, +_3)$ is an algebra.

Practice Problem 1. Define "multiplication" $*_3$ on Z_3 as follows: $[x] *_3 [y] = [x \cdot y]$. Show that $*_3$ is a binary operation on Z_3; that is, show that if $[x] = [x']$ and $[y] = [y']$, then $[x \cdot y] = [x' \cdot y']$. Hence, $(Z_3, *_3)$ is an algebra.

Homomorphisms

Let A and B be the sets in Examples 3 and 2, respectively. Let $f : A \to B$ be defined as follows: $f(a) = f(b) = p$, $f(c) = f(e) = r$, and $f(d) = q$. Form Table 5–3 by applying f to all the entries in Table 5–2.

TABLE 5–3

#	$f(a)$	$f(b)$	$f(c)$	$f(d)$	$f(e)$
$f(a)$	$f(a)$	$f(b)$	$f(e)$	$f(e)$	$f(e)$
$f(b)$	$f(b)$	$f(b)$	$f(e)$	$f(e)$	$f(e)$
$f(c)$	$f(e)$	$f(e)$	$f(c)$	$f(e)$	$f(e)$
$f(d)$	$f(e)$	$f(e)$	$f(e)$	$f(d)$	$f(e)$
$f(e)$	$f(e)$	$f(e)$	$f(e)$	$f(e)$	$f(e)$

Now rewrite Table 5–3, using the function values given for f to obtain Table 5–4.

TABLE 5–4

#	p	p	r	q	r
p	p	p	r	r	r
p	p	p	r	r	r
r	r	r	r	r	r
q	r	r	r	q	r
r	r	r	r	r	r

In spite of the repetitions, Table 5–4 defines an operation # on the set B. The question is, which operation? By examining the entries in the table we see that it defines exactly the same operation on B that Table 5–1 defines. So Tables 5–1, 5–3, and 5–4 are all tables for the meet operation $\wedge$.

Now, let x and y be any two elements of A. Since Table 5–3 is a table for the operation $\wedge$, $f(x) \wedge f(y)$ in that table occupies the same spot as $x \vee y$ occupies in Table 5–2. But the entries in Table 5–3 are obtained from the entries in Table 5–2 by applying the function f. So $f(x) \wedge f(y)$ should equal $f(x \vee y)$. We can easily verify this, for example, when $x = a$ and $y = c$. By Table 5–3, $f(a) \wedge f(c) = f(e)$, since Table 5–3 is a table for $\wedge$. By Table 5–2, $a \vee c = e$, and hence, $f(a \vee c) = f(e)$. Therefore, $f(a) \wedge f(c) = f(e) = f(a \vee c)$. In fact, $f(x) \wedge f(y) = f(x \vee y)$ for all x and y in A.

Definition.   Let $(A, *)$ and $(B, \#)$ be algebras. A function $f : A \to B$ is a **homomorphism** from $(A, *)$ to $(B, \#)$ if, for all $x, y \in A$, $f(x * y) = f(x)\#f(y)$.

We say that a homomorphism **preserves** the **operations** of an algebra.

Example 5

In Example 4, define $f : \mathcal{Z} \to Z_3$ by $f(x) = [x]$. Show that f is a homomorphism from $(\mathcal{Z}, +)$ to $(Z_3, +_3)$.

Solution By the definition of f, $f(x + y) = [x + y]$ and by the definitions of f and $+_3$, $f(x) +_3 f(y) = [x] +_3 [y] = [x + y]$. Hence, $f(x + y) = f(x) +_3 f(y)$, and f is a homomorphism from $(\mathcal{Z}, +)$ to $(Z_3, +_3)$.

Practice Problem 2. Let f be defined as in Example 5 and $*_3$ as in Practice Problem 1. Show that f is a homomorphism from $(\mathcal{Z}, \cdot)$ to $(Z_3, *_3)$.

Example 6

Let $*$ be the operation on $\mathcal{R}$ defined by $x * y = x/(y^2 + 1)$, and let $\#$ be the operation on $B = (-1, \infty)$ defined by $x \# y = \left[\sqrt{|x|}/(y + 1) \right]^2$. Note that $x \# y$ is always nonnegative, and hence, A is closed under $\#$. Then $(\mathcal{R}, *)$ and $(B, \#)$ are algebras. Define $f : \mathcal{R} \to B$ by $f(x) = x^2$. Show that f is a homomorphism from $(\mathcal{R}, *)$ to $(B, \#)$.

Solution

$$f(x * y) = f\left(\frac{x}{y^2 + 1} \right) = \left(\frac{x}{y^2 + 1} \right)^2$$

and

$$f(x) \# f(y) = x^2 \# y^2 = \left(\frac{\sqrt{|x^2|}}{y^2 + 1} \right)^2 = \left(\frac{x}{y^2 + 1} \right)^2$$

Hence, $f(x * y) = f(x) \# f(y)$.

Practice Problem 3. Let $\#$ be the binary operation on $\mathcal{Z}$ defined by $x \# y = (x - 1)(y - 1) + 1$. Define a function $f : \mathcal{Z} \to \mathcal{Z}$ by $f(x) = x^2 + 1$. Show that f is a homomorphism from $(\mathcal{Z}, \cdot)$ to $(\mathcal{Z}, \#)$, where $\cdot$ is the standard multiplication on $\mathcal{Z}$.

In the definition of an algebra, one binary operation on a set is specified. This definition can be generalized to arbitrarily many operations, and they need not be binary. In addition, we may have arbitrarily many **distinguished elements**. A distinguished element is an element in the algebra, singled out, perhaps, for some special properties. In this section, we discuss algebras with at most two binary operations and at most two distinguished elements.

Example 7

$(\mathcal{Z}, +, \cdot, 0, 1)$ is an algebra, since $\mathcal{Z}$ is closed under $+$ and $\cdot$, and $0, 1 \in \mathcal{Z}$. In this example, 0 and 1 are the distinguished elements.

Definition. A function $f : A \to B$ is a **homomorphism** from an algebra $(A, *, \#, a, b)$ to an algebra $(B, \oplus, \odot, c, d)$ if $f(x * y) = f(x) \oplus f(y)$, $f(x \# y) = f(x) \odot f(y)$, $f(a) = c$, and $f(b) = d$.

Notice that a homomorphism, in addition to preserving the operations, takes distinguished elements to corresponding distinguished elements. In the preceding definition, a corresponds to c, and b corresponds to d.

Example 8

Define $f : \mathcal{Z} \to Z_3$ by $f(x) = [x]$. Show that f is a homomorphism from $(\mathcal{Z}, +, \cdot, 0, 1)$ to $(Z_3, +_3, *_3, [0], [1])$.

Solution By Example 5 and Practice Problem 2, we see that $f(x+y) = f(x) +_3 f(y)$ and $f(x \cdot y) = f(x) *_3 f(y)$. In addition, we have $f(0) = [0]$ and $f(1) = [1]$. Hence, f is a homomorphism from $(\mathcal{Z}, +, \cdot, 0, 1)$ to $(Z_3, +_3, *_3, [0], [1])$.

We have defined a homomorphism for algebras of the form $(A, *, \#, a, b)$. A similar definition can be given for algebras of any form.

Isomorphisms

Definition. A bijective function $f : A \to B$ that is a homomorphism from an algebra $(A, *, \#, a, b)$ to an algebra $(B, \oplus, \odot, c, d)$ is called an **isomorphism** from $(A, *, \#, a, b)$ to $(B, \oplus, \odot, c, d)$. When there exists an isomorphism from $(A, *, \#, a, b)$ to $(B, \oplus, \odot, c, d)$, we say that $(A, *, \#, a, b)$ is **isomorphic** to $(B, \oplus, \odot, c, d)$.

A 1–1 function $f : A \to B$ that is a homomorphism from an algebra $(A, *, \#, a, b)$ to an algebra $(B, \oplus, \odot, c, d)$ is called an **embedding** of $(A, *, \#, a, b)$ into $(B, \oplus, \odot, c, d)$.

As with homomorphisms, definitions of isomorphisms and embeddings can be given for algebras of any form.

Example 9

Let $A = \{a, b, c\}$, and define two operations, $*$ and $\#$, on A by Tables 5–5 and 5–6. Then $(A, *, \#, a, b)$ is an algebra.

TABLE 5–5

$*$	a	b	c
a	a	b	c
b	b	c	a
c	c	a	b

TABLE 5–6

$\#$	a	b	c
a	a	a	a
b	a	b	c
c	a	c	b

TABLE 5–7

$+_3$	[0]	[1]	[2]
[0]	[0]	[1]	[2]
[1]	[1]	[2]	[0]
[2]	[2]	[0]	[1]

TABLE 5–8

$*_3$	[0]	[1]	[2]
[0]	[0]	[0]	[0]
[1]	[0]	[1]	[2]
[2]	[0]	[2]	[1]

Next, define the operations $+_3$ and $*_3$ on Z_3 as shown in Tables 5–7 and 5–8, and define $f : A \to Z_3$ by $f(a) = [0]$, $f(b) = [1]$, and $f(c) = [2]$.

By applying f to Tables 5–5 and 5–6, we obtain Tables 5–7 and 5–8. In addition, the distinguished elements are mapped to corresponding distinguished elements. Therefore, f is a homomorphism. Also, by the definition of f, we see that f is a bijection, and hence, f is an isomorphism from $(A, *, \#, a, b)$ to $(Z_3, +_3, \#, *_3, [0], [1])$.

Example 10

Let $A = \{2^n 3^m : n, m \in \mathcal{N}\}$ and $B = \{m + n\sqrt{2} : m, n \in \mathcal{N}\}$. Then A is closed under multiplication since $(2^n 3^m) \cdot (2^p 3^q) = 2^{n+p} 3^{m+q}$. Also, B is closed under addition since $(m + n\sqrt{2}) + (p + q\sqrt{2}) = (m + p) + (n + q)\sqrt{2}$. Hence, $(A, \cdot)$ and $(B, +)$ are algebras.

Now, define f by $f(2^n 3^m) = m + n\sqrt{2}$. To show that f defines a function from A to B, we must show that if $2^n 3^m = 2^p 3^q$, then $m + n\sqrt{2} = q + p\sqrt{2}x$. So let $2^n 3^m = 2^p 3^q$. We show that $n = p$. Suppose that $n \neq p$, say, $p < n$. Then $2^{n-p} 3^m = 3^q$. But the right side of this equation is odd, whereas the left side is even, a contradiction. So we have $n = p$. Therefore, $2^n 3^m = 2^n 3^q$, and hence, $3^m = 3^q$. But then, $m = q$.

To show that f is 1–1, suppose that $f(2^n 3^m) = f(2^p 3^q)$. Then $m + n\sqrt{2} = q + p\sqrt{2}$. We again show that $n = p$. Suppose that $n \neq p$. Solving for $\sqrt{2}$, we obtain $\sqrt{2} = (q - m)/(n - p)$. Now, since we assumed that $n \neq p$, it follows that $\sqrt{2}$ is rational. But this is a contradiction. Hence, $n = p$. So we have $m + n\sqrt{2} = q + n\sqrt{2}$, from which it follows that $m = q$. Therefore, $2^n 3^m = 2^p 3^q$, and hence, f is 1–1.

To show that f is onto, assume that $m + n\sqrt{2} \in B$, where $n, m \in \mathcal{N}$. Then $f(2^n 3^m) = m + n\sqrt{2}$.

Finally, we show that f preserves the operations. We have $f((2^n 3^m) \cdot (2^p 3^q)) = f(2^{n+p} 3^{m+q}) = (m+p) + (n+q)\sqrt{2}$. Also, $f(2^n 3^m) + f(2^p 3^q) = (m+n\sqrt{2}) + (p+q\sqrt{2}) = (m + p) + (n + q)\sqrt{2}$. Hence, $f((2^n 3^m) \cdot (2^p 3^q)) = f(2^n 3^m) + f(2^p 3^q)$. Therefore, f is an isomorphism from $(A, \cdot)$ to $(B, +)$.

Practice Problem 4. Let $A = \{2^n : n \in \mathcal{Z}\}$. Define $f : \mathcal{Z} \to A$ by $f(n) = 2^n$.

a. Show that A is closed under multiplication.

b. Show that f is an isomorphism from $(\mathcal{Z}, +)$ to $(A, \cdot)$. [You may use the fact that if $2^q = 1$, then $q = 0$.]

c. Conclude that f is an embedding of $(\mathcal{Z}, +)$ into $(\mathcal{Q}, \cdot)$.

Homomorphisms and Equivalence Relations

Let $(D, *, a)$ be an algebra with binary operation $*$ and distinguished element a, and let $(C, \#, b)$ be an algebra with binary operation $\#$ and distinguished element b. Suppose that $f : D \to C$ is onto and f is a homomorphism from $(D, *, a)$ to $(C, \#, b)$. Since f is onto, we say that f is a homomorphism from $(D, *, a)$ onto $(C, \#, b)$. In Section 5.1, we defined a relation $\approx$ on D by $x \approx y$ if and only if $f(x) = f(y)$.

We now define $\circledast$ on $D/\approx$ by $[x] \circledast [y] = [x * y]$. The situation here is similar to that in Example 4 and in Practice Problem 1. We need to show that $\circledast$ is an operation on $D/\approx$. This is done next in Theorem 1.

Theorem 1 Let $f : D \to C$ be a homomorphism from $(D, *, a)$ to $(C, \#, b)$. Then

a. $\approx$ is an equivalence relation on D.

b. $(D/\approx, \circledast, [a])$ is an algebra.

c. The function $g : D \to D/\approx$ defined by $g(x) = [x]$ is a homomorphism from $(D, *, a)$ onto $(D/\approx, \circledast, [a])$.

d. If, in addition, f is onto, then there is a unique function $h : D/\approx \to C$ such that h is an isomorphism from $(D/\approx, \circledast, [a])$ to $(C, \#, b)$ and $h \circ g = f$.

Proof. a. This proof is the same as the proof of Theorem 8, part a, in Section 5.1.
b. We must show that $\circledast$ is an operation on $D/\approx$; that is, if $[x] = [x']$ and $[y] = [y']$, then $[x * y] = [x' * y']$. To accomplish this, assume that $[x] = [x']$ and $[y] = [y']$. Then $x \approx x'$ and $y \approx y'$. By the definition of $\approx$, $f(x) = f(x')$ and $f(y) = f(y')$. Since f is a homomorphism, $f(x * y) = f(x) \# f(y) = f(x') \# f(y') = f(x' * y')$. Therefore, $x * y \approx x' * y'$ and $[x * y] = [x' * y']$. Hence, $(D/\approx, \circledast, [a])$ is an algebra.
c. Part of this proof is the same as in Theorem 8, part b, in Section 5.1. That g is homomorphism from $(D, *, a)$ onto $(D/\approx, \circledast, [a])$ is left as an exercise.
d. Define h as in the proof of Theorem 8, part c, in Section 5.1. The proof that h is unique and $h \circ g = f$ is the same as in that theorem. The proof that h is an isomorphism from $(D/\approx, \circledast, [a])$ to $(C, \#, b)$ is left as exercise. $\square$

Theorem 1 can be generalized to algebras of the form $(D, *, \#, a, b)$ or, in fact, to algebras of any form.

Example 11

We illustrate Theorem 1 with algebras of the form $(A, *)$. (See Examples 2 and 3, and Tables 5–1, 5–2, and 5–4.) To obtain Table 5–4, we applied a function f to all the elements in Table 5–2. This function was defined by $f(a) = f(b) = p$, $f(c) = f(e) = r$, and $f(d) = q$. We then observed that f is a homomorphism from the algebra $(A, \vee)$ in Example 2 to the algebra $(B, \wedge)$ in Example 3. For this function, the equivalence classes under the relation $\approx$ are $[a] = \{a, b\}$, $[c] = \{c, e\}$, and $[d] = \{d\}$. Tables 5–9 and 5–10 define the operations $\wedge$ on B and $\circledvee$ on $A/\approx$.

TABLE 5–9			
$\wedge$	p	q	r
p	p	r	r
q	r	q	r
r	r	r	r

TABLE 5–10			
$\circledvee$	$[a]$	$[c]$	$[d]$
$[a]$	$[a]$	$[c]$	$[c]$
$[c]$	$[c]$	$[c]$	$[c]$
$[d]$	$[c]$	$[c]$	$[d]$

Theorem 1 asserts that h is an isomorphism from $(B, \wedge)$ to $(A/\approx, \circledvee)$. The function h has values as follows: $h([a]) = f(a) = p$, $h([c]) = f(c) = r$, and $h([d]) = f(d) = q$. Note that h takes Table 5–10 to Table 5–9 and that h is 1–1. So h is indeed an isomorphism from $(A/\approx, \circledvee)$ to $(B, \wedge)$.

It is customary to use A to denote the algebra $(A, *)$ when it is clearly understood what the operation on A is. Similarly, we use the set of elements to denote an algebra of any form when it is clearly understood what all the operations and distinguished elements are. When this convention is used in a situation where more than one algebra is involved, care must be taken to avoid ambiguity.

Exercise Set 5.3

EXERCISE SET 5.3

1. Determine whether $^\#$ is a unary operation on the given set. If not, give a counterexample.
   (a) $B = \mathcal{R}$; $x^\# = x$
   (b) $B =$ the set of all finite subsets of $\mathcal{N}$; $X^\# = \mathcal{N} - X$, where $X \in B$
   (c) $B = \mathcal{N}$; $x^\# = 1/x$

2. Determine whether $^\#$ is a unary operation on the given set. If not, give a counterexample.
   (a) $B = \{1, 2, 3, 4, 12\}$; $x^\# = 12/x$
   (b) $B = \mathcal{Z}$; $x^\# = -x$
   (c) $B =$ set of finite subsets of $\mathcal{N}$; $X^\# = X - \{0, 1\}$, where $X \in B$

3. Determine whether $*$ is a binary operation on the given set B. If not, give a counterexample.
   (a) $B =$ set of all infinite subsets of $\mathcal{N}$; $X * Y = X \cup Y$
   (b) $B =$ set of all infinite subsets of $\mathcal{N}$; $X * Y = X \cap Y$

4. Determine whether $*$ is a binary operation on the given set B. If not, give a counterexample.
   (a) $B = \mathcal{N}$; $x * y = 2x + 1$
   (b) $B = \mathcal{Z}$; $x * y = x/y$

5. Show that the function $f : P(A) \to P(A)$ defined by $f(X) = A - X$ is an isomorphism from $(P(A), \cap)$ to $P(A), \cup)$.

6. Let $A = \{a, b, c, d, e\}$ and $B = \{p, q, r\}$. The Hasse diagrams in Figures 5–17 and 5–18 define partial orderings on A and B, respectively.

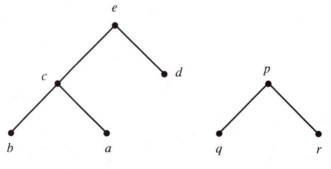

Figure 5–17   Figure 5–18

Using Figure 5–17, construct a table for the operation $\vee$ on A defined by $x \vee y = \text{lub}(\{x, y\})$. Using Figure 5–18, construct a table for the operation $\mathbb{W}$ on B defined by $x \mathbb{W} y = \text{lub}(\{x, y\})$. Define a function $f : A \to B$ by $f(a) = f(b) = f(c) = q$, $f(d) = r$, and $f(e) = p$. Apply f to all the entries in the table for the operation $\vee$ to show that f is a homomorphism from $(A, \vee)$ to $(B, \mathbb{W})$.

7. Using the sets given in Exercise 6, construct a table for the operation $\oslash$ on $A/\approx$ as in Example 11, and verify that the function $h : A/\approx \to B$ defined in the proof of Theorem 1 is in fact an isomorphism from $(A/\approx, \oslash)$ to $(B, \mathbb{W})$.

8. (a) Finish the proof of Theorem 1, part c.
   (b) Finish the proof of Theorem 1, part d.

9. Let $A = \{f : f : \mathcal{Z} \to \mathcal{Z}\}$; that is, A is the set of all function from $\mathcal{Z}$ into $\mathcal{Z}$. Define the function $f + g$ by $(f + g)(x) = f(x) + g(x)$. Then $+$ is an operation on A. (We use the same symbol to denote both addition of integers and addition of functions.) Define a function $\varphi : A \to \mathcal{Z}$ by $\varphi(f) = f(0)$. Show that φ is a homomorphism from $(A, +)$ to $(\mathcal{Z}, +)$.

10. Let $A = \{f : f : [0, 1] \to \mathcal{R} \text{ and } f \text{ is continuous}\}$; that is, A is the set of all continuous function from $[0, 1]$ into $\mathcal{R}$. Define the function $f + g$ by $(f + g)(x) = f(x) + g(x)$. Then $+$ is an operation on A. (We use the same symbol to denote both addition of real numbers and addition of functions.) Define a function $\varphi : A \to \mathcal{R}$ by $\varphi(f) = \int_0^1 f(x)\,dx$. Show that φ is a homomorphism from $(A, +)$ to $(\mathcal{R}, +)$. [You may use the well-known properties of integration.]

11. Show that the function $f : \mathcal{R} \to (0, \infty)$ defined by $f(x) = 2^x$ is an isomorphism from $(\mathcal{R}, +)$ to $((0, \infty), \cdot)$.

12. Show that the function $g : (0, \infty) \to \mathcal{R}$ defined by $g(x) = log_e x$ is an isomorphism from $((0, \infty), \cdot)$ to $(\mathcal{R}, +)$.

13. Define an operation $+$ on $\mathcal{R} \times \mathcal{R}$ by $(x, y) + (u, v) = (x + u, y + v)$ Show that the function $f : \mathcal{R} \times \mathcal{R} \to \mathcal{R} \times \mathcal{R}$ defined by $f(x, y) = (x + y, x - y)$ is an isomorphism from $(\mathcal{R} \times \mathcal{R}, +)$ to $(\mathcal{R} \times \mathcal{R}, +)$.

14. Let $B = \{m + n\sqrt{2} : m, n \in \mathcal{Z}\}$. In Example 10, we showed that B is closed under addition.
    (a) Show that B is closed under multiplication.
    (b) Show that f defined by $f(m + n\sqrt{2}) = m - n\sqrt{2}$ is an isomorphism from $(B, +, \cdot)$ to $(B, +, \cdot)$. (*Hint:* You must show that f actually defines a function. To do this, show that $m + n\sqrt{2} = p + q\sqrt{2}$ implies that $m - n\sqrt{2} = p - q\sqrt{2}$; see Example 10.)

15. Let $A = \{m + n\sqrt{3} : m, n \in \mathcal{Z}\}$ and $B = \{m + n\sqrt{2} : m, n \in \mathcal{Z}\}$. From Exercise 14, B is closed under addition and multiplication.
    (a) Show that A is closed under addition and multiplication.
    (b) Show that f defined by $f(m + n\sqrt{3}) = m + n\sqrt{2}$ is a function from A to B.
    (c) Show that f is a bijection.
    (d) Show that f is not a homomorphism from $(A, +, \cdot)$ to $(B, +, \cdot)$.

16. Let $(A, *, a)$ be an algebra with distinguished element a. Define an operation on $\{p\}$ by $p \# p = p$. Then $(\{p\}, \#, p)$ is an algebra. Show that the function $f : A \to \{p\}$ defined by $f(x) = p$ is a homomorphism from $(A, *, a)$ to $(\{p\}, \#, p)$.

17. Let A be a nonempty set. Define an operation $*$ on A by $x * y = y$. Let $\equiv$ be an equivalence relation on A. Define $\circledast$ by $[x] \circledast [y] = [x * y]$. Show that $\circledast$ defines an operation on $A/\equiv$.

18. Let A and $*$ be defined as in Exercise 17, and let $(B, \#)$ be an algebra. Prove that if $f : A \to B$ is a homomorphism from $(A, *)$ onto $(B, \#)$, then for all $u, v \in B$, $u \# v = v$.

19. Prove: If f is a homomorphism from $(A, *)$ to $(B, \#)$, then $f[A]$ is closed under $\#$.

20. Prove: If f is an embedding of $(A, *)$ into $(B, \#)$, then $(A, *)$ is isomorphic to $(f[A], \#)$.

In Exercises 21 and 22, let A be a set and, for a subset X of A, define a function $\chi_X : A \to \{0, 1\}$, called the **characteristic function,** by

$$\chi_X(x) = \begin{cases} 1 & \text{if } x \in X \\ 0 & \text{if } x \notin X \end{cases}$$

21. Let A be a nonempty set, and let B be the set of all characteristic functions with domain

A. Define the unary operation # on B as follows: The function $\chi_X\#$ is defined by $\chi_X\# = \chi_{A-X}$.

(a) Show that $(B, \#)$ is an algebra.

(b) Show that the function $f : P(A) \to B$ defined by $f(X) = \chi_X$ is an isomorphism from $(P(A),')$ to $(B, \#)$, where $'$ is the complement operation defined by $Y' = A - Y$. (*Hint:* Show that $f(X') = f(X)\#$.)

22. Let A be a nonempty set, and let B be the set of all characteristic functions with domain A. Define two binary operations on B as follows: The function $\vee$ is defined by $(\chi_X \vee \chi_Y)(x) = \max\{\chi_X(x), \chi_{\overline{Y}}(x)\}$; the function $\wedge$ is defined by $(\chi_X \wedge \chi_Y)(x) = \min\{\chi_X(x), \chi_{\overline{Y}}(x)\}$.

(a) Show that $\vee$ and $\wedge$ are both binary operations on B.

(b) Show that the function $f : P(A) \to B$ defined by $f(X) = \chi_X$ is an isomorphism from $(P(A), \cup, \cap)$ to $(B, \vee, \wedge)$.

23. Let f be a homomorphism from $(A, *)$ to $(B, \#)$, and let g be a homomorphism from $(B, \#)$ to $(C, \otimes)$. Prove that $g \circ f$ is a homomorphism from $(A, *)$ to $(C, \otimes)$.

KEY CONCEPTS

Equivalence relation	Least element
Equivalence class	Greatest element
Partition	Strictly less than
Partition induced by an	Strictly greater than
equivalence relation	Minimal element
Relation induced by a partition	Maximal element
Partial ordering, partial order	Lower bound
Partially ordered set, poset	Upper bound
Comparable elements	Greatest lower bound
Linear ordering, linear order	Least upper bound
Standard partial order of $\mathcal{Z}$, of $\mathcal{R}$	Algebra
Set-inclusion partial order	Homomorphism
Hasse diagram	Isomorphism
Extremal elements	Embedding

PROOFS TO EVALUATE

See the instructions following the Proofs to Evaluate in Chapter 2.

1. *Conjecture:* R defined by xRy if and only if $x^2 + y^2 = 1$ or $x = y$ is an equivalence relation on $\mathcal{Z}$.

       Equivalence Relations and Partial Orders   Chap. 5

Argument: Clearly R is reflexive since $x = x$ implies xRx for all integers x. To show that R is symmetric, assume xRy. Then $x^2 + y^2 = 1$ or $x = y$. Hence, $y^2 + x^2 = 1$ or $y = x$, and so yRx. Finally, assume xRz and zRy. There are three cases.

Case 1. Assume $x = z$ and $z = y$. Then $x = y$, and hence, xRy.

Case 2. Assume $x^2 + z^2 = 1$ and $z = y$. Then $x^2 + y^2 = 1$, and again, xRy.

Case 3. Assume $x = z$ and $z^2 + y^2 = 1$. This proof is like the proof of case 2. Therefore, R is transitive.

2. *Conjecture:* If R is an equivalence relation on A and R is a partial ordering on A, then A is empty or A is a singleton set.

   Argument: Suppose $A \neq \emptyset$ and A has two elements x and y. Since R is symmetric, xRy implies yRx. But since R is antisymmetric, xRy and yRx implies $x = y$. Therefore, A is a singleton set.

3. *Conjecture:* Let R be a transitive relation on a set A. If $[x] \cap [y] \neq \emptyset$, then $[x] = [y]$.

   Argument: Since $[x] \cap [y] \neq \emptyset$, let $z \in [x] \cap [y]$. Therefore, xRz and zRy. Since R is transitive, xRy. We show that $[y] \subseteq [x]$. Let $t \in [y]$. Then yRt. But since we have xRy, by transitivity, xRt. Therefore, $t \in [x]$. Similarly, $[x] \subseteq [y]$, and it follows that $[x] = [y]$.

4. *Conjecture:* Let W be a subset of the power set of A. Define a relation R on A by xRy if and only if for some $Y \in W$, $x \in Y$ and $y \in Y$. Then R is transitive.

   Argument: Assume that xRy and yRz. Then for some $Y \in W$, $x \in Y$ and $y \in Y$, and for some $Y \in W$, $y \in Y$ and $z \in Y$. Therefore, $x \in Y$ and $z \in Y$. Hence, xRz.

5. *Conjecture:* Let $(A, \preceq)$ be a poset. If $X \subseteq Y \subseteq A$, then $\text{lub}(X) \preceq \text{lub}(Y)$.

   Argument: Let $x \in X$. Therefore, $x \in Y$, and hence, $x \preceq \text{lub}(Y)$. So $\text{lub}(Y)$ is an upper bound of X. But then, $\text{lub}(X) \preceq \text{lub}(Y)$.

6. *Conjecture:* Let f be an embedding of $(A, *)$ into $(B, \#)$, and let g be an embedding of $(B, \#)$ into $(C, \otimes)$. Then $g \circ f$ is an embedding of $(A, *)$ into $(C, \otimes)$.

   Argument: By Exercise 22 of Section 5.3, $g \circ f$ is a homomorphism from $(A, *)$ to $(C, \otimes)$. Also, by Theorem 3 of Section 4.4, $g \circ f$ is 1–1. Therefore, $g \circ f$ is an embedding of $(A, *)$ into $(C, \otimes)$.

REVIEW EXERCISES

In the following exercises, an extremal element is a greatest, least, maximal, or minimal element.

1. Let $A = \{a, b, c, d\}$ and $R = \{(a, a), (b, b), (c, c), (d, d), (a, b), (b, a)\}$.
   (a) Is R an equivalence relation on A? Justify your answer. If R is an equivalence relation, find A/R.
   (b) Is R a partial ordering on A? Justify your answer. If R is a partial ordering, draw the Hasse diagram and specify any extremal elements of A.

2. Let $A = \{a, b, c, d\}$ and $R = \{(a, a), (b, b), (c, c), (d, d), (a, b)\}$.
   (a) Is R an equivalence relation on A? Justify your answer. If R is an equivalence relation, find A/R.
   (b) Is R a partial ordering on A? Justify your answer. If R is a partial ordering, draw the Hasse diagram and specify any extremal elements of A.

3. Let $A = \{1, 2, 3, 6\}$, and define R on A by xRy if and only if $y - x \in \mathcal{N}$.
   (a) Is R an equivalence relation on A? Justify your answer. If R is an equivalence relation, find A/R.
   (b) Is R a partial ordering on A? Justify your answer. If R is a partial ordering, draw the Hasse diagram and specify any extremal elements of A.

4. Let $A = \{1, 2, 3, 6\}$, and define R on A by xRy if and only if $|y - x| = 3n$ for some $n \in \mathcal{N}$.
   (a) Is R an equivalence relation on A? Justify your answer. If R is an equivalence relation, find A/R.
   (b) Is R a partial ordering on A? Justify your answer. If R is a partial ordering, draw the Hasse diagram and specify any extremal elements of A.

5. Define R on $\mathcal{Z}$ by xRy if and only if $y - x \in \mathcal{N}$.
   (a) Is R an equivalence relation on $\mathcal{Z}$? Justify your answer. If R is an equivalence relation, find $\mathcal{Z}/R$.
   (b) Is R a partial ordering on $\mathcal{Z}$? Justify your answer. If R is a partial ordering, specify any extremal elements of $\mathcal{Z}$.

6. Define R on $\mathcal{Z}$ by xRy if and only if $|y - x| = 3n$ for some $n \in \mathcal{N}$.
   (a) Is R an equivalence relation on $\mathcal{Z}$? Justify your answer. If R is an equivalence relation, find $\mathcal{Z}/R$.
   (b) Is R a partial ordering on $\mathcal{Z}$? Justify your answer. If R is a partial ordering, specify any extremal elements of $\mathcal{Z}$.

7. Let R be an equivalence relation on A. Is R^{-1} an equivalence relation on A? Justify your answer.

8. Let R be a partial ordering on A. Is R^{-1} a partial ordering on A? Justify your answer.

9. Recall that a function is a binary relation and that 1_A is the identity function from A to A.
   (a) Prove that 1_A is an equivalence relation on A.
   (b) Prove that 1_A is a partial ordering on A.

10. Let R be a binary relation on A. What can you say about R if R is both a function from A to A and an equivalence relation on A? Prove your conjecture.

11. Let R be a binary relation on A. What can you say about R if R is both a function from A to A and a partial ordering on A? Prove your conjecture.

12. Let R be a binary relation on A. What can you say about R if R is both an equivalence relation on A and a partial ordering on A? Prove your conjecture.

13. Define a relation T on $\mathcal{R} \times \mathcal{R}$ by $(x, y)T(z, w)$ if and only if $x^2 + y^2 = z^2 + w^2$.
   (a) Prove that T is an equivalence relation.
   (b) Specify the elements in the equivalence classes $[(0, 0)]$ and $[(1, 0)]$.
   (c) Let $(a, b) \in \mathcal{R} \times \mathcal{R}$. Specify the elements in $[(a, b)]$.

14. Let r be a real number and $C_r = \{(x, y) : x^2 + y^2 = r\}$. Consider the collection $\Pi = \{C_r : r \geq 0\}$.

    (a) Prove that Π is a partition of $\mathcal{R} \times \mathcal{R}$.

    (b) Prove that $(\mathcal{R} \times \mathcal{R})/\Pi$ is the same equivalence relation as T in Exercise 13.

15. Define a relation T on $\mathcal{N} \times \mathcal{N}$ by $(u, v)T(z, w)$ if and only if $u + w = v + z$.

    (a) Prove that T is an equivalence relation on $\mathcal{N} \times \mathcal{N}$.

    (b) Find a bijective function $f : (\mathcal{N} \times \mathcal{N})/T \rightarrow \mathcal{Z}$. [Remember to show that f is well defined.]

    (c) Prove that f is bijective.

16. In Exercise 11 of Section 5.1, we defined a relation R on $\mathcal{Z} \times (\mathcal{N} - \{0\})$ by $(u, v)R(z, w)$ if and only if $w \cdot u = z \cdot v$ and proved that R is an equivalence relation.

    (a) Find a bijective function $f : (\mathcal{Z} \times (\mathcal{N} - \{0\}))/R \rightarrow \mathcal{Q}$. [Remember to show that f is well defined.]

    (b) Prove that f is bijective.

17. Consider the relation T Exercise 15. Let $+$ and $\cdot$ be the standard addition and multiplication on $\mathcal{N}$. Define an "addition" $\oplus$ and a "multiplication" $\odot$ on $(\mathcal{N} \times \mathcal{N})/T$ by $[(a, b)] \oplus [(c, d)] = [(a + c, b + d)]$ and $[(a, b)] \odot [(c, d)] = [(ac + bd, ad + bc)]$.

    (a) Verify that $((\mathcal{N} \times \mathcal{N})/T, \oplus, \odot)$ is an algebra.

    (b) Show that $((\mathcal{N} \times \mathcal{N})/T, \oplus, \odot, [(0, 0)], [(1, 0)])$ is isomorphic to $(\mathcal{Z}, +, \cdot, 0, 1)$, where $+$ and $\cdot$ are the standard addition and multiplication on $\mathcal{Z}$.

18. Consider the relation R in Exercise 16. Let $+$ and $\cdot$ be the standard addition and multiplication on $\mathcal{Z}$. Define an "addition" $\oplus$ and a "multiplication" $\odot$ on $(\mathcal{Z} \times (\mathcal{N} - \{0\}))/R$ by $[(a, b)] \oplus [(c, d)] = [(ad + bc, bd)]$ and $[(a, b)] \odot [(c, d)] = [(ac, bd)]$.

    (a) Verify that $((\mathcal{Z} \times (\mathcal{N} - \{0\}))/R, \oplus, \odot)$ is an algebra.

    (b) Show that $((\mathcal{Z} \times (\mathcal{N} - \{0\}))/R, \oplus, \odot, [(0, 1)], [(1, 1)])$ is isomorphic to $(\mathcal{Q}, +, \cdot, 0, 1)$, where $+$ and $\cdot$ are the standard addition and multiplication on $\mathcal{Q}$.

19. Let f be an isomorphism from $(A, *)$ to $(B, \#)$, and let g be an isomorphism from $(B, \#)$ to $(C, \otimes)$. Prove that $g \circ f$ is an isomorphism from $(A, *)$ to $(C, \otimes)$.

20. Let $\mathbf{A}$ be a set of algebras of the same form. Define a relation on $\mathbf{A}$ by $(A, *) \simeq (B, \#)$ if and only if $(A, *)$ is isomorphic to $(B, \#)$. Prove that $\simeq$ is an equivalence relation on $\mathbf{A}$.

SUPPLEMENTARY EXERCISES

Let Φ be the collection of partitions of a set A. For Exercises 1–5, consider the relation $\preceq$ on Φ defined by $\Pi_1 \preceq \Pi_2$ if and only if every member of Π_1 is a subset of some member of Π_2.

1. Prove that $\preceq$ is a partial order of Φ.

2. Let $A = \{a, b, c\}$.

    (a) Find Φ. (*Hint:* There are five partitions of A.)

    (b) Specify $\preceq$.

3. Let R and S be equivalence relations on A. Prove that $R \subseteq S$ if and only if $A/R \preceq A/S$.

4. Let Π_1 and Π_2 be partitions of set A. Use the results of Exercise 3 above and Theorem 7 of Section 5.1 to show that $\Pi_1 \preceq \Pi_2$ if and only if $A/\Pi_1 \subseteq A/\Pi_2$.

5. Use the results of Exercise 4 above and Theorem 7 of Section 5.1 to construct an alternative proof of Exercise 1.

6. Let $(A, \preceq)$ be a poset such that every subset of A has a least upper bound in A. Note that lub($\emptyset$) is the least element in A. Prove that if $f : A \rightarrow A$ is a function such that for all $x, y \in A$, $x \preceq y$ implies $f(x) \preceq f(y)$, then there is a $z_0 \in A$ such that $f(z_0) = z_0$. (z_0 is called a **fixed point** for the function f.) (*Hint:* Let $B = \{x : x \in A \text{ and } x \preceq f(x)\}$, and let $z_0 = \text{lub}(B)$. Show that $f(z_0)$ is an upper bound of B, and hence, that $z_0 \preceq f(z_0)$. Then, by the property of f, $f(z_0) \preceq f(f(z_0))$. Therefore, $f(z_0) \in B$, and it follows that $f(z_0) \preceq z_0$.)

7. Let $f : P(A) \rightarrow P(A)$ be a function such that $X \subseteq Y$ implies $f(X) \subseteq f(Y)$. Use the results of Exercise 6 to show that there is a $Z_0 \in P(A)$ such that $f(Z_0) = Z_0$. (This exercise will be used to prove the Schröder-Bernstein Theorem in the Supplementary Exercises of Chapter 6.)

8. Let $(A, \preceq)$ and $(B, \propto)$ be posets. A function $f : A \rightarrow B$ is **order preserving** if, for all $x, y \in A$, $x \preceq y$ if and only if $f(x) \propto f(y)$. Assume that f is a bijection and is order preserving. Define the following operations on A: $x \vee y = \text{lub}(\{x, y\})$, $x \wedge y = \text{glb}(\{x, y\})$. (Here, the lub and the glb are in terms of the partial ordering $\preceq$.) Define the following operations on B: $u \, \mathbb{W} \, v = \text{lub}(\{u, v\})$, $u \, \mathbb{A} \, v = \text{glb}(\{u, v\})$. (Here, the lub and the glb are in terms of the partial ordering $\propto$.) Show that f is an isomorphism from $(A, \vee, \wedge)$ to $(B, \mathbb{W}, \mathbb{A})$

SOLUTIONS TO PRACTICE PROBLEMS

Section 5.1

1. *Proof:* R is reflexive, since $x - x = 0 \in \mathcal{Z}$. To show that R is symmetric, let xRy. Then $x - y \in \mathcal{Z}$. Therefore, $y - x = -(x - y) \in \mathcal{Z}$, and hence, yRx. Finally, to show that R is transitive, assume xRy and yRz. Then $x - y, y - z \in \mathcal{Z}$. Hence, $x - z = (x - y) + (y - z) \in \mathcal{Z}$. Therefore, xRz.

2. a. $x \in [0]$ if and only if $0 - x \in \mathcal{Z}$ if and only if $-x \in \mathcal{Z}$ if and only if $x \in \mathcal{Z}$. Therefore, $[0] = \mathcal{Z}$.

   b. $x \in [\frac{1}{2}]$ if and only if $\frac{1}{2} - x \in \mathcal{Z}$ if and only if $x - \frac{1}{2} \in \mathcal{Z}$ if and only if $x - \frac{1}{2} = k$, for some $k \in \mathcal{Z}$ if and only if $x = (2k + 1)/2$ for some $k \in \mathcal{Z}$. Hence, $[\frac{1}{2}] = \{(2k + 1)/2 : k \in \mathcal{Z}\} = \{\ldots, -\frac{3}{2}, -\frac{1}{2}, \frac{1}{2}, \frac{3}{2}, \frac{5}{2}, \ldots\}$.

   c. $x \in [\sqrt{2}]$ if and only if $\sqrt{2} - x \in \mathcal{Z}$ if and only if $x - \sqrt{2} \in \mathcal{Z}$ if and only if $x - \sqrt{2} = k$, for some $k \in \mathcal{Z}$ if and only if $x = k + \sqrt{2}$, for some $k \in \mathcal{Z}$. Therefore, $[\sqrt{2}] = \{k + \sqrt{2} : k \in \mathcal{Z}\} = \{\ldots, -2 + \sqrt{2}, -1 + \sqrt{2}, \sqrt{2}, 1 + \sqrt{2}, 2 + \sqrt{2}, \ldots\}$.

   d. By the same method as in the first three parts, it is easily seen that $[3] = [0] = \mathcal{Z}$.

3. *Proof:* R is reflexive, since $m - m = 0 = 4 \cdot 0$ and $0 \in \mathcal{Z}$. R is symmetric, for if $n - m = 4k$ then $m - n = 4(-k)$. R is transitive, since if mRn and nRp, then $m - n = 4k$ and $n - p = 4j$ for some $k, j \in \mathcal{Z}$. But $m - p = (n - p) + (m - n) = 4j + 4k = 4(j + k)$. Hence, $m - p$ is divisible by 4, and therefore, mRp.

   The equivalence classes are $\{1, 5, 9\}$, $\{2, 6, 10\}$, $\{3, 7\}$, and $\{4, 8\}$.

4. (a) The equivalence classes are the set of even integers and the set of odd integers.
   (b) There are three equivalence classes: $[0] = \{\ldots, -9, -6, -3, 0, 3, 6, 9, \ldots\}$, $[1] = \{\ldots, -8, -5, -2, 1, 4, 7, \ldots\}$, $[2] = \{\ldots, -7, -4, -1, 2, 5, 8, \ldots\}$.

5. To find $D/\approx$, we find $f^{-1}[\{x\}]$ for $x \in \{0, 1, 2, 3\}$. $[0] = f^{-1}[\{0\}] = \{x : f(x) \in \{0\}\}$ $= \{x : \lfloor x \rfloor = 0\} = \{x : 0 \le x < 1\} = [0, 1)$ and $[1] = f^{-1}[\{1\}] = \{x : f(x) \in \{1\}\} =$ $\{x : \lfloor x \rfloor = 1\} = \{x : 1 \le x < 2\} = [1, 2)$. Similarly, $[2] = f^{-1}[\{2\}] = [2, 3)$ and $[3] = f^{-1}[\{3\}] = [3, 4)$. Therefore, $[0, 4)/\approx = \{[0, 1), [1, 2), [2, 3), [3, 4)\}$. A table for h is

$[x]$	$[0, 1)$	$[1, 2)$	$[2, 3)$	$[3, 4)$
$h([x])$	0	1	2	3

Section 5.2

1. **(a)** For example, $\{1, 2, 3\}$ and $\{4\}$ are not comparable.

    (b) Both $\emptyset$ and $\mathcal{Z}$ are comparable to every other set in $P(\mathcal{Z})$, and these are the only two such sets.

2.

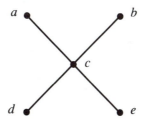

3. $\mathcal{Z}$ with the standard order has neither a maximal nor a minimal element.

4. **(a)** The elements a, c and d are lower bounds of $\{d, f\}$; f is the only upper bound of $\{d, f\}$.

    (b) The element c is the only lower bound of $\{c, d, e\}$; there are no upper bounds of $\{c, d, e\}$.

    (c) There are no lower bounds of $\{a, b\}$; there are no upper bounds of $\{a, b\}$.

5. **(a)** The greatest lower bound of $\{d, f\}$ is d; the least upper bound of $\{d, f\}$ is f.

    (b) The greatest lower bound of $\{c, d, e\}$ is c; there is no least upper bound of $\{c, d, e\}$.

    (c) There is no greatest lower bound of $\{a, b\}$; there is no least upper bound of $\{a, b\}$.

6. **(a)** $\text{glb}(B) = \pi - 1$. Note that $\text{glb}(B) \in B$.

    (b) $\text{lub}(B) = \pi$. Note that $\text{lub}(B) \notin B$.

Section 5.3

1. Let $[x] = [x']$ and $[y] = [y']$. Then $x - x' = 3k$ and $y - y' = 3j$. We multiply the first equation by y and the second equation by x' to obtain $xy - x'y = 3yk$ and $x'y - x'y' = 3x'j$. Adding the last two equations yields $xy - x'y' = 3(yk + x'j)$. Therefore, $xy \equiv_3 x'y'$ and $[xy] = [x'y']$.

2. $f(x \cdot y) = [x \cdot y] = [x] *_3 [y] = f(x) *_3 f(y)$.

3. $f(x \cdot y) = (xy)^2 + 1$, $f(x) \# f(y) = (x^2 + 1) \# (y^2 + 1) = ((x^2 + 1) - 1)((y^2 + 1) - 1) + 1 = x^2 y^2 + 1 = (xy)^2 + 1$.

Solutions to Practice Problems

4.   (a) $2^n 2^m = 2^{n+m} \in A$.

   (b) f is 1–1: Assume that $f(n) = f(m)$. Then $2^n = 2^m$. Assume that $n \neq m$, say, $m < n$. Then $2^{n-m} = 1$, and hence, $n - m = 0$. So $n = m$.

   f is onto: Let $2^n \in A$, where $n \in \mathcal{Z}$. Then $f(n) = 2^n$. Also, f preserves operations since $f(n + m) = 2^{n+m} = 2^n \cdot 2^m = f(n) \cdot f(m)$.

   (c) f is an embedding of $(\mathcal{Z}, +)$ into $(\mathcal{Q}, \cdot)$ by part b, since $A \subseteq \mathcal{Q}$.

6

Cardinality

In this chapter, we discuss the relative sizes of sets. Consider the sets $\{a, b, c\}$, $\{1, 2, 3\}$, and $\{\pi, e, i\}$. They are certainly not equal; in fact, they are pairwise disjoint. But they do have something in common: Each set has three elements; that is, they are of "size three." On the other hand, the sets $\{a, b, c\}$ and $\{2, 4, 6, 8, 10\}$ are not of the "same size." In Section 6.1, we define an equivalence relation $\approx$ on sets which yields a rigorous method for comparing their sizes.

The four sets discussed in the preceding paragraph are all finite. However, certain sets, such as $\mathcal{N}$ and $\mathcal{R}$, are not finite. An interesting question is, can we nonetheless compare their relative sizes? In this chapter, we address this decidedly unintuitive issue.

All missing proofs and missing solutions to examples are exercises. In constructing a proof, any previous definition or result may be used.

6.1 THE CARDINALITY OF FINITE SETS

Suppose we want to determine, without counting, whether the number of people in a room is the same as the number of chairs in the room. One solution is to ask everyone to sit in a chair such that each person uses only one chair and no two people sit in the same chair. If there are no chairs left unoccupied and no one is left without a chair, then we know that the number of people and the number of chairs is the same. What we have actually done is to define a bijective function f from the set of people to the set of chairs, namely, $(p, c) \in f$ if and only if c is the chair in which person p sits. The function f is well defined since each person uses only one chair, f is 1–1 since two different people do not sit in the same chair, and f is onto since no chairs are left unoccupied.

The set of people and the set of chairs in the preceding discussion are, of course, finite. To show that any two arbitrary sets have the same number of elements, we use the same idea: we construct a bijective function from one set to the other.

As usual, we will assume that U is a universal set and all sets are subsets of U. We define a relation $\approx$ on $P(U)$ as follows.

Definition. A is **equinumerous to** B, $(A \approx B)$ if there exists a bijective function $f : A \to B$.

If there is no bijective function $f : A \to B$, we write $A \not\approx B$.

Example 1

Prove that $\{a, b, c\}$ is equinumerous to $\{0, 1, 2\}$.

Solution We need to find a bijective function $g : \{a, b, c\} \to \{0, 1, 2\}$. One such function is given by $g(a) = 0$, $g(b) = 1$, and $g(c) = 2$ (see Figure 6–1).

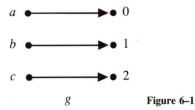

g **Figure 6–1**

Theorem 1 $A \approx \emptyset$ if and only if $A = \emptyset$.

Theorem 2 The relation $\approx$ is an equivalence relation on $P(U)$.

Theorem 3 $A \times \{b\} \approx A$.

Theorem 4 If $A_1 \approx A_2$, $B_1 \approx B_2$, $A_1 \cap B_1 = \emptyset$, and $A_2 \cap B_2 = \emptyset$, then $A_1 \cup B_1 \approx A_2 \cup B_2$.
Hint for the proof: The assumptions $A_1 \approx A_2$ and $B_1 \approx B_2$ imply the existence of bijective functions from A_1 to A_2 and B_1 to B_2. Use these bijections to construct a bijection from $A_1 \cup B_1$ to $A_2 \cup B_2$. ☐

For $k \in \mathcal{N} = \{0, 1, 2, 3, ...\}$, we write $N_k = \{j : j \in \mathcal{N} \text{ and } j < k\} = \{0, 1, 2, ..., k - 1\}$. Observe that $N_0 = \emptyset$; in fact, $N_k = \emptyset$ if and only if $k = 0$. Note also that $N_{k+1} = N_k \cup \{k\}$.

Lemma Let $f : A \to B$ be onto, $x \in A$, and $y \in B$. Then there is a function $g : A \to B$ such that g is onto and $g(x) = y$. In addition, if f is 1–1, then so is g.

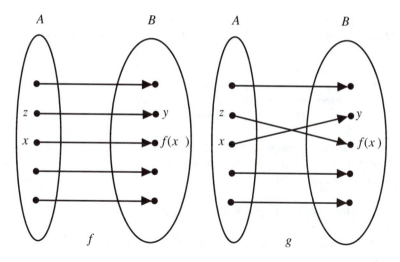

Figure 6–2

Proof. Since f is onto, there is a $z \in A$ such that $f(z) = y$. The idea is to define g so that it agrees with f for elements of A that are not equal to x or z and then interchange the values y and $f(x)$ (see Figure 6–2). Specifically, we define $g : A \to B$ by

$$g(t) = \begin{cases} f(t) & \text{if } t \neq x, z \\ f(x) & \text{if } t = z \\ y & \text{if } t = x \end{cases}$$

It is easy to verify that g is onto, and if f is 1–1, then so is g. $\qquad\square$

Theorem 5   If $N_n \approx N_m$ then $n = m$.

Proof. The proof is by mathematical induction on m.
Basis step $(m = 0)$: If $N_n \approx \emptyset$, then $n = 0$.
Induction step: First, for all $n \in \mathcal{N}$, if $N_n \approx N_k$, then $n = k$. Now assume $N_n \approx N_{k+1}$. Then there exists a bijection $f : N_{k+1} \to N_n$. We wish to show that $n = k + 1$. Since $k \in N_{k+1}$, $f(k) \in N_n$, and hence, $N_n \neq \emptyset$. Therefore, $n \neq 0$ and hence $n = p + 1$ for some $p \in \mathcal{N}$. So $f : N_{k+1} \to N_{p+1}$ is a bijection. By the previous lemma, there is a bijection $g : N_{k+1} \to N_{p+1}$ such that $g(k) = p$. Define $h : N_k \to N_p$ by $h(x) = g(x)$ for $x \in N_k$. Show that h is a bijection and, hence, that $N_k \approx N_p$. By the induction hypothesis, $k = p$, and it follows that $n = p + 1 = k + 1$. $\qquad\square$

Definition.   A set X is **finite** if X is equinumerous to N_k for some $k \in \mathcal{N}$. When $X \approx N_k$, we say that the **cardinality** of X is k and write $\mathrm{Card}(X) = k$.

Note that by Theorem 5, there is at most one k such that $X \approx N_k$, and hence, $\mathrm{Card}(X)$ is well defined. Also, by definition, $\mathrm{Card}(A) = k$ if and only if A is equinumerous to N_k.

In Example 1, we proved that the set $\{a, b, c\}$ is finite and that the cardinality of $\{a, b, c\}$ is three; that is, $\text{Card}(\{a, b, c\}) = 3$. In order to prove that a set A is finite, either we must show that $A \approx N_k$ for some k by finding a bijection, or we can use a previously proved theorem.

Theorem 6   Let A be finite. Then $\text{Card}(A) = \text{Card}(B)$ if and only if $A \approx B$.
Hint for the proof of ($\Rightarrow$): Let $\text{Card}(A) = m = \text{Card}(B)$, where $m \in \mathcal{N}$. $\square$

The next theorem is both intuitively obvious and useful in subsequent proofs. It says that when a new element is adjoined to a finite set, the cardinality of the resulting set is increased by one.

Theorem 7   If A is finite and $x \notin A$, then $A \cup \{x\}$ is finite and $\text{Card}(A \cup \{x\}) = \text{Card}(A) + 1$.
Hint for the proof: Let $A \approx N_k$, and define a function $f : A \cup \{x\} \to N_{k+1}$ which is a bijection. $\square$

Corollary.   If B is finite and $x \in B$, then $B - \{x\}$ is finite and $\text{Card}(B - \{x\}) = \text{Card}(B) - 1$.
Hint for the proof: Let $A = B - \{x\}$, and then use Theorem 7. $\square$

Theorem 8   If A and B are disjoint finite sets, then $A \cup B$ is a finite set and $\text{Card}(A \cup B) = \text{Card}(A) + \text{Card}(B)$.
Hint for the proof: Let $\text{Card}(A) = m$ and $\text{Card}(B) = n$. Fix m and use mathematical induction on n. $\square$

Theorem 9   If A and B are finite sets, then $A \times B$ is a finite set and $\text{Card}(A \times B) = \text{Card}(A) \cdot \text{Card}(B)$.
Hint for the proof: Let $\text{Card}(A) = m$ and $\text{Card}(B) = n$. Fix m and use mathematical induction on n. For the induction step, note that $A \times B = (A \times (B - \{x\})) \cup (A \times \{x\})$ where $x \in B$. $\square$

Theorem 10   If B is finite and $A \subseteq B$, then A is finite and $\text{Card}(A) \leq \text{Card}(B)$.

Proof. We prove the equivalent theorem, if $\text{Card}(B) = n$ and $A \subseteq B$, then A is finite and $\text{Card}(A) \leq n$. The proof is by mathematical induction on n. The basis step ($n = 0$) is easy.
    Induction hypothesis: If $\text{Card}(D) = k$ and $C \subseteq D$, then C is finite and $\text{Card}(C) \leq k$.
    Let $\text{Card}(B) = k + 1$ and $A \subseteq B$. If $A = B$, then the conclusion follows easily. So assume $A \subset B$. Then there exists an $x \in B - A$. Hence, $A \subseteq B - \{x\}$ and $\text{Card}(B - \{x\}) = k$. Now use the corollary to Theorem 7 and the induction hypothesis to obtain the desired conclusion. $\square$

Corollary.   If A and B are finite sets, then $A \cup B$ is a finite set.
Hint for the proof: Note that $A \cup B = A \cup (B - A)$. Now use Theorem 8. $\square$

Theorem 11   If B is finite and is a proper subset of A, then $B \not\approx A$.

Hint for the proof: If A is not finite, then $B \not\approx A$. So assume that A is finite. Let $x \in A - B$. Then $B \subseteq A - \{x\}$. Use the corollary to Theorem 7 and Theorem 10 to show that $\text{Card}(B) < \text{Card}(A)$.   □

Corollary.   If B is a subset of a finite set A and $B \approx A$, then $B = A$.

Theorem 12   (Pigeonhole Principle) If C is finite and there exists a 1–1 function $f : D \to C$, then D is finite and $\text{Card}(D) \leq \text{Card}(C)$.

Hint for the proof: Note that $f : D \to \text{ran}(f)$ is a bijection, and use Theorem 10.   □

The pigeonhole principle can be restated as the following logically equivalent statement.

Pigeonhole Principle (restated):   If C is finite and $\text{Card}(D) > \text{Card}(C)$, then any function $f : D \to C$ is not 1–1.

The pigeonhole principle receives its name from the fact that if there are more objects than there are pigeonholes, then at least one pigeonhole contains two or more objects. Let P be a set of objects and H be a set of pigeonholes, where $\text{Card}(P) > \text{Card}(H)$. Suppose every object in P is in some pigeonhole from H. Now, if we define a function f from P to H by $f(p)$ is the pigeonhole containing object p, then $f : P \to H$ is not 1–1. This means that there are two different objects p and q such that $f(p) = f(q)$; that is, p and q share the same pigeonhole.

The pigeonhole principle has many important applications in the branch of mathematics called combinatorics. The following example can be found in *Applied Combinatorics* by Alan Tucker (see references).

Example 2

Show there are at least two people living in New York City with the same number of hairs on their head.

Solution   We use the pigeonhole principle. There are more than 7 million people living in New York City, but it is known that the maximum number of hairs on any human head is less than 200,000.

Theorem 13   If B is a finite set and $f : B \to A$ is onto, then there is a function $g : A \to B$ such that g is 1–1.

Proof.   Since B is finite, there is a bijection $h : N_k \to B$ for some k. We wish to define g so that $g(x)$ is a specific element of $f^{-1}[\{x\}]$. Accordingly, define $g : A \to B$ by $g(x) = h(j)$, where j is the least integer such that $h(j) \in f^{-1}[\{x\}]$. We then have $g(x) \in f^{-1}[\{x\}]$; that is, $f(g(x)) = x$ for all x. To show that g is 1–1, assume $g(u) = g(v)$. Then $u = f(g(u)) = f(g(v)) = v$.   □

Theorem 14   If B is a finite set and $f : B \to A$ is onto, then A is finite and $\text{Card}(A) \leq \text{Card}(B)$.

Proof. By Theorem 13, there is a function $g : A \to B$ such that g is 1–1. The desired result then follows from the pigeonhole principle. $\qquad\square$

Theorem 15   If A is finite, $B \subseteq A$, and $g : B \to A$ is onto, then $B = A$.

Hint for the proof: Use Theorems 10 and 14 and the corollary to Theorem 11. $\qquad\square$

EXERCISE SET 6.1

1. Prove Theorem 1.
2. Prove Theorem 2.
3. By Theorem 2, $\approx$ is an equivalence relation on $P(U)$. Describe the elements of the equivalence class $[N_2]$.
4. (a) Prove that $(0, 1) \approx (-\pi, \pi)$.
   (b) Prove that $(0, 1) \approx \mathcal{R}$.
5. Prove Theorem 3.
6. Prove that $A \times B \approx B \times A$.
7. Prove Theorem 4.
8. (a) Prove $A \approx A \times N_1$.
   (b) Find a set B such that $B \not\approx B \times N_2$.
9. Prove $\mathcal{N} \approx \mathcal{N} \times N_2$.
10. Prove Theorem 6.
11. Prove Theorem 7.
12. Prove the corollary to Theorem 7.
13. Prove Theorem 8.
14. *Prove:* If A and B are finite sets, then $\operatorname{Card}(A \cup B) = \operatorname{Card}(A) + \operatorname{Card}(B) - \operatorname{Card}(A \cap B)$.
15. Prove Theorem 9.
16. Prove the corollary to Theorem 10.
17. Prove Theorem 11.
18. Prove the corollary to Theorem 11.
19. Prove Theorem 12.
20. Prove Theorem 15.
21. *Prove:* If $f : N_k \to N_k$ is onto, then f is 1–1.
22. *Prove:* If $f : N_k \to N_k$ is 1–1, then f is onto.
23. *Prove:* If $A \approx B$ and $C \approx D$, then $A \times C \approx B \times D$.

6.2 THE CARDINALITY OF INFINITE SETS

Definition.   A set is **infinite** if it is not finite.

The following two theorems are very useful for proving that a given set is infinite.

Theorem 1   If A is infinite and A is equinumerous to B, then B is infinite.
Hint for the proof: Assume that B is not infinite, and use a theorem from Section 6.1 about finite sets.   □

Theorem 2   If there is a proper subset D of A such that $D \approx A$, then A is infinite.

Hint for the proof: Assume that A is not infinite.   □

Example 1

Prove that the set $\mathcal{N}$ of nonnegative integers is infinite.

Solution   Let $A = \{1, 2, 3, ...\}$. Define $f : \mathcal{N} \to A$ by $f(x) = x + 1$. It is easy to show that f is a bijection. Therefore, $\mathcal{N}$ is infinite, since A is a proper subset of $\mathcal{N}$ and $A \approx \mathcal{N}$.

Theorem 3   If A is a subset of B and A is infinite, then B is infinite.

Example 2

a. The set, $\mathcal{Z}$, of integers is infinite.

b. The set, $\mathcal{R}$, of real numbers is infinite.

Definition.   A set D is **denumerable** if $D \approx \mathcal{N}$. A set is **countable** if it is finite or denumerable. A set is **uncountable** if it is not countable.

Example 3

a. $\mathcal{N}$ is denumerable.

b. $\mathcal{Z}$ is denumerable.

Indexing Denumerable Sets

If D is denumerable, there is a bijection $d : \mathcal{N} \to D$. Since d is onto, $D = \{d(x) : x \in \mathcal{N}\} = \{d(0), d(1), d(2), ...\}$. Since d is 1–1, $d(0), d(1), d(2), ...$ are distinct. If we write d_i for $d(i)$, then $D = \{d_0, d_1, d_2, ...\}$. When a denumerable set D is written as $D = \{d_0, d_1, d_2, ...\}$, we say that D is **indexed** by $\mathcal{N}$. Furthermore, a set D is denumerable if and only if D can be indexed by $\mathcal{N}$.

Theorem 4   If D is denumerable, then D is infinite.

Theorem 5   If D is denumerable and $A \subseteq D$, then A is countable.

Proof. Let D be denumerable and $A \subseteq D$. Then $D = \{d_0, d_1, d_2, \ldots\}$ and $A \subseteq \{d_0, d_1, d_2, \ldots\}$. If A is finite, then A is countable. So assume A is infinite. We index A using a recursive definition.

First, pick $d_{i_0} \in A$, where i_0 is the least of the subscripts on elements in A. Next, assume the elements $d_{i_0}, d_{i_1}, \ldots, d_{i_k}$ have been chosen. Then, since A is infinite, $A - \{d_{i_0}, d_{i_1}, \ldots, d_{i_k}\}$ is nonempty. Now, pick $d_{i_{k+1}} \in A - \{d_{i_0}, d_{i_1}, \ldots, d_{i_k}\}$, where i_{k+1} is the least of the subscripts on elements in $A - \{d_{i_0}, d_{i_1}, \ldots, d_{i_k}\}$. Clearly, the elements of $\{d_{i_0}, d_{i_1}, d_{i_2}, \ldots\}$ are distinct, and $\{d_{i_0}, d_{i_1}, d_{i_2}, \ldots\} \subseteq A$. The proof that the set $\{d_{i_0}, d_{i_1}, d_{i_2}, \ldots\}$ includes all the elements of A follows from the method of choosing the elements d_{i_u}. At each stage of the process, an element is chosen from A with the least subscript of those that have not already been chosen. From this fact, it follows that all the elements of A are in $\{d_{i_0}, d_{i_1}, d_{i_2}, \ldots\}$. (See Exercise 23 for a more detailed proof.) Therefore, A is indexed by $\mathcal{N}$, and hence, $A \approx \mathcal{N}$. $\qquad\square$

Corollary. If A is a subset of a countable set C, then A is countable.

Theorem 6 If A is a countable set and B is a denumerable set, then $A \cup B$ is a denumerable set.

Hint for the proof: Let $C = A - B$. Then C is countable, $A \cup B = C \cup B$, and $C \cap B = \emptyset$. There are two cases: (1) C is finite and (2) C is denumerable. $\qquad\square$

Theorem 7 If E is infinite, D is denumerable, and $E \subseteq D$, then E is denumerable.

Corollary. If D is denumerable and $f : A \to D$ is 1–1, then A is countable. In addition, if A is infinite, then A is denumerable.

Hint for the proof: $f[A] \subseteq D$ and $f[A] \approx A$. $\qquad\square$

Corollary. If A and B are denumerable, then $A \times B$ is denumerable.

Proof. First we show that $\mathcal{N} \times \mathcal{N}$ is denumerable. Define $f : \mathcal{N} \times \mathcal{N} \to \mathcal{N}$ by $f(m, n) = 2^m 3^n$. We show that f is 1–1. Let $f(m, n) = f(p, q)$ and $m \neq p$, say, $m > p$. Then $2^m 3^n = 2^p 3^q$; that is, $2^{m-p} 3^n = 3^q$. But the left side of this equation is even and the right side is odd. This contradiction yields $m = p$. Hence, $3^n = 3^q$, and it follows that $n = q$.

By theorem 3 of Section 6.1, $\mathcal{N} \times \{0\} \approx \mathcal{N}$, and hence, $\mathcal{N} \times \{0\}$ is infinite. Since $\mathcal{N} \times \{0\} \subseteq \mathcal{N} \times \mathcal{N}$, $\mathcal{N} \times \mathcal{N}$ is also infinite by Theorem 3. But then, by the previous corollary, $\mathcal{N} \times \mathcal{N}$ is denumerable.

Now let $A \approx \mathcal{N}$ and $B \approx \mathcal{N}$. Then $A \times B \approx \mathcal{N} \times \mathcal{N}$, and hence, $A \times B$ is denumerable. $\qquad\square$

For a more constructive proof of this corollary, see Exercise 24.

Example 4

$\mathcal{N} \times (\mathcal{Z} - \{0\})$ is denumerable.

Theorem 8   If D is a denumerable set and $f : D \rightarrow A$ is onto, then there is a $g : A \rightarrow D$ such that g is 1–1.

Proof. Let D be denumerable and let $h : \mathcal{N} \rightarrow D$ be a bijection. The remainder of the proof is similar to the proof of Theorem 13 in Section 6.1. □

Theorem 9   If D is denumerable and $f : D \rightarrow A$ is onto, then A is countable. In addition, if A is infinite, then A is denumerable.

Example 5

The set $\mathcal{Q}$ of rational numbers is denumerable.

Hint for the solution: First note that every rational number can be written in the standard form x/y, where x is a nonnegative integer and y is a nonzero integer. Define $f : \mathcal{N} \times \left(\mathcal{Z} - \{0\} \right) \rightarrow \mathcal{Q}$ by $f(x, y) = x/y$.

Theorem 10   If F is uncountable and $F \approx G$, then G is uncountable.

Theorem 11   The interval $(0, 1)$ is uncountable.

Proof. The proof is by contradiction. Suppose $(0, 1)$ is countable. Now, the interval $(0, 1)$ contains an infinite set—for example, $\{.1, .11, .111, \ldots\}$. Hence, the set $(0, 1)$ must be denumerable. So, $(0, 1) = \{d_0, d_1, d_2, \ldots\}$. Each element d_i is a decimal number between 0 and 1. Hence

$$d_0 = .a_{00}a_{01}a_{02}a_{03} \ldots$$

$$d_1 = .a_{10}a_{11}a_{12}a_{13} \ldots$$

$$d_2 = .a_{20}a_{21}a_{22}a_{23} \ldots$$

$$\vdots$$

where a_{ij}, $i, j = 0, 1, 2, 3, \ldots$, is a digit.

We will obtain a contradiction by constructing a decimal number d between 0 and 1 that is not an element of $\{d_0, d_1, d_2, \ldots\}$. Define $d = .b_0 b_1 b_2 b_3 \ldots$ such that $b_i \neq a_{ii}$ for all i. For example, we let $b_i = 3$ if $a_{ii} = 5$ and $b_i = 5$ otherwise. To see that $d \notin \{d_0, d_1, d_2, \ldots\}$, suppose $d = d_j$ for some j. Then $b_j = a_{jj}$, which is impossible by the way d is defined. □

In the proof of Theorem 11, the number d was defined in such manner that for all i, the digit in its ith place is different from the ith diagonal digit in the array of digits making up the decimal numbers $d_0, d_1, d_2, \ldots$. This method is called the **Cantor diagonalization method**, after Georg Cantor, the 19th-century German mathematician who devised the technique. Cantor diagonalization is an important method of proof and will surface again in the Supplementary Exercises.

Theorem 12   The set of real numbers is uncountable.

EXERCISE SET 6.2

1. Prove Theorem 1.

2. Prove Theorem 2.

3. Prove Theorem 3.

4. Either prove or disprove the following conjectures.
   (a) $\{1/2^n : n \in \mathcal{N}\}$ is denumerable.
   (b) $\{\cos \pi n : n \in \mathcal{N}\}$ is denumerable.

5. Either prove or disprove the following conjectures.
   (a) The set of primes is denumerable.
   (b) $N_3 \times \mathcal{N}$ is denumerable.

6. Prove Theorem 4.

7. Prove the corollary to Theorem 5.

8. Prove Theorem 6.

9. Prove Theorem 7.

10. Prove the first corollary to Theorem 7.

11. Prove Theorem 8.

12. Prove Theorem 9.

13. *Prove:* If A is denumerable and B is finite, then $A - B$ is denumerable.

14. Prove Theorem 10.

15. Prove Theorem 12.

16. *Prove:* If A is uncountable and B is countable, then $A - B$ is uncountable.

17. *Prove:* $N_m \times \mathcal{N}$ is denumerable for all positive integers m.

18. *Prove:* If A and B are countable, then $A \times B$ is countable.

19. *Prove:* If Ω is a finite collection of countable sets, then $\cup \Omega$ is countable.

20. *Prove:* Every interval of real numbers is uncountable.

21. *Prove:* If A is uncountable and $A \subseteq B$, then B is uncountable.

22. Prove that the set of irrational numbers is uncountable.

23. Refer to the proof of Theorem 5. Supply the details in the following outline of the proof that $A \subseteq \{di_0, di_1, di_2, \ldots\}$.
    Outline of the proof: Suppose not. Let j_0 be the least subscript such that $dj_0 \in A - \{d_{i_0}, d_{i_1}, d_{i_2}, \ldots\}$. Take the least subscript i_u on an element in $\{d_{i_0}, d_{i_1}, d_{i_2}, \ldots\}$ such that $j_0 < i_u$. Show that $i_u = j_0$.

24. In the proof of the second corollary to Theorem 7, it was proved that $\mathcal{N} \times \mathcal{N}$ is denumerable. Fill in the details of the following alternative proof. Define $f : \mathcal{N} \times \mathcal{N} \to \mathcal{N}$ by $f(m, n) = m + (1/2)(m+n)(m+n+1)$. Prove that f is bijective. (*Hint:* First draw a picture of $\mathcal{N} \times \mathcal{N}$, and label the points by the function value at that point. For example, $f(0, 2) = 3$, $f(1, 1) = 4$, and $f(2, 0) = 5$.)

25. In this exercise we give one way of writing $\mathcal{N}$ as a denumerable union of pairwise disjoint denumerable sets. Refer to Exercise 24 above. Let $A_m = \{f(m, n) : n \in \mathcal{N}\}$, and define $\Omega = \{A_m : m \in \mathcal{N}\}$. (It is helpful to draw a picture. Draw a rectangular array with A_0 as the first row, A_1 as the second row, etc.) (*Hint:* For the proofs that follow, use the result of Exercise 24.)

(a) Show that each A_m is denumerable.

(b) Prove that Ω is a partition of $\mathcal{N}$.

6.3 TWO WAYS OF LOOKING AT INFINITY

Historically, there were two definitions of an infinite set. The word "infinite" means "not finite," and in Section 6.2 we defined an infinite set as a set that is not finite. Another definition, given by Richard Dedekind (1831–1916) in 1888, is as follows.

Definition. A set B is **Dedekind infinite** if B includes a proper subset D such that $D \approx B$.

We use the term "Dedekind infinite" to distinguish between the two types of infinite sets. In what follows, we will determine whether sets such as N_k, $\mathcal{N}$, and $\mathcal{R}$ are Dedekind infinite. Also, we examine whether a Dedekind infinite set is infinite (not finite) as well, and whether an infinite set is Dedekind infinite.

Theorem 1 If A is Dedekind infinite, then A is infinite.

Proof. This is just a restatement of Theorem 2 in Section 6.2. $\qquad\square$

At this stage, we know that there are finite sets and infinite (nonfinite) sets. Furthermore, every Dedekind infinite set is also infinite (see Figure 6–3). Example 1 following shows that there exist Dedekind infinite sets.

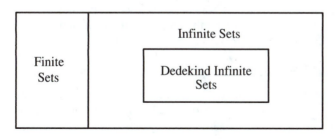

Figure 6–3 Types of sets

Example 1

Show that $\mathcal{N}$ is Dedekind infinite.

Solution Let $E = \{0, 2, 4, 6, \ldots\}$. Then the function $f : E \to \mathcal{N}$ defined by $f(n) = n/2$ is bijective. Hence, $E \approx \mathcal{N}$. Note that the solution to Example 1 of Section 6.2 is also a solution to this example.

Theorem 2 If B is denumerable, then B is Dedekind infinite.

Example 2

Show that for all nonnegative integers k, N_k is not Dedekind infinite.

Solution The result follows from Theorem 11 in Section 6.1.

Theorem 3   If $A \subseteq B$ and A is Dedekind infinite, then B is Dedekind infinite.

Corollary.   Every set with a denumerable subset is Dedekind infinite.

Example 3

Show that $\mathcal{R}$ is Dedekind infinite.

Solution   This follows from the preceding corollary.

Figure 6–4 illustrates the relationship between infinite sets and Dedekind infinite sets, as we are thus far aware. The question remains, Are all infinite sets also Dedekind infinite? In other words, can we prove the converse of Theorem 1?

Sets:	Finite	Infinite	
		Dedekind Infinite	Other?
	e.g., N_k	e.g., $\mathcal{N}, \mathcal{R}$	

Figure 6–4

The corollary to Theorem 3 is a key result, since it provides a tool for proving that a given set is Dedekind infinite. If we can prove that every infinite set includes a denumerable subset, then we have proved the converse of Theorem 1. First we prove that every Dedekind infinite set includes a denumerable subset.

Theorem 4   If A is Dedekind infinite, then A includes a denumerable subset.

Proof.   Let A be Dedekind infinite. Then there exist a set $D \subseteq A$ with $D \neq A$ and a bijective function $f : A \rightarrow D$. Let $a_0 \in A - D$. We use definition by recursion to construct a set $B = \{a_0, a_1, a_2, \ldots\}$, where $a_{k+1} = f(a_k)$ for $k = 0, 1, 2, \ldots$. To prove that B is denumerable, we need to show that the elements of B are distinct. We thus prove that $a_m = a_n$ implies $m = n$.

Suppose $a_j = a_i$ for some $j \neq i$, say, $j < i$. Let j be the least integer such that $a_j = a_i$ and $j < i$. Clearly, $j \neq 0$ since $a_0 \in A - D$ and $a_i \in D$ for $i = 1, 2, 3, \ldots$. But then, since $f(a_{j-1}) = a_j = a_i = f(a_{i-1})$ and f is bijective, we have $a_{j-1} = a_{i-1}$, contrary to the supposition that j is the least integer such that a_j equals another element of B.

Since B is indexed by $\mathcal{N}$, B is denumerable.   □

Theorem 5   If D is infinite, then D includes a denumerable subset.

Proof.   Let D be infinite. The procedure for constructing a denumerable subset E of D appears quite simple: (See The Axiom of Choice below.) Pick $a_0 \in D$, then pick

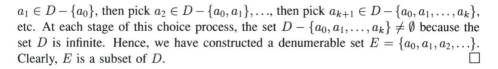

$a_1 \in D - \{a_0\}$, then pick $a_2 \in D - \{a_0, a_1\}, \ldots$, then pick $a_{k+1} \in D - \{a_0, a_1, \ldots, a_k\}$, etc. At each stage of this choice process, the set $D - \{a_0, a_1, \ldots, a_k\} \neq \emptyset$ because the set D is infinite. Hence, we have constructed a denumerable set $E = \{a_0, a_1, a_2, \ldots\}$. Clearly, E is a subset of D. $\qquad\square$

Corollary.   Every infinite set is Dedekind infinite.

The Axiom of Choice

As mentioned in the proof of Theorem 5, constructing a denumerable subset of an infinite set appears straightforward and simple. However, appearances are misleading. First, let us compare the proof of Theorem 5 with the proof of Theorem 4.

The difference between the two proofs lies in how the choice of each a_{k+1} was made. In the proof of Theorem 4, each a_{k+1} is the image of a_k by a given function f. In other words, the choice at each stage of the process is made by an existing function. On the other hand, the use of the word "pick" to indicate each choice of a_i in the proof of Theorem 5 implies the existence of a choice function, that is, a function that chooses an element from each set in a given collection of nonempty sets. But how do we know that such a function exists? In the case of a finite number of choices, there is no difficulty: The usual laws of set theory imply the existence of such a choice function. In the case of an infinite number of choices, however, the usual laws of set theory are inadequate, and another axiom is needed. This axiom is called the *axiom of choice*.

Axiom of Choice   Let Ω be an arbitrary collection of nonempty sets. There is a function $f : \Omega \to \cup\Omega$ such that $f(A) \in A$ for each $A \in \Omega$.

Note that the function f, held to exist by the axiom of choice, simultaneously chooses an element from each set in Ω. Any function $f : \Omega \to \cup\Omega$ such that $f(A) \in A$ for each $A \in \Omega$ is called a **choice function** for Ω.

Now compare the proofs of Theorems 4 and 5 again. The proof of Theorem 4 does not require the axiom of choice because a choice function exists. However, in the proof of Theorem 5, the hypotheses of the theorem do not seem to suggest the existence of a choice function. Hence, since the collection of nonempty sets is infinite, the axiom of choice was used to obtain a choice function.

Proof of Theorem 5 Using the Axiom of Choice: Let Ω be the collection of nonempty subsets of D. By the axiom of choice, there exists a choice function f for Ω. Hence, for each nonempty subset A of D, $f(A) \in A$; in other words, f chooses an element of A.

Now we define the set $E = \{a_0, a_1, a_2, \ldots\}$ by recursion as follows. Let $a_0 = f(D)$ and $a_{k+1} = f\big(D - \{a_0, a_1, \ldots, a_k\}\big)$. Since D is infinite, the sets $D - \{a_0, a_1, \ldots, a_k\} \neq \emptyset$ and consequently, are sets in Ω. By a proof similar to that of Theorem 4, we can then show that E is denumerable. $\qquad\square$

In the proof of Theorem 13 in Section 6.1, a choice of an element in $f^{-1}\big[\{x\}\big]$ was needed. The axiom of choice was not used in this proof since $f^{-1}\big[\{x\}\big]$ is a subset of a finite set B. That is, since there is a bijection $h : N_k \to B$, we had a method for choosing an element of $f^{-1}[\{x\}]$. We chose $h(j)$, where j is the least integer such that $h(j) \in f^{-1}[\{x\}]$. A similar situation occurs in the proof of Theorem 8 in Section 6.2.

The existence and use of a choice function in proofs can be very subtle. Recognizing that a choice function is being used is sometimes difficult. Moreover, it can be tricky deciding whether the axiom of choice is needed to assert the existence of a choice function. After being convinced that the axiom of choice may be implicit in proofs requiring a choice function, some people start to see it in arguments where it is not needed.

Given that a choice function is required, how do we know whether the axiom of choice is needed? A nice example due to Bertrand Russell concerning pairs of shoes and pairs of socks is helpful.

Suppose we have an infinite collection of pairs of socks and we wish to choose one sock from each pair. The existence of such a choice function requires the axiom of choice, since there are an infinite number of choices to be made and there is no way to distinguish one sock from another in each pair. But suppose, instead, that we have an infinite collection of pairs of shoes and we wish to choose one shoe from each pair. In this case, the axiom of choice is not needed because each pair has a right shoe and a left shoe. We can choose, for example, the left shoe from each pair. In other words, a choice function exists. Furthermore, if the number of pairs of socks is finite, then a choice function can be constructed without the axiom of choice:

Following are several situations in which a choice function exists without the axiom of choice:

1. A finite collection of arbitrary sets.
2. The collection of nonempty subsets of $\mathcal{N}$. In this case, each set in the collection has a least element (by the well-ordering property), and so a choice function exists.
3. The collection of closed intervals $[a, b]$ in $\mathcal{R}$. Here, each interval has a right end point (or left end point, or midpoint, etc.).

The axiom of choice has a history of controversy. It was initially viewed with suspicion because it asserts the existence of a function without specifying (or constructing) that function. Today, most mathematicians accept the axiom of choice as an essential axiom of set theory. However, there are still a few mathematicians, often called constructivists, who limit their attention to mathematics that can be derived with a restricted form of logic and set theory, which excludes the axiom of choice. In any case, it is common for mathematicians to explicitly mention any use made of the axiom.

EXERCISE SET 6.3

Do not use the corollary to Theorem 5 for any proof in this exercise set.

1. Prove Theorem 2.

2. Prove Theorem 3.

3. Prove the corollary to Theorem 3.

4. *Prove:* If D is a Dedekind infinite set, then $D - \{a\}$ is Dedekind infinite.

5. *Prove:* If D is a Dedekind infinite set, then $D - F$ is Dedekind infinite if F is finite.

6. *Prove:* If D is a Dedekind infinite set and $A \approx D$, then A is Dedekind infinite.

7. Prove that the interval $(0, 1)$ is Dedekind infinite.

8. *Prove:* If B is infinite, then $B - \{a\}$ is infinite.

9. *Prove:* If B is infinite and F is finite, then $B - F$ is infinite.

10. Use the axiom of choice to prove that for any set B, there is a function $g : P(B) - \{\emptyset\} \to B$ such that $g(A) \in A$ for every nonempty subset A of B.

11. Use the axiom of choice to prove that for any relation R, there exists a function $f \subseteq R$ such that dom(f) = dom(R).

12. Let A be a finite set and Ω be the collection of all nonempty subsets of A. Find a choice function for Ω without using the axiom of choice.

Prove the theorems stated in Exercises 13 and 14. Determine whether the axiom of choice is used in your proof.

13. A function $f : D \to C$ is onto if and only if there exists a function $g : C \to D$ such that $f \circ g = 1_C$. (*Hint:* See Theorem 13 in Section 6.1 and Theorem 8 in Section 6.2.)

14. A function $f : D \to C$ is 1–1 if and only if there exists a function $g : C \to D$ such that $g \circ f = 1_D$.

15. *Prove:* If Υ is a countable collection of countable sets, then $\cup \Upsilon$ is countable. (*Hint for the case* $\Upsilon = \{B_m : m \in \mathcal{N}\}$, *where each B_m is denumerable:* By Exercise 25 of Section 6.2, $\mathcal{N} = \cup\{A_m : m \in \mathcal{N}\}$, and the sets A_m are denumerable and pairwise disjoint. Use the axiom of choice to obtain a collection of functions $\{f_m : m \in \mathcal{N}$ and $f_m : A_m \to B_m$ is bijective$\}$, and then use the functions f_m to define an onto function $h : \mathcal{N} \to \cup \Upsilon$.)

KEY CONCEPTS

The relation $\approx$, A is equinumerous to B	Countable
	Uncountable
Finite set	Indexed by $\mathcal{N}$
Cardinality	Cantor diagonalization method
Pigeonhole principle	Dedekind infinite
Infinite set	Axiom of choice
Denumerable	Choice function

REVIEW EXERCISES

1. Prove that $(0, 1] \approx (0, 1)$. (*Hint:* Define a function f such that $f(1) = \frac{1}{2}, f(\frac{1}{2}) = \frac{1}{3}, f(\frac{1}{3}) = \frac{1}{4}, \ldots$, and for all other elements x in $(0, 1]$, $f(x) = x$.)

2. *Prove:* If $A \approx B$, then $P(A) \approx P(B)$.

3. *Prove:* If A is infinite and $x \notin A$, then $A \cup \{x\} \approx A$. (*Hint:* A includes a denumerable set. Use the idea in Exercise 1 to construct a bijection $f : A \cup \{x\} \to A$.)

4. *Prove:* If A is infinite, then $A \times N_m \approx A$ for $m = 1, 2, 3, \ldots$.

5. Without using the corollary to Theorem 5 in Section 6.3, prove that if $x \notin A$ and $A \cup \{x\} \approx A$, then A is Dedekind infinite.

6. If A is denumerable and B is finite, then $A \cup B$ is denumerable. (*Hint:* Let $\text{Card}(B) = n$. Use mathematical induction on n.)

7. Let Π be a partition of a set A. Use the axiom of choice to prove that there is a $B \subseteq A$ such that $B \cap X$ has exactly one element for every $X \in \Pi$.

8. Let A be a denumerable set and Ω be the collection of all nonempty subsets of A. Find a choice function for Ω without using the axiom of choice.

SUPPLEMENTARY EXERCISES

Define 2^A to be the set of all functions $f : A \to \{0, 1\}$.

1. Prove that $\text{Card}(2^{N_k}) = 2^k$.

2. Prove that $2^A \approx P(A)$. (*Hint:* For each subset $X \subseteq A$, consider the characteristic function $\chi_X : A \to \{0, 1\}$ defined by $\chi_X(z) = 0$ if $z \notin X$ and $\chi_X(z) = 1$ if $z \in X$.)

3. Prove that $A \not\approx 2^A$ for any set A. (*Hint:* Suppose $A \approx 2^A$ and obtain a bijective function $f : A \to 2^A$. Let $f_a = f(a)$, and note that $f_a \in 2^A$. Use the Cantor diagonalization method to find a function $d \in 2^A$ such that $d \neq f_a$ for all $a \in A$.)

4. Prove that $A \not\approx P(A)$ for any set A. (You may use Exercise 3 to prove this, but a direct proof is instructive.)

5. Prove that $P(\mathcal{N})$ is uncountable. (*Hint:* See Exercise 4.)

Define a relation $\preceq$ on sets by $A \preceq B$ if there exists a 1–1 function $f : A \to B$. Moreover, $A \prec B$ if $A \preceq B$ and $A \not\approx B$.

6. **(a)** Prove that $\preceq$ is reflexive.
   (b) Prove that $\preceq$ is transitive.

7. *Prove:* If $A \preceq B$ and $B \preceq A$, then $A \approx B$. (This is known as the **Schröder-Bernstein Theorem.** It says that the relation $\preceq$ has a property that closely resembles the antisymmetric property, and thus, $\preceq$ is almost a partial order.)
   (*Hint:* Assume $f : A \to B$ and $g : B \to A$ are bijections. Define a function $h : P(A) \to P(A)$ by $h(X) = A - g\big[B - f[X]\big]$. Show that $X \subseteq Y$ implies that $h(X) \subseteq h(Y)$, and use Supplementary Exercise 7 in Chapter 5 to conclude that there is a $Z_0 \in P(A)$ such that $h(Z_0) = Z_0$. This means that $g\big[B - f[Z_0]\big] = A - Z_0$. Now define a function $k : A \to B$ by

$$k(x) = \begin{cases} f(x) & \text{if } x \in Z_0 \\ g^{-1}(x) & \text{if } x \in A - Z_0 \end{cases}$$

and show that k is a bijection.)

8. *Prove:* If $A \subseteq B$, then $A \preceq B$.

9. *Prove:* If $m \leq n$, then $N_m \preceq N_n$.

10. *Prove:* If A is finite, then $A \prec \mathcal{N}$.

11. Prove that $A \prec P(A)$. (*Hint:* See Exercise 4.)

12. *Prove:* If R is an equivalence relation on A, then $A/R \preceq A$.

7

Boolean Algebra

The mathematical structure now called Boolean algebra was formulated by George Boole (1815–1864) in his 1854 publication, "An Investigation of the Laws of Thought." Boolean algebra has been a valuable tool in the study of logic and, because of its applicability to the design of logic circuits, has been very useful in computer science.

7.1 INTRODUCTION TO BOOLEAN ALGEBRA

One of the most familiar examples of a Boolean algebra is from set theory. Consider the power set $P(U)$ of a set U. Set union ($\cup$) and set intersection ($\cap$) are binary operations on $P(U)$, and set complementation ($'$) is a unary operation on $P(U)$. Recall that the following laws hold for all X, Y, and Z in $P(U)$:

1a. $X \cup Y = Y \cup X$	1b. $X \cap Y = Y \cap X$
2a. $(X \cup Y) \cup Z = X \cup (Y \cup Z)$	2b. $(X \cap Y) \cap Z = X \cap (Y \cap Z)$
3a. $X \cup (Y \cap Z) = (X \cup Y) \cap (X \cup Z)$	3b. $X \cap (Y \cup Z) = (X \cap Y) \cup (X \cap Z)$
4a. $X \cup \emptyset = X$	4b. $X \cap U = X$
5a. $X \cup X' = U$	5b. $X \cap X' = \emptyset$

The algebra $(P(U), \cup, \cap, ', \emptyset, U)$ is an example of a Boolean algebra. Using this algebra as a model, we formally state the definition of a Boolean algebra.

206

> **Definition.** A **Boolean Algebra** $(B, +, \cdot, ', 0, 1)$ consists of a nonempty set B, binary operations $+$ and $\cdot$ on B, a unary operation $'$ on B, and distinguished elements 0 and 1 in B such that the following axioms hold for all x, y, and z in B:
>
> 1a. $x + y = y + x$ 1b. $x \cdot y = y \cdot x$
>
> 2a. $(x + y) + z = x + (y + z)$ 2b. $(x \cdot y) \cdot z = x \cdot (y \cdot z)$
>
> 3a. $x + (y \cdot z) = (x + y) \cdot (x + z)$ 3b. $x \cdot (y + z) = x \cdot y + x \cdot z$
>
> 4a. $x + 0 = x$ 4b. $x \cdot 1 = x$
>
> 5a. $x + x' = 1$ 5b. $x \cdot x' = 0$

Since $+$ and $\cdot$ are binary operations on B, $x + y$ and $x \cdot y$ are in B when x and y are in B. Similarly, because $'$ is a unary operation on B, $x' \in B$ when $x \in B$.

Each of the preceding axioms has a descriptive name. Axiom 1 is called the **commutative law,** Axiom 2 the **associative law,** Axiom 3 the **distributive law,** Axiom 4 the **identity law,** and Axiom 5 the **complement law.** Axioms 4a and 4b define the elements 0 and 1; by Theorem 5 following, 0 and 1 are unique. Axioms 5a and 5b define x' for $x \in B$. We call x' the **complement** of x; by Theorem 3 following, the complement of an element is unique.

Note that the axioms are given in pairs (for example, 3a and 3b). Each statement in a pair is the dual of the other statement in the pair. The **dual** of a statement is obtained by simultaneously replacing $+$ by $\cdot$, $\cdot$ by $+$, 0 by 1, and 1 by 0. In a Boolean algebra, if a statement is true, then its dual is also true. This is called the **principle of duality.**

The 0 and 1 in the Boolean algebra $(B, +, \cdot, ', 0, 1,)$ are called the **zero element** and the **one element,** respectively. The 0 and 1 can be considered the generic zero and one elements of a Boolean algebra. Each example of a Boolean algebra will have a specific zero and one, but the specific zero and one elements will usually be designated with symbols other than 0 and 1. The operations $+$, $\cdot$, and $'$ are called **sum, product,** and **complement,** respectively.

When the context is clear and there is no danger of confusion, we sometimes write B instead of $(B, +, \cdot, ', 0, 1,)$ to denote a Boolean algebra.

The operations for a Boolean algebra follow a precedence hierarchy. Parenthesized expressions are applied first, as in any algebraic procedure. Among the operations $+$, $\cdot$, and $'$, the unary operation $'$ has the highest precedence. Next is the binary operation $\cdot$, followed by $+$. Finally, the equality sign in an equation has the lowest precedence. Parentheses are used for clarity, but can be omitted in many cases by using the precedence hierarchy. So $(x \cdot y) + z$ can be written as $x \cdot y + z$, and $z \cdot (y') + x$ written as $z \cdot y' + x$. For simplicity, we will usually write $x \cdot y$ as xy. Thus, $x \cdot (y' + x \cdot z) = x(y' + xz)$. Similarly, the distributive law 3a may be written as $x + yz = (x + y)(x + z)$.

Now we consider the important, two-element Boolean algebra.

Example 1

Consider $(B, +, \cdot, ', 0, 1)$ with $B = \{0, 1\}$, where the operations $+$, $\cdot$, and $'$ are defined in Table 7–1.

TABLE 7-1

$0 + 0 = 0$	$0 \cdot 0 = 0$	$0' = 1$
$0 + 1 = 1$	$0 \cdot 1 = 0$	$1' = 0$
$1 + 0 = 1$	$1 \cdot 0 = 0$	
$1 + 1 = 1$	$1 \cdot 1 = 1$	

With these operations, the system $(B, +, \cdot, ', 0, 1)$ is a Boolean algebra. In other words, the axioms 1a–5a and 1b–5b hold for all elements in B. The zero element is 0 and the one element is 1.

Let us verify Axiom 3b, $x \cdot (y + z) = x \cdot y + x \cdot z$, for $x = 0$, $y = 1$, and $z = 1$. The left side is

$$
\begin{aligned}
x \cdot (y + z) &= 0 \cdot (1 + 1) && \text{(substitution)} \\
&= 0 \cdot 1 && (1 + 1 = 1) \\
&= 0 && (0 \cdot 1 = 0)
\end{aligned}
$$

The right side is

$$
\begin{aligned}
x \cdot y + x \cdot z &= 0 \cdot 1 + 0 \cdot 1 && \text{(substitution)} \\
&= 0 + 0 && (0 \cdot 1 = 0) \\
&= 0 && (0 + 0 = 0)
\end{aligned}
$$

Hence, $x \cdot (y + z) = x \cdot y + x \cdot z$ for the special case when $x = 0, y = 1$, and $z = 1$.

It is possible, though tedious, to verify Axiom 3b and, in fact, all the other axioms for all the elements in $B = \{0, 1\}$. We will skip the details here.

The two-element Boolean algebra is related to propositional logic. If we consider 0 to be the truth value FALSE and 1 to be the truth value TRUE, then we see that $+$ is OR, $\cdot$ is AND, and $'$ is NOT.

In the following, if A is a finite set, then **Card(A)** is the number of elements in A (see Chapter 6).

In the Boolean algebra $(P(U), \cup, \cap, ', \emptyset, U)$, the empty set $\emptyset$ is the zero element and the universal set U is the one element. From this example, we see that it is easy to construct a Boolean algebra $(P(U), \cup, \cap, ', \emptyset, U)$ from a universal set U. If U is finite and $\text{Card}(U) = n$, then $P(U)$ has 2^n elements. Thus, the system $(P(U), \cup, \cap, ', \emptyset, U)$ is a Boolean algebra with 2^n elements.

There are many elementary theorems in Boolean algebra. We will state and prove some of the most useful ones. Other proofs are left as exercises. The theorems can be proved from the axioms for Boolean algebra. Once we have proved them, we will have useful facts that are true for every example of a Boolean algebra.

Parts a and b in many of the following theorems are duals of each other. Since by the principle of duality, the dual of a theorem is a theorem, we need only prove one of the two. However, it may be instructive to prove the dual of a theorem that has been proved. We usually refer to the axioms and theorems of Boolean algebra as **Boolean algebra laws.**

In what follows, we assume that $(B, +, \cdot, ', 0, 1)$ is a Boolean algebra and x and y are arbitrary elements of B.

Theorem 1   (Idempotent Laws)

a. $x + x = x$

b. $x \cdot x = x$

Proof. We prove that $x + x = x$ for all $x \in B$. Let x be an arbitrary element of B. Then:

$$
\begin{aligned}
x + x &= (x + x) \cdot 1 && \text{(Identity law 4b)} \\
&= (x + x) \cdot (x + x') && \text{(Complement law 5a)} \\
&= x + xx' && \text{(Distributive law 3a)} \\
&= x + 0 && \text{(Complement law 5b)} \\
&= x && \text{(Identity law 4a)} \qquad \square
\end{aligned}
$$

Theorem 2   (Identity Laws)

a. $x + 1 = 1$

b. $x \cdot 0 = 0$

Hint for proof of part a: $x + 1 = x + (x + x')$ $\qquad \square$

The next theorem is useful for proving subsequent theorems of Boolean algebra.

Theorem 3   (Unique Complement Law) For each element $x \in B$, x' is the unique element in B that satisfies the properties $x + x' = 1$ and $x \cdot x' = 0$.

Proof. We need to prove that if $x + y = 1$ and $x \cdot y = 0$, then $y = x'$. Let $x + y = 1$ and $x \cdot y = 0$. Then $y = y \cdot 1 = y(x + x') = yx + yx' = xy + x'y = 0 + x'y = xx' + x'y = x'x + x'y = x'(x + y) = x'1 = x'$. $\qquad \square$

Theorem 4   (Involution Laws)

a. $1' = 0$

b. $0' = 1$

c. $(x')' = x$

Proof. We prove a, using the unique complement law:

$$
\begin{aligned}
1 + 0 &= 1 \text{ and } 1 \cdot 0 = 0 \cdot 1 = 0 && \text{(Axioms 4a, 1b, 4b)} \\
\text{But } 1 + 1' &= 1 \text{ and } 1 \cdot 1' = 0. && \text{(Axioms 5a, 5b)} \\
\text{Hence } 1' &= 0. && \text{(Unique complement law)} \qquad \square
\end{aligned}
$$

If we consider Axioms 4a and 4b carefully, we see that they guarantee the existence of a zero element and a one element. But they say nothing about uniqueness—in other words, are 0 and 1 the *only* elements satisfying Axioms 4a and 4b? Furthermore, they say nothing about whether 0 and 1 are distinct. The next theorem clarifies these issues.

Theorem 5

a. 0 is the unique element in B such that $x + 0 = x$ for all $x \in B$.

b. 1 is the unique element in B such that $x \cdot 1 = x$ for all $x \in B$.

c. 0 and 1 are distinct if B has more than one element.

Theorems 6, 7, and 8 following are simplification laws.

Theorem 6

a. $x + x \cdot y = x$

b. $x \cdot (x + y) = x$

Hint for proof of part a: $x + xy = x \cdot 1 + xy$ □

Theorem 7

a. $x \cdot y + x \cdot y' = x$

b. $(x + y) \cdot (x + y') = x$

Theorem 8

a. $x + y \cdot x' = x + y$

b. $x \cdot (y + x') = x \cdot y$

Hint for proof: Use the distributive law. □

By the associative law $(x+y)+z = x+(y+z)$, we can write $x+y+z$ unambiguously to mean either $(x + y) + z$ or $x + (y + z)$. Similarly, $xyz = (x \cdot y) \cdot z = x \cdot (y \cdot z)$. In general, the terms in the sum or product of any number of variables can be associated unambiguously in any way. For example, $xyzw = x \cdot (yzw) = (xy)(zw) = (xyz) \cdot w$.

Theorem 9   (De Morgan's Laws)

a. $(x + y)' = x' \cdot y'$

b. $(x \cdot y)' = x' + y'$

Proof. We prove $x' + y' = (x \cdot y)'$ using the unique complement law. First,
$$xy + (x' + y') = (xy + x') + y' = (x' + xy) + y' = (x' + x)(x' + y) + y'$$
$$= 1 \cdot (x' + y) + y' = (x' + y) + y' = x' + (y + y') = x' + 1 = 1.$$
Next,
$$xy(x' + y') = xyx' + xyy' = yxx' + xyy' = y \cdot 0 + y \cdot 0 = 0 + 0 = 0.$$
Since $xy + (x' + y') = 1$ and $xy(x' + y') = 0$, $x' + y' = (xy)'$, by the unique complement law. □

After all the preceding hard work proving the various theorems, it is worth stating explicitly what has been accomplished: We now know that all the theorems proved are true of any Boolean algebra; hence, it is not necessary to verify any of these theorems for any specific Boolean algebra.

So far, we have considered the two-element Boolean algebra and the Boolean algebra consisting of subsets of a given set. There are many other interesting Boolean algebras. Example 2 presents one based on elementary concepts of number theory. In particular, we will need the concept of the greatest common divisor (gcd) and the least common multiple (lcm) of two positive integers (see Chapter 8).

Example 2

Let $B = \{1, 3, 4, 12\}$. Define $x' = 12/x$ (where $/$ is ordinary division), $x + y = \text{lcm}(x, y)$, and $x \cdot y = \text{gcd}(x, y)$. It is not difficult to verify that $'$ defines a unary operation on B, while both $+$ and $\cdot$ define binary operations on B. With patience and a little elementary number theory, we can verify that the system $(B, +, \cdot, ', 1, 12)$ is a Boolean algebra. Note that for this Boolean algebra, the zero element is 1 and the one element is 12.

The next example gives two systems that are very similar to Example 2. However, neither of them is a Boolean algebra.

Example 3

Define $x' = 12/x$, $x + y = \text{lcm}(x, y)$, and $x \cdot y = \text{gcd}(x, y)$.

a. Let $B = \{1, 3, 12\}$. Then $(B, +, \cdot, ', 1, 12)$ is not a Boolean algebra, because $'$ is not a unary operation on B. To see this, note that the complement of 3 is $3' = \frac{12}{3} = 4$. But $4 \notin B$; hence, the operation $'$ is not a unary operation on B.

b. Let $B = \{1, 2, 3, 4, 6, 12\}$, and consider $(B, +, \cdot, ', 1, 12)$. In this case, $'$ is a unary operation on B, while $+$ and $\cdot$ are both binary operations on B. Hence, $(B, +, \cdot, ', 1, 12)$ is an algebra. Now, the element 2 is not the zero element of the system, since $3 + 2 \neq 3$. However, $2 \cdot 2' = 2 \cdot 6 = \text{gcd}(2, 6) = 2$. But this violates Axiom 5b. Hence, the algebra $(B, +, \cdot, ', 1, 12)$ is not a Boolean algebra.

EXERCISE SET 7.1

1. Determine whether $*$ is a binary operation on the given set B. If not, give a counterexample.
   (a) $B = \{1, 2, 3, 4, 12\}$; $x * y = \text{lcm}(x, y)$
   (b) $B = \{1, 2, 3, 4, 12\}$; $x * y = \text{gcd}(x, y)$

2. Determine whether $*$ is a binary operation on the given set B. If not, give a counterexample.
   (a) $B = $ set of all infinite subsets of $\mathcal{N}$; $X * Y = X \cup Y$
   (b) $B = $ set of all infinite subsets of $\mathcal{N}$; $X * Y = X \cap Y$

3. Write the dual of each Boolean equation or statement.
   (a) $xy' + x' = x' + y'$
   (b) $(x + 0) \cdot 1 + x' = 1$
   (c) If $x \cdot y = 1$, then $x = 1$ and $y = 1$.

4. Write the dual of each Boolean equation or statement.
   (a) If $x + y = y + z$ and $x' + y = x' + z$, then $y = z$.
   (b) If $x + y = y$, then $x \cdot y' = 1$.

5. Let $x' = 24/x$, $x + y = \text{lcm}(x, y)$, and $x \cdot y = \text{gcd}(x, y)$. For each of the given sets B, determine whether $(B, +, \cdot, ', 1, 24)$ is a Boolean algebra.
   (a) $B = \{1, 2, 3, 12, 24\}$

(b) $B = \{1, 2, 3, 4, 6, 8, 12, 24\}$

6. Let $x' = 24/x$, $x + y = \mathrm{lcm}(x, y)$, and $x \cdot y = \gcd(x, y)$. For each of the given sets B, determine whether $(B, +, \cdot, ', 1, 24)$ is a Boolean algebra.

   (a) $B = \{1, 2, 3, 8, 12, 24\}$
   (b) $B = \{1, 3, 8, 24\}$

7. Prove Theorem 1b.

8. Prove Theorem 2.

9. **(a)** Prove Theorem 4b.
   (b) Prove Theorem 4c.

10. Prove Theorem 5.

For a given positive integer n, let D_n be the set of all positive divisors of n. For example, $D_{24} = \{1, 2, 3, 4, 6, 8, 12, 24\}$.

11. For a given positive integer n, let $B = D_n$, and define $x' = n/x$, $x + y = \mathrm{lcm}(x, y)$, and $x \cdot y = \gcd(x, y)$. For each of the given values of n, determine whether $(D_n, +, \cdot, ', 1, n)$ is a Boolean algebra.

    (a) $n = 1$
    (b) $n = 2$
    (c) $n = 12$

12. For a given positive integer n, let $B = D_n$, and define $x' = n/x$, $x + y = \mathrm{lcm}(x, y)$, and $x \cdot y = \gcd(x, y)$. For each of the given values of n, determine whether $(D_n, +, \cdot, ', 1, n)$ is a Boolean algebra.

    (a) $n = 4$
    (b) $n = 15$
    (c) $n = 30$

13. **(a.)** Do Exercise 12 for D_{40} and D_{42}.
    (b.) Find a property of n that is a necessary and sufficient condition for D_n to form a Boolean algebra. Prove your conjecture.

14. Prove Theorem 6.

15. Prove Theorem 7.

16. Prove Theorem 8.

17. Prove De Morgan's law, $(x \cdot y)' = x' + y'$, using Theorems 9a and 4c.

18. Use Boolean algebra laws to simplify the following.
    (a) $(x + y)(x' + y)$
    (b) $(xy)(x' + y)$

19. Use Boolean algebra laws to simplify the following.
    (a) $(x + 0)(y + 1)$
    (b) $[(xy)z + (xy)z'] + xy'$

20. **(a)** Prove the extended De Morgan's law $(x + y + z)' = x'y'z'$.
    (b) State the dual of this version of De Morgan's law.

21. **(a)** Prove the extended distributive law $w + xyz = (w + x)(w + y)(w + z)$.
    (b) State the dual of this version of the distributive law.

22. Use Boolean algebra laws to simplify $xy + xy' + x'y + x'y'$.

23. Use Boolean algebra laws to simplify $xy + (xy)'z + z'$.

24. Use Boolean algebra laws to simplify $xyz + xy'z + xy'z' + xyz'$.

7.2 THE STRUCTURE OF A BOOLEAN ALGEBRA

In Section 7.1, we stated and proved the most important theorems pertaining to the algebraic structure of a Boolean algebra. But we are still not able to answer easily the question "Is this a Boolean algebra?" when faced with an arbitrary algebraic structure. To answer this question for certain structures, we need to prove a categorization theorem, which in turn requires that we consider a partial ordering on a Boolean algebra. Accordingly, we define a relation $\leq$ on a Boolean algebra as follows.

Definition. $x \leq y$ if $x \cdot y = x$.

To illustrate the foregoing relation in a specific Boolean algebra, we examine $(P(U), \cup, \cap,', \emptyset, U)$. For this Boolean algebra, the condition $x \cdot y = x$ is $X \cap Y = X$, which is equivalent to $X \subseteq Y$. So the relation analogous to $\leq$ for $(P(U), \cup, \cap,', \emptyset, U)$ is the subset relation.

Theorem 1

a. $0 \leq x$

b. $x \leq 1$

Theorem 2   The three conditions $x \cdot y = x$, $x + y = y$, and $x \cdot y' = 0$ are logically equivalent. Hence, these conditions are all equivalent to $x \leq y$.

Outline of the proof: It is sufficient, because of the transitivity of the implication, to prove three implications (the proofs are left as exercises):

1. If $x \cdot y' = 0$, then $x \cdot y = x$.
2. If $x \cdot y = x$, then $y = x + y$.
3. If $x + y = y$, then $x \cdot y' = 0$.

So $x \leq y$ is logically equivalent to any of the three conditions $x \cdot y = x$, $x + y = y$, $x \cdot y' = 0$. $\square$

Theorem 3

a. $x \cdot y \leq y$

b. $y \leq x + y$

For another example of $\leq$ in a Boolean algebra, consider the Boolean algebra $(B, +, \cdot, 1, 12)$ introduced in Example 2 of Section 7.1, with $B = \{1, 3, 4, 12\}$, $x + y = \text{lcm}(x, y)$, $x \cdot y = \gcd(x, y)$, and $x' = 12/x$. In this Boolean algebra, the condition $x \cdot y = x$ is $\gcd(x, y) = x$, which holds if and only if x divides y. So in this Boolean algebra, $x \leq y$ means that x divides y. Think of this example and the subset ordering while considering the following theorems.

The next three theorems prove that the relation $\leq$ is a partial ordering.

Theorem 4 $x \leq x$.

Theorem 5   If $x \leq y$ and $y \leq x$, then $x = y$.

Theorem 6   If $x \leq y$ and $y \leq z$, then $x \leq z$.

Theorems 4, 5, and 6 in effect state that $\leq$ is reflexive, antisymmetric, and transitive. Hence, $\leq$ is a partial ordering. When we wish to emphasize the partial ordering $\leq$, we write $(B, +, \cdot, ', 0, 1, \leq)$.

Theorem 7   Let $x \leq y$ and $w \leq z$. Then:

a. $x + w \leq y + z$.

b. $xw \leq yz$.

Whenever $x \leq y$ and $x \neq y$, we say that x is **strictly less than** y and write $x < y$.

Now consider the Boolean algebra $(D_{30}, +, \cdot, 1, 30)$, with $D_{30} = \{1, 2, 3, 5, 6, 10, 15, 30\}$, $x + y = \mathrm{lcm}(x, y)$, $x \cdot y = \gcd(x, y)$, and $x' = 30/x$. Each element $x \in D_{30}$ can be written $x = 2^i 3^j 5^k$, where i, j, $k = 0$ or 1. For example, $5 = 2^0 3^0 5^1$ and $6 = 2^1 3^1 5^0$. In other words, all the elements of this Boolean algebra are built from the elements 2, 3, and 5. Similarly, for the Boolean algebra $(P(U), \cup, \cap, ', \emptyset, U)$, all elements (the subsets of U) can be written as a union of some collection of singleton sets. For example, with $U = \{a, b, c, d\}$, the subset $\{a, c, d\} = \{a\} \cup \{c\} \cup \{d\}$. Of course, the empty set is the union of *no* singleton sets; this is analogous to $1 = 2^0 3^0 5^0$ in D_{30}.

The primes 2, 3, and 5 in D_{30} and the singletons in U act as building blocks. Because of that, they are called the atoms of the Boolean algebras to which they belong.

Definition.   An element a in $(B, +, \cdot, ', 0, 1, \leq)$ is an **atom** if $a \neq 0$ and for all $x \in B$, if $x < a$, then $x = 0$.

Thus, a is an atom if and only if no nonzero element is strictly less than a; that is, a is a minimal element in the set of nonzero elements.

Theorem 8   Let a_1 and a_2 be atoms. If $a_1 \leq a_2$, then $a_1 = a_2$.

Theorem 9   An element a is an atom if and only if for all x, $a \cdot x = a$ or $a \cdot x = 0$.

Proof. ($\Rightarrow$) Let a be an atom, and suppose $a \cdot x \neq a$. Then since $a \cdot x \leq a$, it follows immediately that $a \cdot x = 0$.

($\Leftarrow$) Suppose $a \cdot x = a$ or $a \cdot x = 0$ for all x. To prove that a is an atom, we may assume that x is arbitrary with $x \leq a$ and $x \neq a$. But $x \leq a$ implies $x \cdot a = x$. So by our first assumption, $x = a$ or $x = 0$. Hence, $x = 0$, since $x \neq a$. Therefore, a is an atom. $\square$

Theorem 10   Let a be an atom and x, y be any element of B. Then:

a. $a \leq x + y$ if and only if $a \leq x$ or $a \leq y$.

b. $a \leq xy$ if and only if $a \leq x$ and $a \leq y$.

c. $a \leq x$ if and only if $a \not\leq x'$.

Definition.   For $x \in B$, **Atom[x]**$= \{a : a$ is an atom of B and $a \leq x\}$.

Atom$[x]$ is thus composed of all the atoms less than or equal to x. We denote the set of all atoms in B by $\mathbf{W}$.

Example 1

a. Consider the Boolean algebra $(D_{30}, +, \cdot, 1, 30)$. Clearly, Atom$[15] = \{3, 5\}$ and Atom$[30] = \{2, 3, 5\} = W$. Also, $15 = 3 + 5$ and $30 = 2 + 3 + 5$ in this Boolean algebra.

b. Let $U = \{a, b, c\}$ in the Boolean algebra $(P(U), \cup, \cap, ', \emptyset, U)$. Then Atom$[\{a, c\}] = \big\{\{a\}, \{c\}\big\}$ and $W = \big\{\{a\}, \{b\}, \{c\}\big\}$.

Theorem 11

a. Atom$[0] = \emptyset$.

b. Atom$[1] = W$.

c. If a is an atom, then Atom$[a] = \{a\}$.

Theorem 12

a. Atom$[x + y] = $ Atom$[x] \cup$ Atom$[y]$.

b. Atom$[xy] = $ Atom$[x] \cap$ Atom$[y]$.

c. Atom$[x'] = W - $ Atom$[x]$.

d. $x \leq y$ if and only if Atom$[x] \subseteq$ Atom$[y]$.

We say that $(B, +, \cdot, ', 0, 1, \leq)$ is a **finite Boolean algebra** if B is a finite set. As usual, we write Card(B) for the number of elements in a finite Boolean algebra.

Definition.   A Boolean algebra B is **atomistic** if for every nonzero element x in B, there exists an atom a such that $a \leq x$.

Theorem 13   Every finite Boolean algebra is atomistic.

Proof. Let B be a finite Boolean algebra. Pick any nonzero element x. We must prove that there is an atom a such that $a \leq x$.

If x is an atom, we are done. If x is not an atom, then there exists an element $x_1 \neq 0$ such that $x_1 < x$. Similarly, if x_1 is not an atom, there exists an element $x_2 \neq 0$ such that $x_2 < x_1$. Continuing in this way, we have $x > x_1 > x_2 > \ldots$. But B is finite, so this process must terminate, say, at x_k. We thus have $x > x_1 > x_2 > \ldots > x_k > 0$.

So there is an element $x_k \le x$ with no nonzero element strictly less than x_k. Therefore, x_k is the atom we are seeking. $\quad\square$

What about the converse of Theorem 13? See Exercise 26 for an atomistic Boolean algebra that is not finite.

Every atomistic Boolean algebra satisfies two properties that make it possible to describe its structure.

Theorem 14   Let B be an atomistic Boolean algebra. Then:
a. Atom$[x] = \emptyset$ if and only if $x = 0$.
b. Atom$[x] = $ Atom$[y]$ if and only if $x = y$.

Proof of part b ($\Rightarrow$): Let Atom$[x] = $ Atom$[y]$. Then Atom$[x] \cap (W - $ Atom$[y]) = \emptyset$. Hence, Atom$[xy'] = $ Atom$[x] \cap $ Atom$[y'] = \emptyset$. Therefore, $xy' = 0$, by part a, and similarly, $x'y = 0$. Hence, $x \le y$ and $y \le x$, and we have $x = y$.

Recall that W is the set of all atoms in a Boolean algebra B. So $P(W)$ is the collection of all subsets of atoms in B.

Corollary.   Let B be an atomistic Boolean algebra. Then there is an embedding of $(B, +, \cdot,', 0, 1, \le)$ into $(P(W), \cup, \cap,', \emptyset, W, \subseteq)$.

Proof. Define $f : B \to P(W)$ by $f(x) = $ Atom$[x]$. Then f is a well-defined 1–1 function, by Theorem 14b. We next show that f is a homomorphism. We use Theorem 12 to prove that f preserves operations and order. For example, $f(xy) = $ Atom$[xy] = $ Atom$[x] \cap $ Atom$[y] = f(x) \cap f(y)$. Note that in the Boolean algebra $P(W)$, the complement operation $'$ is defined by $X' = W - X$. To say that f preserves complements means that $f(x') = f(x)'$, where the complement on the left is in B and the complement on the right is in $P(W)$. This translates to $f(x') = W - f(x)$, which follows from Theorem 12c. Also, by Theorem 11, $f(0) = \emptyset$ and $f(1) = W$. $\quad\square$

The corollary to Theorem 14 tells us that every atomistic Boolean algebra can be embedded by a 1–1 function into a collection of sets. So an atomistic Boolean algebra must "look like" a collection of sets. As we will see in the next corollary, the situation is even more completely described for finite Boolean algebras. The following example illustrates these ideas.

Example 2

Consider the Boolean algebra $B = (D_{30}, +, \cdot, 1, 30)$. Then $W = \{2, 3, 5\}$. The function $f : B \to P(W)$ defined by $f(x) = $ Atom$[x]$ is shown in Table 7.2.

TABLE 7–2

x	$f(x) = \text{Atom}[x]$
1	$\{\emptyset\}$
2	$\{2\}$
3	$\{3\}$
5	$\{5\}$
6	$\{2, 3\}$
10	$\{2, 5\}$
15	$\{3, 5\}$
30	$\{2, 3, 5\}$

The preceding corollaries and the next corollary are illustrated by the following examples.

a. $f(2 + 3) = f(6) = \{2, 3\}$. $f(2) \cup f(3) = \{2\} \cup \{3\} = \{2, 3\}$.
Therefore, $f(2 + 3) = f(2) \cup f(3)$.
b. $f(10 \cdot 15) = f(5) = \{5\}$. $f(10) \cap f(15) = \{2, 5\} \cap \{3, 5\} = \{5\}$.
Therefore, $f(10 \cdot 15) = f(10) \cap f(15)$.
c. $f(10)' = f(3) = \{3\}$. $f(10)' = \{2, 3, 5\} - \{2, 5\} = \{3\}$.
Therefore, $f(10') = f(10)'$.

Corollary. Let B be a finite Boolean algebra. Then $(B, +, \cdot, ', 0, 1, \leq)$ is isomorphic to $(P(W), \cup, \cap, ', \emptyset, W, \subseteq)$.

Proof. As in the previous corollary, define $f : B \to P(W)$ by $f(x) = \text{Atom}[x]$. By Theorem 13 and the previous corollary, we only need to prove that f is onto. Accordingly, let $Y \subseteq W$ be a finite collection of atoms and y be the sum of the atoms in Y. We leave it as an exercise to show that $\text{Atom}[y] = Y$ (see Exercise 25). Since W is finite, it then follows that for every set $Y \in P(W)$, $Y = \text{Atom}[y]$ for some $y \in B$. Therefore, f is onto. $\qquad\square$

The following proposition was proved in Chapter 6.

Proposition. Let $\text{Card}(A) = m$ and $\text{Card}(B) = n$. If there exists a bijective function $f : A \to B$, then $m = n$.

Corollary. Let $\text{Card}(W) = k$ in the finite Boolean algebra B; that is, B has exactly k atoms. Then $\text{Card}(B) = 2^k$.

Proof. By the second corollary to Theorem 14, there is a bijection from a finite Boolean algebra to the power set of W. But $\text{Card}(P(W)) = 2^k$. Therefore, $\text{Card}(B) = 2^k$, by the previous proposition. $\qquad\square$

As an example of how to apply the preceding corollary, consider $D_{36} = \{1, 2, 3, 4, 6, 9, 12, 18, 36\}$. Is it possible to form a Boolean algebra using $B = D_{36}$? The answer is no, since the cardinality of D_{36} is 9, and $9 \neq 2^n$ for any n.

Exercise Set 7.2   **217**

EXERCISE SET 7.2

1. Prove Theorem 1.
2. Prove Theorem 2.
3. Prove Theorem 3.
4. *Prove:* If $z \leq y$ and $z \leq y'$, then $z = 0$.
5. *Prove:* If $1 \leq y$, then $y = 1$.
6. *Prove:* If $x \leq y$, then $y' \leq x'$.
7. Prove that $x \leq y$ if and only if $x' + y = 1$.
8. Prove Theorem 4.
9. Prove Theorem 5.
10. Prove Theorem 6.
11. Prove Theorem 7.
12. Here is another proof that x' is unique. Fill in the missing steps of the proof. Let $y \cdot x = 0$ and $y + x = 1$.
    (a) $y = y \cdot 1 = y \cdot (x + x') = \ldots = y \cdot x'$. Hence, $y = y \cdot x'$, and therefore, $y \leq x'$.
    (b) $y = y + 0 = y + x \cdot x' = \ldots = y + x'$. Hence, $y = y + x'$, and therefore, $x' \leq y$. It follows that $y = x'$.
13. Prove Theorem 8.
14. Prove Theorem 10.
15. Prove Theorem 11.
16. Prove Theorem 12.
17. Finish the proof of Theorem 14.
18. For each of the following sets B, decide whether $(B, +, \cdot, 0, 1)$ is a Boolean algebra. In each case, justify your answer.
    (a.) $B = D_{25}$
    (b.) $B = D_{26}$
19. For each of the following sets B, decide whether $(B, +, \cdot, 0, 1)$ is a Boolean algebra. In each case, justify your answer.
    (a) $B = D_{32}$
    (b) $B = D_{31}$
20. For the Boolean algebra $(D_{42}, +, \cdot, ', 1, 42)$, find $P(W)$ and specify the bijective function given in the second corollary to Theorem 14.
21. For the Boolean algebra $(D_{105}, +, \cdot, ', 1, 105)$, find $P(W)$ and specify the bijective function given in the second corollary to Theorem 14.
22. For the Boolean algebra $(P(U), \cup, \cap, ', \emptyset, U)$, verify that $W = \big\{ \{x\} : x \in U \big\}$ and specify the 1–1 function given in the first corollary to Theorem 14.
23. As we have seen, for certain positive integers n, we can form a Boolean algebra $(D_n, +, \cdot, ', 1, n)$, while for others (for example, $n = 36$), we cannot. Here, as usual, we are assuming that $+$ is the lcm, $\cdot$ is the gcd, and $x' = n/x$. Prove that $(D_n, +, \cdot, ', 1, n)$ is a Boolean algebra if and only if there is no integer $m > 1$ such that m^2 divides n.
24. Let a_1 and a_2 be atoms. Prove that $a_1 \neq a_2$ if and only if $a_1 \cdot a_2 = 0$.

25. Let $Y \subseteq W$ be a finite collection of atoms and y be the sum of the atoms in Y. Prove that Atom$[y] = Y$.

26. This exercise defines an atomistic Boolean algebra that is not finite. Let S be the set of all subsets A of $\mathcal{N}$, where A is finite or A' is finite.
    (a) Prove that $(S, \cup, \cap, ', \emptyset, \mathcal{N})$ is a Boolean algebra. (*Hint:* You need only verify that $\cup$ and $\cap$ are binary operations on S and $'$ is a unary operation on S.)
    (b) Prove that the Boolean algebra in part a is atomistic.

27. For the atomistic Boolean algebra defined in Exercise 26:
    (a) Find $P(W)$, and specify the 1–1 function given in the first corollary to Theorem 14.
    (b) Show that the 1–1 function specified in part a is not bijective.

KEY CONCEPTS

Boolean algebra	Boolean algebra laws
Complement	Partial ordering for a Boolean algebra
Dual	Atom
Principle of Duality	Atom$[x]$
Zero element and one element	Atomistic Boolean algebra
Axioms for Boolean algebra	Finite Boolean algebra

REVIEW EXERCISES

1. Let $B = \{0, 1\}$. Define the binary operations $+$ and $\cdot$ by $x + y = \max(x, y)$ and $x \cdot y = \min(x, y)$, respectively. Define the unary operation $'$ by $0' = 1$ and $1' = 0$. It can be shown that $(B, +, \cdot, ', 0, 1)$ is a Boolean algebra. Verify that axioms 3a and 4b for a Boolean algebra hold by checking all possible cases. [Max(x, y) is the maximum value of the elements in the set $\{x, y\}$, for example, $\max(0, 1) = 1$; $\min(0, 1) = 0$.]

2. Give an example of a Boolean algebra $(B, +, \cdot, ', 0, 1)$, where B has exactly eight elements. Specify B, $+$, $\cdot$, $'$, 0, and 1 for this Boolean algebra.

3. Can you give an example of a Boolean algebra $(B, +, \cdot, ', 0, 1)$, where B has exactly 10 elements? Explain.

4. Let $(B, +, \cdot, ', 0, 1)$ be a Boolean algebra and $x, y, z \in B$. Prove or disprove the following conjecture: If $x \neq 0$ and $x \cdot y = x \cdot z$, then $y = z$.

5. Let $(B, +, \cdot, ', 0, 1)$ be a Boolean algebra and $x, y, z \in B$. Prove or disprove the following conjecture: If $x \neq 1$ and $x + y = x + z$, then $y = z$.

6. Define a relation R on a finite Boolean algebra by xRy if and only if Card(Atom$[x]$) = Card(Atom$[y]$).
    (a) Prove that R is an equivalence relation.
    (b) Specify the equivalence classes with respect to R for the Boolean algebra $(D_{30}, \text{lcm}, \text{gcd}, ', 1, 30)$, where $x' = 30/x$.

7. Use Boolean algebra laws to simplify $x(x' + y) + z + y$.

8. Use Boolean algebra laws to simplify $(x + xy' + z')' + xz$.

9. Use Boolean algebra laws to prove that $x'w + x'y' + yz' + x'z = x' + yz'$.

10. Use Boolean algebra laws to prove that $(x'yw + x'z'w + x'y'w' + yz' + x'z)' = x(y' + z)$. (*Hint:* You may wish to use the result of Exercise 9.)

11. Find an atomistic Boolean algebra that is not finite and which is not the same as the Boolean algebra described in Exercise 26 of Section 7.2. (*Hint:* Consider Exercise 22 of Section 7.2.)

12. For $x, y \in B$, define $x \ominus y = x'y + xy'$. Recall that $A \triangle B = (A - B) \cup (B - A)$. Prove that $\text{Atom}[x \ominus y] = \text{Atom}[x] \triangle \text{Atom}[y]$.

13. Let that B be a finite Boolean algebra. Prove that every element x in B is the sum of atoms in $\text{Atom}[x]$.

14. In this exercise do not use the last corollary in Section 7.2 or any equivalent statement. Let B be a finite Boolean algebra.
    (a) *Prove:* If $0 = 1$, then $\text{Card}(B) = 1$.
    (b) *Prove:* If $x = x'$ for some $x \in B$, then $\text{Card}(B) = 1$.
    (c) *Prove:* If $\text{Card}(B) > 1$, then $\text{Card}(B)$ is even.

SUPPLEMENTARY EXERCISES

We have seen that every atomistic Boolean algebra can be embedded by a 1–1 function into a collection of sets. This is, in fact, true for any Boolean algebra. To prove this, we need the concept of a filter on a Boolean algebra B. F is a **filter** on a Boolean algebra B if $F \subseteq B$ and F satisfies the following conditions:

a. $F \neq \emptyset$.

b. If $x \in F$ and $y \in F$, then $x \cdot y \in F$.

c. If $x \in F$ and $y \in B$, then $x + y \in F$.

1. Let $H \subseteq X$ and $F = \{A : H \subseteq A \text{ and } A \subseteq X\}$. Show that F is a filter on the Boolean algebra $P(X)$.

2. Show that $F \subseteq B$ is a filter on a Boolean algebra B if and only if
    (a) $F \neq \emptyset$.
    (b) If $x \in F$ and $y \in F$, then $x \cdot y \in F$.
    (c) If $x \in F$, $y \in B$, and $x \leq y$, then $\in F$.

3. Let F be a filter on a Boolean algebra B. Prove that $F = B$ if and only if $0 \in F$.

U is an **ultrafilter** on a Boolean algebra B if U is a filter on B, $0 \notin U$, and for every $x \in B$, $x \in F$ or $x' \in F$.

4. Let $y \in X$ and $U = \{A : y \in A \text{ and } A \subseteq X\}$. Prove that U is an ultrafilter on the Boolean algebra $P(X)$.

In Exercises 5–7, let U be an ultrafilter on a Boolean algebra B, and suppose $x \in B$ and $y \in B$.

5. Prove that $x \cdot y \in U$ if and only if $x \in U$ and $y \in U$. (Here we only need that U is a filter.)

6. *Prove:* $x + y \in U$ if and only if $x \in U$ or $y \in U$.

7. *Prove:* $x' \in U$ if and only if $x \notin U$.

In the rest of the exercises, we use the ultrafilter theorem: Let F be a filter on a Boolean algebra B such that $F \neq B$. Then there is an ultrafilter U on B such that $F \subseteq U$. The proof of this theorem is beyond the scope of the text.

8. Prove that $F = \{t : t \in B$ and $x \leq t\}$ is a filter on B and that $x \in F$.

9. *Prove:* If $x \neq 0$, then there is an ultrafilter U on B such that $x \in U$.

Define Filter$[x] = \{U : U$ is an ultrafilter on B and $x \in U\}$; that is, Filter$[x]$ is the set of ultrafilters on B which contain the element x. Let $V = \{U : U$ is an ultrafilter on $B\}$; that is, V is the set of all ultrafilters on B.

10. *Prove:* Filter$[x + y] = $ Filter$[x] \cup$ Filter$[y]$.

11. *Prove:* Filter$[x \cdot y] = $ Filter$[x] \cap$ Filter$[y]$.

12. *Prove:* Filter$[x'] = V -$ Filter$[x]$.

13. *Prove:* Filter$[1] = V$.

14. *Prove:* Filter$[x] = \emptyset$ if and only if $x = 0$.

15. *Prove:* If Filter$[x] = $ Filter$[y]$, then $x = y$.

Hence, as in the case of atomistic Boolean algebras, we can define a 1–1 function $f : B \rightarrow P(V)$ by $f(x) = $ Filter$[x]$. Using Exercises 10–12, it is then easy to prove that f preserves operations, and, by Exercises 13 and 14, $f(0) = \emptyset$ and $f(1) = V$. Hence, f is an embedding. This tells us that every Boolean algebra can be embedded into a collection of sets.

8

The Integers

Many of the concepts and theorems of abstract algebra depend on facts concerning the arithmetic of the integers. Furthermore, the algebraic structures studied in abstract algebra can be better understood after the integers and their underlying structure are well understood. Even some of the techniques used in abstract algebra are similar to those we will be using to prove theorems concerning the integers.

Our main goal in this chapter is to describe the system of integers. In order to do that, we first list axioms for an algebraic system called an integral domain. The integers form an integral domain. We then use the axioms of an integral domain to deduce properties of integral domains in general and the integers in particular.

Many other algebras having very little resemblance to the integers are also integral domains. To characterize the integers, we will add some axioms to the integral domain axioms and then prove that any algebra satisfying the combined set of axioms is isomorphic to the integers. The last three sections of the chapter are an introduction to—perhaps the oldest branch of mathematics—number theory. Chapter 8 might be subtitled: "A Prerequisite to Abstract Algebra."

8.1 INTEGRAL DOMAINS

Definition.   An **integral domain** $(D, +, \cdot, , 0, 1)$ consists of a nonempty set D, binary operations $+$ and $\cdot$ on D, and distinguished elements 0 and 1 in D such that the following axioms hold for all x, y, and z in D:

1a. $x + y = y + x$ 1b. $x \cdot y = y \cdot x$ (Commutative laws)

2a. $(x + y) + z = x + (y + z)$ 2b. $(x \cdot y) \cdot z = x \cdot (y \cdot z)$ (Associative laws)

3. $x \cdot (y + z) = x \cdot y + x \cdot z$ (Distributive law)

4a. $x + 0 = x$

4b. $x \cdot 1 = x$

5.   For every element $x \in D$, there exists an element $w \in D$ such that $x + w = 0$.

                                                      (Additive inverse law)

6.   If $x \cdot y = 0$, then $x = 0$ or $y = 0$.

The elements 0 and 1 are called the **zero** and **one** elements, or the **additive identity** and **multiplicative identity,** respectively. Theorem 2 states that these elements are unique. It is an exercise to prove that $x \cdot 0 = 0$ for all x (see Exercise 4).

If D contains at least two elements, then $0 \neq 1$. For if $0 = 1$, then for all x, $x = x \cdot 1 = x \cdot 0 = 0$.

From now on, we write xy for $x \cdot y$.

The set of integers $\mathcal{Z}$ with the standard addition and multiplication is an integral domain. Both $\mathcal{Q}$ and $\mathcal{R}$ are also integral domains, as is the algebra Z_3 (that is, the algebra $(Z_3, +_3, *_3, [0], [1])$ defined in Section 5.3). In Section 8.5, Theorem 4 states that Z_d is an integral domain if and only if d is a prime.

Axioms 1–6 are sufficient for deriving many of the algebraic properties of an integral domain.

Theorem 1   If $x + y = x + z$, then $y = z$.

Proof. Assume that $x + y = x + z$. Then, by Axiom 5, there is a w such that $x + w = 0$. Now, $w + (x + y) = w + (x + z)$. By Axiom 2a, $(w + x) + y = (w + x) + z$. So $0 + y = 0 + z$, and by Axioms 1a and 4a, $y = z$.   □

Theorem 2

   a. 0 is unique; that is, if $x + u = x$ for all x, then $u = 0$.

   b. 1 is unique; that is, if $xv = x$ for all x, then $v = 1$.

   c. Additive inverses are unique; that is, for a given x, if $x + w = 0$ and $x + u = 0$, then $w = u$.

Proof. We prove part b. Suppose there is an element u such that $xu = x$ for all x. Then, in particular, $1u = 1$. But $u1 = u$ by Axiom 4b. Hence, $u = u1 = 1u = 1$.   □

By Theorem 2c, the **additive inverse** of x is well defined. We write the additive inverse of x as $-x$. We then have $x + (-x) = 0$.

The properties for an integral domain allow us to solve equations such as $x + a = b$. From elementary algebra, the solution is $x = b - a$. Formally, we can justify this solution by using Axiom 5 as follows. Assume $x + a = b$. Then $(x + a) + (-a) = b + (-a)$. By Axioms 2a, 4a, and 5, we then obtain $x = b + (-a)$. For simplicity, we write $b - a$ for $b + (-a)$. This defines another binary operation, **subtraction,** for an integral domain.

So the solution of $x + a = b$ and the definition of subtraction both depend on the existence of an additive inverse for elements of an integral domain.

Besides their algebraic properties, the integers satisfy order properties. So to completely characterize the integers by a list of axioms, the axioms will have to include those for a partial order. We need three more axioms to define an ordered integral domain.

An **ordered integral domain** is an integral domain D with a subset D^+ such that the following axioms are satisfied:

7. The subset D^+ is closed under addition on D.
8. The subset D^+ is closed under multiplication on D.
9. For all $x \in D$, exactly one of the following holds:

$$x = 0, x \in D^+, \text{ or } -x \in D^+$$

For example, when the integral domain is the rational numbers, D^+ is the set of positive rational numbers.

Linear Ordering on D

We define a partial ordering on an ordered integral domain as follows.

Definition.   For $x, y \in D$, $x \leq y$ if $y = x$ or $y - x \in D^+$.

To prove Theorems 3–6 following, you may use the results of Exercises 2–7.

Theorem 3   The relation $\leq$ is a linear ordering on D.

As usual, $x < y$ stands for $x \leq y$ and $x \neq y$. When $x < y$, we say that x is **strictly less than** y. Note that $0 < x$ if and only if $x \in D^+$.

Corollary.   (Trichotomy Law) For $x, y \in D$, exactly one of the following holds: (1) $x < y$; (2) $x = y$; (3) $y < x$.

By the trichotomy law, for every element x in an ordered integral domain, either $x < 0$ or $x = 0$ or $0 < x$. If $x > 0$, we say that x is **positive;** if $x < 0$, we say that x is **negative.** Note that $x \in D^+$ if and only if x is positive. An element x in $\mathcal{Z}$ is said to be a **positive integer** if $x > 0$; x is a **negative integer** if $x < 0$.

Theorem 4
a. $x \leq y$ if and only if $x + w \leq y + w$.
b. For $w > 0$, $x \leq y$ if and only if $xw \leq yw$.

Theorem 5   For $w < 0$, $x \leq y$ if and only if $yw \leq xw$.

Corollary.   If $x \neq 0$, then xx is positive.

Proof. Assume $x \neq 0$. If $x > 0$, we use Theorem 4b to conclude that $xx > 0$. If $x < 0$, we conclude that $xx > 0$ by Theorem 5. $\square$

Since $(1)(1) = 1$ and $1 \neq 0$, 1 is positive. Moreover, $1 + 1$ is positive, $1 + 1 + 1$ is positive, etc. To make these ideas more precise, we need a definition and some notation.

Definition.   Let $a \in D$.
a. $0 \cdot a$ is the additive identity in the integral domain, where $0 \in \mathcal{N}$.
b. $(n + 1) \cdot a = n \cdot a + a$, where $n \in \mathcal{N}$.
c. If m is a negative integer, then $m \cdot a = (-m) \cdot (-a)$.

In particular, if 1 is the multiplicative identity in an integral domain, then $n \cdot 1$ designates $1 + 1 + \ldots + 1$ (n times) when n is a positive integer. For $0 \in \mathcal{N}, 0 \cdot 1$ designates the additive identity in the integral domain; if n is a negative integer, $n \cdot 1$ designates $(-1) + (-1) + \ldots + (-1)$ (n times).

A note of caution: The same symbols are used to denote the additive and multiplicative identities in $\mathcal{Z}$ as are used in an integral domain D. Also, the same notation is used for the respective binary operations. Moreover, the multiplication sign used in the preceding definition is not a binary operation in either $\mathcal{Z}$ or an arbitrary integral domain. For this reason, we omit the multiplication sign for any integral domain and write xy or mn. Confusion can easily be avoided by noting the context.

It is an exercise to prove that $n \cdot 1$ is positive for all positive integers n (see Exercise 13). In other words, $\{n \cdot 1 : n \text{ a positive integer}\} \subseteq D^+$. Moreover, $(n + 1) \cdot 1 > n \cdot 1$. Similarly, $n \cdot 1$ is negative for all negative integers n.

Theorem 6   Let $m, n \in \mathcal{Z}$ and $a, b \in D$. Then
a. $m \cdot a + n \cdot a = (m + n) \cdot a$ b. $m \cdot (a + b) = m \cdot a + m \cdot b$
c. $m \cdot (n \cdot a) = (mn) \cdot a$ d. $m \cdot (ab) = (m \cdot a)b = a(m \cdot b)$
e. $(m \cdot a)(n \cdot b) = (mn) \cdot (ab)$

Proof. We prove part a. Let $m \in \mathcal{Z}$ be fixed. We prove that $m \cdot a + n \cdot a = (m + n) \cdot a$ for all $n \in \mathcal{N}$ by mathematical induction. For the basis step, note that $m \cdot a + 0 \cdot a = m \cdot a + 0 = m \cdot a = (m + 0) \cdot a$.

Assume $m \cdot a + k \cdot a = (m + k) \cdot a$. Then $m \cdot a + (k + 1) \cdot a = m \cdot a + (k \cdot a + a) = (m \cdot a + k \cdot a) + a = (m + k) \cdot a + a = [(m + k) + 1] \cdot a = [m + (k + 1)] \cdot a$.

For n a negative integer, we consider two cases. *Case 1.* Assume $m \in \mathcal{N}$. Then $n \cdot a + m \cdot a = (n + m) \cdot a$ by what was just proved. Hence, $m \cdot a + n \cdot a = (m + n) \cdot a$, by the commutative law for addition in an integral domain.

Case 2. Assume m is negative. Then both $-m$ and $-n$ are positive integers. Hence, $(-n) \cdot (-a) + (-m) \cdot (-a) = [(-n) + (-m)] \cdot (-a)$, by what was just proved. Therefore, $n \cdot a + m \cdot a = (n + m) \cdot a$. $\square$

Since the integers, the rational numbers, and the real numbers are all ordered integral domains, Axioms 1–9 are not quite enough to characterize the integers. We need one more axiom.

A **well-ordered integral domain** is an ordered integral domain D such that

10. Every nonempty subset of D^+ has a least element.

Axiom 10 is called the **well-ordering property** for D^+. So an ordered integral domain is a well-ordered integral domain if D^+ satisfies the well-ordering property.

Theorem 7   There is no element x in a well-ordered integral domain such that $0 < x < 1$.

Proof. Suppose $\{x : 0 < x < 1\}$ is nonempty. Then by the well-ordering property for D^+, $\{x : 0 < x < 1\}$ has a least element u. Hence, $0 < u < 1$, and by Theorem 4b, $0 < uu < u < 1$. But this contradicts the fact that u is least. $\square$

Note that by Theorem 7, it is easy to see that neither the rational numbers nor the real numbers is a well-ordered integral domain.

Lemma   If D is a well-ordered integral domain, then:
a. $\{n \cdot 1 : n$ is a positive integer$\} = D^+$
b. $\{n \cdot 1 : n$ is an integer$\} = D$.

Proof. a. By the remarks preceding Theorem 6, we only need to prove that $D^+ \subseteq \{n \cdot 1 : n$ is a positive integer$\}$. We do a proof by contradiction. Suppose $D^+ - \{n \cdot 1 : n$ is a positive integer$\} \neq \emptyset$. Let v be the least element of $D^+ - \{n \cdot 1 : n$ a positive integer$\}$. By Theorem 7, 1 is the least element of D^+. Since $1 \in \{n \cdot 1 : n$ a positive integer$\}$, $1 < v$. Hence, $v - 1$ is positive and $v - 1 < v$. So $v - 1 \notin D^+ - \{n \cdot 1 : n$ a positive integer$\}$. In other words, $v - 1 = m \cdot 1$ for some positive integer m. But then, $v = m \cdot 1 + 1 = (m + 1) \cdot 1$, and we have a contradiction.
The proof of part b is an exercise. $\square$

We are now in a position to prove that every well-ordered integral domain "looks like" $\mathcal{Z}$.

Theorem 8   If D is a well-ordered integral domain, then there is an order-preserving isomorphism $h : \mathcal{Z} \to D$.

Proof. Define h by $h(m) = m \cdot 1$. Then:
1. h is a function with domain $\mathcal{Z}$, by the definition of $m \cdot 1$.
2. h is onto by the previous lemma.
3. h is 1–1. (Assume $h(m) = h(n)$ and $m \neq n$. If $n > m$, then $n - m$ is a positive integer, and $(n - m) \cdot 1 = n \cdot 1 - m \cdot 1 = h(n) - h(m) = 0$. This is a contradiction, since $0 \notin D^+$. If $m > n$, we obtain the same contradiction.)
4. $h(m + n) = (m + n) \cdot 1 = m \cdot 1 + n \cdot 1$ and $h(mn) = (mn) \cdot 1 = (m \cdot 1)(n \cdot 1)$.
5. If $m < n$, then $n - m$ is a positive integer and $h(n - m) = (n - m) \cdot 1$ is an element of D^+. Also, $h(n) - h(m) = h(n - m)$. Hence, $h(m) < h(n)$. $\square$

EXERCISE SET 8.1

1. Finish the proof of Theorem 2.

In Exercises 2–8 use Axioms 1–6 for an integral domain D to prove the given statements.

2. $-0 = 0$

3. $-(-x) = x$

4. $x(0) = 0$. (*Hint:* $x(0) = x(0 + 0)$.)

5. **Cancellation law:** If $ax = ay$ and $a \neq 0$, then $x = y$. (*Hint:* First show that $ax = ay$ implies $a(x - y) = 0$.)

6. $x(y - z) = xz - xy$

7. $(-x)y = x(-y) = -(xy)$

8. $(-x)(-y) = xy$

9. Prove Theorem 3.

10. Prove the trichotomy law.

11. Prove Theorem 4.

12. Prove Theorem 5.

13. Assume that 1 is the multiplicative identity in D. Use mathematical induction to prove that $n \cdot 1$ is positive for all positive integers n.

14. Assume that 1 is the multiplicative identity in D. Prove that $n \cdot 1$ is negative for all negative integers n.

15. Prove Theorem 6b.

16. Prove Theorem 6c.

17. Prove Theorem 6d.

18. Prove Theorem 6e.

19. Prove part b of the lemma preceding Theorem 8.

8.2 DIVISIBILITY IN THE INTEGERS

All the axioms for a well-ordered integral domain and all the theorems we proved from those axioms are properties of the integers. In this section, we concentrate on the integral domain $\mathcal{Z}$.

Because there is no multiplicative inverse for every element of $\mathcal{Z}$, it is not always possible to solve an equation like $x \cdot a = b$ using the properties of $\mathcal{Z}$. In other words, there is no binary operation called division for $\mathcal{Z}$. Of course, for special cases, such as $a = 6$ and $d = 3$, we can certainly say that d "divides evenly into" a. But 3 is a factor of 6 (or, 6 is a multiple of 3). In general, it is not possible to take just any pair of elements $a, d \in \mathcal{Z}$ and obtain a/d in $\mathcal{Z}$.

Definition. Let $a, d \in \mathcal{Z}$, with $d \neq 0$. Then the integer d **divides** a if there exists an $m \in \mathcal{Z}$ such that $a = m \cdot d$.

The integer a is a **multiple** of d when d divides a. When d divides a, we write $d|a$. If $d|a$, d is called a **factor** of a or a **divisor** of a. Note that since we wish to discuss divisibility in the context of the integers, writing a/d or $a \div d$ is never allowed. For example, since $6 = 2 \cdot 3$, we may write $2|6$ or $3|6$, but we do not write $6/2$ or $6/3$.

Theorem 1   If $xy = 1$, then $x = y = 1$ or $x = y = -1$.

Corollary.   The integer 1 has exactly one positive factor, namely, itself.

Theorem 2   Let $a, d \in \mathcal{Z}$, with $d \neq 0$. Then:
a. $d|0$ b. $1|a$ c. $d|d$

Proof. Part a. $0 = 0 \cdot d$. Therefore, $d|0$. The proofs of parts b and c are left as exercises. □

Theorem 3   Let $a, b, d \in \mathcal{Z}$, with $d \neq 0$. Then:
a. If $d|a$ and $d|b$, then $d|(a + b)$.
b. If $d|a$, then $d|a \cdot b$.

Proof. Part b. Assume $d|a$. Then $a = m \cdot d$ for some $m \in \mathcal{Z}$. Hence, $a \cdot b = (m \cdot d) \cdot b = (m \cdot b) \cdot d$. Therefore, $d|a \cdot b$, since $m \cdot b \in \mathcal{Z}$. □

Theorem 4   Let $b, d, e \in \mathcal{Z}$, with $d \neq 0$ and $e \neq 0$. Then:
a. If $d|e$ and $e|b$, then $d|b$.
b. If $d|e$ and $e|d$, then $d = e$ or $d = -e$.

Proof. Part b. Assume $d|e$ and $e|d$. Then $e = md$ and $d = ne$ for some integers m and n. Hence, $e = mne$, and we have $mn = 1$. So $m|1$, and therefore, $m = 1$ or $m = -1$. If $m = 1$, then $d = e$. If $m = -1$, then $d = -e$. □

As usual, $|x|$ designates the absolute value of x.

Theorem 5   Let $d, e \in \mathcal{Z}$, with $d \neq 0$ and $e \neq 0$. If $d|e$, then $|d| \leq |e|$.

Theorem 6   (**Division Algorithm** for $\mathcal{Z}$) Let $a, d \in \mathcal{Z}$, with $d \neq 0$. Then there exist unique elements $q, r \in \mathcal{Z}$ such that $a = q \cdot d + r$ and $0 \leq r < |d|$, where $|d|$ is the absolute value of d.
Outline of the proof: Let $a, d \in \mathcal{Z}$ with $d \neq 0$, and consider the set $A = \{a - x \cdot d : x \in \mathcal{Z}$ and $a - x \cdot d \in \mathcal{N}\}$. By the well-ordering principle, A has a least element r. Since $r \in A$, $r = a - q \cdot d$ for some $q \in \mathcal{Z}$.
1. Prove that $0 \leq r < |d|$.
2. Prove that r and q are unique. (*Hint:* Assume there exist integers r' and q' such that $a = q' \cdot d + r'$ and $0 \leq r' < |d|$, and show that $r' = r$ and $q' = q$. You may use the results of Exercises 15 and 16.) □

Note that no divisor d can be zero. After applying the division algorithm to integers a and d, we say that a is **divided by** d. The elements q and r are respectively called the **quotient** and **remainder** of the division.

The division algorithm is a formal statement of what we ordinarily call division. For example, if we use division to divide 72 by 5, we get a quotient of 14 and a remainder of 2. That is, $72 = 14 \cdot 5 + 2$. Here we have $a = 72, d = 5$, $q = 14$, and $r = 2$. It is also clear, using the notation of the division algorithm, that $d|a$ if and only if $r = 0$.

EXERCISE SET 8.2

1. Prove Theorem 1.

2. Prove the corollary to Theorem 1.

3.   (a) Prove Theorem 2b.
      (b) Prove Theorem 2c.

4. Prove Theorem 3a.

5. Prove Theorem 4a.

6. Prove Theorem 5.

7. Use the division algorithm to find q and r such that $a = q \cdot d + r$ and $0 \leq r < |d|$ for each pair of integers a, d.
    (a) $a = 17, d = 23$
    (b) $a = 23, d = -3$

8. Use the division algorithm to find q and r such that $a = q \cdot d + r$ and $0 \leq r < |d|$ for each pair of integers a, d.
    (a) $a = 67, d = 4$
    (b) $a = -23, d = 5$

9. Use mathematical induction to prove that $n^2 + n$ is divisible by 2 for $n = 0, 1, 2, \ldots$.

10. Use mathematical induction to prove that $n^3 + 2 \cdot n$ is divisible by 3 for $n = 0, 1, 2, \ldots$.

11. Use mathematical induction to prove that $n^4 + 3 \cdot n^2$ is divisible by 4 for $n = 0, 1, 2, \ldots$.

12. Prove: $3|(7^n - 4^n)$ for $n = 0, 1, 2, \ldots$.

We define **exponentiation** on $\mathcal{Z}$, by recursion, as follows: $x^0 = 1$ and $x^{n+1} = x^n \cdot x$ for $n \in \mathcal{N}$.

13. Let x and y be integers. Prove that for all natural numbers n and m:
    (a) $x^n x^m = x^{n+m}$
    (b) $(x^n)^m = x^{nm}$
    (c) $(xy)^n = x^n y^n$

14. Let x and y be arbitrary unequal integers. Prove that $(x - y)|(x^n - y^n)$ for $n = 0, 1, 2, \ldots$.

15. *Prove:* If $0 \leq a < d$ and $0 \leq b < d$, then $|b - a| < d$.

16. *Prove:* If $0 \leq a < d$ and $d|a$, then $a = 0$.

17. Finish the proof of the division algorithm.

Exercises 18 and 19 refer to the division algorithm.

18. Let d and k be positive integers. Suppose q is the quotient and r is the remainder when a is divided by d. Prove that q is the quotient and kr is the remainder when ka is divided by kd.

19. Let d and k be positive integers. Suppose q is the quotient when a is divided by d and s is the quotient when q is divided by k. Prove that s is the quotient when a is divided by kd.

8.3 THE EUCLIDEAN ALGORITHM AND THE GREATEST COMMON DIVISOR

Definition.   Let a, b, and g be positive integers. A **greatest common divisor** g of a and b satisfies the following properties:
(a) $g|a$ and $g|b$;   (b) For all m, if $m|a$ and $m|b$, then $m|g$

Property (a) states that g is a **common divisor** of a and b. Property (b) states that a greatest common divisor of a and b is divisible by every other common divisor. In the definition of g, property (b) makes g the *greatest* common divisor because $m|g$ implies $m \le g$.

For small integers a and b, such as 18 and 48, or 18 and 496, it is easy to find a greatest common divisor.

Example 1

a. 6 is a greatest common divisor of 18 and 48.

b. 2 is a greatest common divisor of 18 and 496.

For larger integers, such as 2,765 and 145,299, the problem of finding a greatest common divisor appears quite difficult. Moreover, how do we know these integers even have a greatest common divisor? Also, if they do have a greatest common divisor, is it unique? Thus, we see that there are three basic questions concerning greatest common divisors: Given positive integers a and b,

1. Does a greatest common divisor exist?

2. Is there at most one greatest common divisor?

3. How do we find a greatest common divisor?

Theorem 1   (Uniqueness) Given positive integers a and b, there is at most one greatest common divisor. That is, let g and d be positive integers satisfying the following properties:

a. $g|a$ and $g|b$.   b. For all m, if $m|a$ and $m|b$, then $m|g$.
c. $d|a$ and $d|b$.   d. For all m, if $m|a$ and $m|b$, then $m|d$.
Then $g = d$.

Theorem 2   (Existence) Suppose a and b are positive integers. Then there is a common divisor d of a and b of the form $d = ax + by$, where x and y are integers. Furthermore, d is the greatest common divisor of a and b.

Proof. The proof is by strong induction on $n = a + b$. For the basis step, take $n = 2$. Then $a = 1$ and $b = 1$. Choose $x = 1$ and $y = 0$. Then $d = 1$ is the greatest common divisor of a and b.

Induction step: Assume the theorem is true for any positive integers c and f such that $c + f \leq k$. Suppose $a + b = k + 1$. If $a = b$, we choose $x = 1$ and $y = 0$. Then $d = a$ is the greatest common divisor of a and b. Without loss of generality, we may assume $b < a$. Then $a - b$ and b are positive integers, and $(a - b) + b \leq k$. By the induction hypothesis, there are integers x and y such that $d = (a - b)x + by$ is a common divisor of $a - b$ and b. Then $d = ax + b(y - x)$. Since d is a common divisor of $(a - b)$ and b, d divides $(a - b) + b = a$. Hence, d is a common divisor of a and b.

Now, any common divisor g of a and b must also divide $(a - b)$ and b. Hence, $g \mid d$, and it follows that d is the greatest common divisor of a and b. $\square$

By Theorems 1 and 2, the greatest common divisor g of a and b is well defined. We write **gcd**(a, b) for g.

Corollary. Let $g = \gcd(a, b)$. Then there exist integers x and y such that $w = ax + by$ if and only if $g \mid w$.
Hints for the proof:
($\Rightarrow$) This part follows easily from Theorem 3 of Section 8.2.
($\Leftarrow$) This follows from Theorem 2. $\square$

There is a systematic method for finding $\gcd(a, b)$. Let us find $\gcd(18, 496)$ by using the division algorithm repeatedly (the right-hand side of each line explains algebraically what is going on in that line):

$$496 = 18 \cdot 27 + 10 \qquad b = aq_1 + r_1, 0 < r_1 < a$$
$$18 = 10 \cdot 1 + 8 \qquad a = r_1 q_2 + r_2, 0 < r_2 < r_1$$
$$10 = 8 \cdot 1 + 2 \qquad r_1 = r_2 q_3 + r_3, 0 < r_3 < r_2$$
$$8 = 2 \cdot 4 + 0 \qquad r_2 = r_3 q_4 + 0$$

The last nonzero remainder is $\gcd(18, 496)$. Hence, $\gcd(18, 496) = 2$. Note that we first divided 496 by 18. If, instead, we first divided 18 by 496, we would have obtained $18 = 496 \cdot 0 + 18$. So the next step would be to divide 496 by 18, and the remaining steps would be identical to those given. Hence, we need not be concerned whether $a \leq b$ or $b \leq a$.

Let us prove that the method illustrated is an algorithm for finding the greatest common divisor of two positive integers. First we prove a preliminary theorem.

Theorem 3 Let a and d be positive integers. If $a = d \cdot q + r$ and $0 \leq r < d$, then $\gcd(a, d) = \gcd(d, r)$.

Proof. Assume $a = qd + r$ and $0 \leq r < d$. Let $v = \gcd(a, d)$ and $u = \gcd(d, r)$. Then $u \mid a$, since $u \mid d$, $u \mid r$, and $a = qd + r$. Hence, $u \mid v$, since $u \mid d$ and $v = \gcd(a, d)$. Also,

$v|r$, since $v|a$, $v|d$, and $r = a - qd$. So $v|u$, since $v|d$ and $u = \gcd(d, r)$. Therefore, $v = u$, since $u|v$ and $v|u$. $\qquad\square$

Theorem 4 (Euclidean Algorithm) Let a and b be positive integers such that a does not divide b. Apply the division algorithm repeatedly as follows:

$$b = aq_1 + r_1, \qquad\qquad 0 < r_1 < a$$
$$a = r_1 q_2 + r_2, \qquad\qquad 0 < r_2 < r_1$$
$$r_1 = r_2 q_3 + r_3, \qquad\qquad 0 < r_3 < r_2$$
$$\vdots \qquad\qquad\qquad\qquad \vdots$$
$$r_{m-2} = r_{m-1} q_m + r_m, \qquad\qquad 0 < r_m < r_{m-1}$$
$$r_{m-1} = r_m q_{m+1}$$

Then $\gcd(a, b) = r_m$.

Proof. Since $a > r_1 > r_2 > \dots$, the remainder must eventually be zero; say, $r_{m+1} = 0$. Since r_m is a factor of r_{m-1}, $\gcd(r_{m-1}, r_m) = r_m$. So by repeated use of Theorem 2, $\gcd(a, b) = \gcd(a, r_1) = \gcd(r_1, r_2) = \dots = \gcd(r_{m-1}, r_m) = r_m$. $\qquad\square$

Note that if $a|b$, then the Euclidean algorithm does not apply. However, it is easy to verify that $a|b$ implies that $\gcd(a, b) = a$.

Example 2

Let $a = 46$ and $b = 64$.
a. Apply the Euclidean algorithm to find $\gcd(46, 64)$.

Solution

$$64 = 46 \cdot 1 + 18$$
$$46 = 18 \cdot 2 + 10$$
$$18 = 10 \cdot 1 + 8$$
$$10 = 8 \cdot 1 + 2$$
$$8 = 2 \cdot 4$$

Hence, $\gcd(46, 64) = 2$.
b. Use the foregoing work to find integers m and n such that $2 = 46 \cdot m + 64 \cdot n$.

Solution By part a, we see that

$$18 = 64 - 46 \cdot 1$$
$$10 = 46 - 18 \cdot 2$$
$$8 = 18 - 10 \cdot 1$$
$$2 = 10 - 8 \cdot 1$$

Working backwards and substituting, we obtain

$$2 = 10 - (18 - 10 \cdot 1) \cdot 1 = 2 \cdot 10 - 18 = 2 \cdot (46 - 18 \cdot 2) - 18$$
$$= 2 \cdot 46 - 5 \cdot 18 = 2 \cdot 46 - 5 \cdot (64 - 46 \cdot 1) = 7 \cdot 46 + (-5) \cdot 64.$$

Hence, $m = 7$ and $n = -5$ are two possible values.

EXERCISE SET 8.3

1. Prove Theorem 1.

2. Find the gcd(48, 18) by using the Euclidean algorithm.

3. Prove that if a and b are positive integers such that $a|b$, then $\gcd(a, b) = a$.

4. Prove the corollary to Theorem 2.

In Exercises 5–9, use the Euclidean algorithm to find $\gcd(a, b)$, and then find m, n such that $ma + nb = \gcd(a, b)$.

5. $a = 155, b = 20$

6. $a = 29, b = 11$

7. $a = 32, b = 76$

8. $a = 93, b = 496$

9. $a = 2,765, b = 145,299$

10. Use a theorem from this section to justify your answer to each of the following questions.
    (a) Can you find integers x, y such that $35x + 49y = 3$?
    (b) Can you find integers x, y such that $35x + 49y = 21$?

11. Use a theorem from this section to justify your answer to each of the following questions.
    (a) Can you find integers x and y such that $42x + 66y = 3$?
    (b) Can you find integers x and y such that $47x + 11y = 1$?

12. (a) Find gcd(10, 33) and m, n such that $10m + 33n = \gcd(10, 33)$.
    (b) Suppose you have two unmarked beakers that will hold exactly 10 cc and 33 cc, respectively. Describe how you would measure out exactly 1 cc of liquid using only the two beakers.

13. Suppose you have two unmarked beakers that will hold exactly 10 cc and 48 cc, respectively. Can you measure out exactly 1 cc of liquid using only the two beakers? Use a theorem from this section to justify your answer.

14. Suppose all the common divisors of a and b are also common divisors of c and d, and vice versa. Prove that $\gcd(a, b) = \gcd(c, d)$.

15. *Prove:* $\gcd(a, b) = 1$ if and only if there exist integers m and n such that $1 = am + bn$.

16. *Prove:* If $\gcd(ab, c) = 1$, then $\gcd(a, c) = 1$ and $\gcd(b, c) = 1$.

17. Suppose $\gcd(a, b) = d$, and write $a = cd$ and $b = fd$. Prove that $\gcd(c, f) = 1$.

18. *Prove:* If $\gcd(a, b) = 1$ and $b|ac$, then $b|c$.

19. *Prove:* If $\gcd(a, b) = d$ and $b|ag$, then $b|dg$. (*Hint:* Use Exercises 17 and 18.)

20. *Prove:* If $\gcd(a, b) = 1$ and $ad|c$ and $bd|c$, then $abd|c$.

21. Suppose $\gcd(a, b) = 1$ and $a > b$. Determine the possible values of $\gcd(a + b, a - b)$. Justify your answer.

22. Suppose a is odd and b is even, or vice versa, and $a > b$. Prove that $\gcd(a, b) = \gcd(a + b, a - b)$.

8.4 FACTORIZATION

If p is a positive integer, then 1 and p are factors of p.

Definition. *A positive integer p is a **prime** if p has exactly two positive factors.*

Since 1 has only one positive factor, 1 is not a prime. A **composite** is an integer greater than 1 which is not a prime. If b is a composite, then b has a factor other than 1 and b. Also, if a prime p is a factor of a prime q then $p = q$.

Example 1

The primes between 1 and 20 inclusive are 2, 3, 5, 7, 11, 13, 17, and 19.

If we consider the numbers between 2 and 20 that are composites, we see that all of them have prime factors. This is a special case of the following theorem.

Theorem 1 Every integer $n \geq 2$ has a prime factor.

Proof. The proof is by strong induction (see the supplementary exercises in Chapter 2).

Induction Hypothesis: Assume that every integer less than k (but greater than or equal to 2) has a prime factor.

If k is prime, then it has a prime factor. If k is not prime, then by definition, k has a factor m such that $1 < m < k$. But then, by the induction hypothesis, m has a prime factor which is also a prime factor of k. $\qquad\square$

Corollary. Every integer $n \geq 2$ is the product of primes.

Hint for proof: Use strong induction. $\qquad\square$

From Theorem 1 and the prime factorization theorem (Theorem 4), to be proved shortly, the prime numbers are considered the "building blocks" of the integers. How many primes are there? Euclid proved over 20 centuries ago that the number of primes is infinite. Euclid's theorem was also proved in the review exercises of Chapter 2. The proof depends on the fact that if $p_0, p_1, \ldots, p_k$ are distinct primes, then $1 + p_0 p_1 \cdots p_k$ has a prime factor different from any of $p_0, p_1, \ldots, p_k$. The proof of this is left as an exercise.

The prime factorization theorem is also called the **fundamental theorem of arithmetic.** In order to prove it, we need several additional facts about prime factors.

Theorem 2 Let a be a positive integer and p be a prime. If p is not a factor of a, then there exist $m, n \in \mathcal{Z}$ such that $1 = a \cdot m + p \cdot n$.

Hint for the proof: See the corollary to Theorem 2 in Section 8.3. $\qquad\square$

Theorem 3   Let p be a prime and a, b be positive integers. If p is a factor of $a \cdot b$, then p is a factor of a or p is a factor of b.

Proof. Suppose $p | a \cdot b$ and p is not a factor of b. Then, by Theorem 2, $1 = m \cdot b + n \cdot p$ for some integers m, n. Multiplying both sides by a, we obtain $a = m \cdot (ab) + (na) \cdot p$. It then follows by Theorem 3 in Section 8.2 that $p | a$. □

Note that Theorem 3 is not true if p is not a prime. For example, 6 is a factor of $24 = 3 \cdot 8$, but 6 is not a factor of either 3 or 8.

Corollary.   Let p be a prime and a_i be positive integers for $i = 1, 2, \ldots, n$. If p is a factor of the product $a_1 \cdot a_2 \cdots a_n$, then p is a factor of at least one of the a_i's for $i = 1, 2, \ldots, n$.

Hint for the proof: Use mathematical induction on n. □

We are now ready to prove the important **prime factorization theorem.**

Theorem 4   (Prime Factorization Theorem) Every integer $a > 1$ has a unique factorization into primes. That is, there exist primes $p_1, p_2, \ldots, p_m$ with $p_1 \leq p_2 \leq \cdots \leq p_n$ such that $a = p_1 p_2 \cdots p_n$, and if $a = q_1 q_2 \cdots q_m$, where $q_1, q_2, \ldots, q_m$ are primes with $q_1 \leq q_2 \leq \cdots \leq q_n$, then $n = m$ and $p_1 = q_1, p_2 = q_2, \ldots, p_n = q_m$.

Proof. We assume in this proof that all products of primes are written in nondescending order; that is, in the product $p_1 p_2 \cdots p_n, p_1 \leq p_2 \leq \cdots \leq p_n$. By the corollary to Theorem 1, a is a product of primes. To show uniqueness, let p_i and q_j be prime. We prove, by induction on n, that if $p_1 p_2 \cdots p_n = q_1 q_2 \cdots q_m$, then $n = m$ and $p_1 = q_1, p_2 = q_2, \ldots, p_n = q_m$.

Basis step: Assume that $p_1 = q_1 q_2 \cdots q_m$. By the corollary to Theorem 3, $p_1 = q_i$ for some i. So $p_1 \leq q_m$ since $q_i \leq q_m$. Clearly, $q_m \leq p_1$, and hence, $p_1 = q_m$. Therefore, $1 = q_1 q_2 \cdots q_{m-1}$, and hence, $q_1 = q_2 = \cdots = q_{m-1} = 1$. Now, since each q_i is a prime, $q_i \neq 1$. So we must have $m = 1$.

Induction step: Assume that if $p_1 p_2 \cdots p_k = q_1 q_2 \cdots q_m$, then $k = m$ and $p_1 = q_1, p_2 = q_2, \ldots, p_k = q_m$. Let $p_1 p_2 \cdots p_{k+1} = q_1 q_2 \cdots q_m$. Then $p_{k+1} = q_i$ for some i. Since $q_i \leq q_m$, it follows that $p_{k+1} \leq q_m$. Similarly, $q_m \leq p_{k+1}$. Hence, $p_{k+1} = q_m$. It then follows that $p_1 p_2 \cdots p_k = q_1 q_2 \cdots q_{m-1}$. By the induction hypothesis, $k = m - 1$ and $p_1 = q_1, p_2 = q_2, \ldots, p_k = q_{m-1}$. Therefore, $k + 1 = m$ and $p_1 = q_1, p_2 = q_2, \ldots, p_{k+1} = q_m$. □

As a consequence of the prime factorization theorem, if n is a positive integer, then each prime in the prime factorization of n^2 occurs an even number of times.

The following example gives a nice application of the prime factorization theorem.

Example 2

Prove that $\sqrt{3}$ is irrational.

Proof. We use a proof by contradiction. Suppose $\sqrt{3}$ is a rational number. Then $\sqrt{3} = m/n$, where m, n are positive integers with no common factors. Multiplying both sides by n and then squaring, we get $3 \cdot n^2 = m^2$. Now, the left side of this equation must have an odd number of 3's in its prime factorization, while the right side must have an even number of 3's in its prime factorization. But by the prime factorization theorem, the number of 3's must be the same on both sides. So we have a contradiction, and it follows that $\sqrt{3}$ is irrational.

We can use the prime factorization theorem to find the greatest common divisor of two integers.

Example 3

Find $\gcd(60, 100)$.

Solution   The prime factorizations of the two numbers are $60 = 2^2 \cdot 3 \cdot 5$ and $100 = 2^2 \cdot 5^2 = 2^2 \cdot 3^0 \cdot 5^2$. We take the minimum power of each prime factor to obtain $\gcd(60, 100) = 2^2 \cdot 3^0 \cdot 5$.

In general, we have the following theorem.

Theorem 5   If $a = p_1^{k_1} \cdot p_2^{k_2} \cdots p_m^{k_m}$ and $b = p_1^{j_1} \cdot p_2^{j_2} \cdots p_m^{j_m}$, where the p_i are distinct primes, then we have: $\gcd(a, b) = p_1^{v_1} \cdot p_2^{v_2} \cdots p_m^{v_m}$, where $v_i = \min(k_i, j_i)$

Least Common Multiple

Let a and b be positive integers. If we write the sum of two fractions $1/a + 1/b$ as a single fraction, so that the numerator and denominator have no factor in common, the denominator will be the least common multiple of a and b.

Example 4

Let $a = 46$ and $b = 64$. Then

$$\frac{1}{a} + \frac{1}{b} = \frac{1}{2 \cdot 23} + \frac{1}{2 \cdot 32} = \frac{32 + 23}{2 \cdot 23 \cdot 32}$$

Definition.   Let a, b, and v be positive integers. A **least common multiple** v of a and b satisfies the following properties: (a) $a|v$ and $b|v$; (b) for all m, if $a|c$ and $b|c$, then $v|c$.

Property (a) of the definition says that v is a **common multiple** of a and b. Property (b) says that v is a *least* common multiple of a and b.

As with the greatest common divisor, we need to prove that the least common multiple exists and, if it exists, then it is unique.

Theorem 6   The least common multiple of two positive integers exists.

Hint for the proof: Let a and b be positive integers. Then $a = p_1^{k_1} p_2^{k_2} \cdots p_m^{k_m}$ and $b = p_1^{j_1} p_2^{j_2} \cdots p_m^{j_m}$, where the p_i are distinct primes. Prove that $p_1^{u_1} p_2^{u_2} \cdots p_m^{u_m}$, where $u_i = \max(k_i, j_i)$ is a least common multiple of a and b. $\qquad\square$

Theorem 7   The least common multiple of two positive integers is unique.

Hint for the proof: See Theorem 1 of Section 8.3.   □

When v is the least common multiple of a and b, we write $v = \textbf{lcm}(a, b)$. To find lcm(60, 100), we take the maximum power of each prime factor to obtain lcm(60, 100) = $2^2 \cdot 3 \cdot 5^2$.

From Example 4, lcm(46, 64) = $2 \cdot 23 \cdot 32$. We have already shown that gcd(46, 64) = 2. Note that $46 \cdot 64 = (2 \cdot 23) \cdot (2 \cdot 32) = 2 \cdot (2 \cdot 23 \cdot 32) = \text{gcd}(46, 64) \cdot \text{lcm}(46, 64)$. This is no accident, as we shall see in Theorem 8.

Theorem 8   Let a and b be positive integers. Then $ab = \text{gcd}(a, b) \cdot \text{lcm}(a, b)$.

Hint for the proof: See Theorem 5 and the hint for the proof to Theorem 6.   □

Theorem 8 provides a nice method for calculating lcm(a, b) from the product $a \cdot b$ and the greatest common divisor gcd(a, b), because

$$\text{lcm}(a, b) = \frac{ab}{\text{gcd}(a, b)}$$

Example 5

Refer to Example 1 in Section 8.3.
  a. Let $a = 18$ and $b = 496$. Then gcd(18, 496) = 2. So lcm(18, 496) = $(18 \cdot 496)/2 = 9 \cdot 496 = 4,464$.
  b. Let $a = 18$ and $b = 48$. Then gcd(18, 48) = 6. So lcm(18, 48) = $(18 \cdot 48)/6 = 3 \cdot 48 = 144$.

EXERCISE SET 8.4

1. Prove the corollary to Theorem 1.
2. Prove Theorem 2.
3. Let $a = 20$ and $p = 13$. Find $m, n \in \mathcal{Z}$ such that $1 = a \cdot m + p \cdot n$.
4. Prove the corollary to Theorem 3.
5. Write each of the following integers as a product of powers of primes.
   (a) 210   **(b)** 1,024
6. Write each of the following integers as a product of powers of primes.
   (a) 426   **(b)** 8,128
7. Let $p_0, p_1, \ldots, p_k$ be primes. Prove that $1 + p_0 p_1 \cdots p_k$ has a prime factor different from any of $p_0, p_1, \ldots, p_k$.
8. *Prove:* $\sqrt{2}$ is irrational.
9. Let p be any odd prime. Prove that $\sqrt{p}$ is irrational.
10. *Prove:* $\sqrt[3]{2}$ is irrational.

11. *Prove:* $\sqrt[3]{p}$ is irrational, where p is prime.

12. *Prove:* $\log_2 5$ is irrational, where $\log_2 5 = y$ if and only if $2^y = 5$.

13. Let p be any odd prime. Prove that $\log_2 p$ is irrational, where $\log_2 p = y$ if and only if $2^y = p$.

14. Let n be a positive integer. Prove that if $2^n - 1$ is a prime, then n is a prime.

In Exercises 15–19, find lcm(a, b) for each of the pairs a, b.

15. $a = 40$, $b = 68$

16. $a = 175$, $b = 450$

17. $a = 196$, $b = 302$

18. $a = 47$, $b = 90$

19. $a = 2,765$, $b = 145,299$

20.   (a) Find lcm$(21, 96)$ using the prime factorization theorem.
    (b) Find gcd$(21, 96)$ using the Euclidean algorithm.
    (c) Verify that $21 \cdot 96 = $ gcd$(21, 96) \cdot$ lcm$(21, 96)$.

21.   (a) Find lcm$(45, 225)$ using the prime factorization theorem.
    (b) Find gcd$(45, 225)$ using the Euclidean algorithm.
    (c) Verify that $45 \cdot 225 = $ gcd$(45, 225) \cdot$ lcm$(45, 225)$.

22. *Prove:* lcm$(b, c) = bc$ if and only if gcd$(b, c) = 1$.

23. *Prove:* lcm$(a, ab) = ab$.

24. Prove Theorem 5.

25. Prove Theorem 6.

26. Prove Theorem 7.

27. Prove Theorem 8.

8.5 INTEGERS MOD d

Recall a definition given originally in Section 5.1.

Definition.   Let m and n be integers and d be some positive integer. Then $n \equiv_d m$ if $n - m$ is divisible by d.

We read $n \equiv_d m$ as "n is **congruent** to m **modulo** d." In many mathematics books, $n \equiv_d m$ is also written as $n \equiv m \pmod{d}$.

Example 1

Let $d = 3$.
    a. $17 \equiv_2 2$, since $17 - 2$ is a multiple of 3.
    b. $-5 \equiv_3 1$, since $-5 - 1$ is a multiple of 3.

Note that $k \equiv_3 0$ whenever k is divisible by 3. In this case, we say k is 0 mod 3.

For more complex examples, the division algorithm is useful.

Example 2

Find r such that $517 \equiv_7 r$ and $0 \le r < 7$.

Solution   Use the division algorithm to get $517 = 7 \cdot 73 + 6$. Hence, $517 \equiv_7 6$.

In Section 5.1, we proved that the relation $\equiv_d$ is an equivalence relation on $\mathcal{Z}$. Therefore, $\equiv_d$ partitions the integers into equivalence classes. We use the division algorithm to characterize these equivalence classes.

It follows from the division algorithm for the integers that for each positive integer d and any integer n, there is a unique r with the properties that $n \equiv_d r$ and $0 \le r < d$. So for a given integer d, every integer n satisfies one of the following: $n \equiv_d 0, n \equiv_d 1, \ldots, n \equiv_d d - 1$. Hence, the equivalence classes for d are $[0], [1], [2], \ldots, [d-1]$.

Example 3

a. For $d = 2, [0]$ contains all the even integers and $[1]$ contains all the odd integers. In other words, $n \equiv_2 0$ means that n is even and $n \equiv_2 1$ means that n is odd.

b. For any integer n, $n \equiv_3 0$ or $n \equiv_3 1$ or $n \equiv_3 2$. Hence, the equivalence classes for $d = 3$ are $[0], [1], [2]$.

In general, when $n \equiv_d r$ and $0 \le r < d$, we say that n is r **mod** d. So every integer is either 0 mod 3, 1 mod 3, or 2 mod 3. In Example 2, we found that 517 is 6 mod 7.

Modular Arithmetic

Let $Z_d = \{[0], [1], \ldots, [d-1]\}$ be the set of equivalence classes. For simplicity, we often write $[x]$ as x. So $Z_d = \{0, 1, 2, \ldots, d-1\}$.

Now let $d = 12$, and consider the clockface shown in Figure 8–1. We have designated noon (or midnight) as 0, since 12 is 0 mod 12.

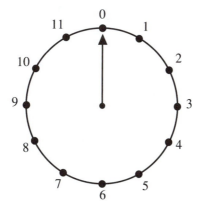

Figure 8–1

Imagine advancing the hand seven hours from 0 to 7. Then advance it nine more hours from 7. The hand is now pointing to 4. This is an example of modular addition: $7 + 9 = 4$ mod 12.

To make this more precise, we will shortly define addition mod d, written $+_d$, on Z_d in the following way: $[x] +_d [y] = [x + y]$. For the example of $7 + 9$, we have $[7] +_{12} [9] = [7 + 9] = [16] = [4]$. Before giving a formal definition, we need a theorem.

Theorem 1 Let $[x]$, $[u]$, $[y]$, and $[v]$ be elements of Z_d. If $[x] = [u]$ and $[y] = [v]$, then $[x + y] = [u + v]$.

Proof. Let $[x] = [u]$ and $[y] = [v]$. Then $x - u$ is a multiple of d, and $y - v$ is also a multiple of d. Hence, $x = u + kd$ and $y = v + jd$ for some integers k and j. Adding, we obtain $x + y = u + v + d(k + j)$. Hence, $(x + y) - (u + v)$ is a multiple of d. So $[x + y] = [u + v]$. $\square$

Definition. **Addition mod d,** written $+_d$, is defined on Z_d as
$$[x] +_d [y] = [x + y]$$

Corollary. Addition mod d is a binary operation on Z_d.

Proof. It is easy to see that $[x] +_d [y] \in Z_d$ if $[x], [y] \in Z_d$. What we need to prove is that the evaluation of $[x] +_d [y]$ does not depend on the choice of element in the equivalence classes $[x]$ or $[y]$. But, by Theorem 1, $[x] = [u]$ and $[y] = [v]$ implies $[x] +_d [y] = [u] +_d [v]$. $\square$

We will write $[x] +_d [y]$ more simply as $x +_d y$. When this is done, it is understood that the result is written as an integer mod d, for example, $3 +_5 3 = 1$.

Table 8–1 defines addition for Z_5.

TABLE 8–1

$+_5$	0	1	2	3	4
0	0	1	2	3	4
1	1	2	3	4	0
2	2	3	4	0	1
3	3	4	0	1	2
4	4	0	1	2	3

From the table, we see that $4 +_5 2 = 1$, $3 +_5 1 = 4$, etc., and it is plain that $+_5$ is a binary operation. By the corollary to Theorem 1, in general, $+_d$ is a binary operation on Z_d.

We define multiplication mod d on Z_d after stating a theorem that guarantees that it is a binary operation on Z_d.

Theorem 2 Let $[x]$, $[u]$, $[y]$, and $[v]$ be elements of Z_d. If $[x] = [u]$ and $[y] = [v]$, then $[xy] = [uv]$.
Hint for the proof: Consider the proof of Theorem 1. $\square$

Definition. **Multiplication mod** d, written $*_d$, is defined on Z_d as

$$[x] *_d [y] = [x \cdot y]$$

By Theorem 2, $*_d$ is a binary operation on Z_d.

Notational simplifications similar to those for addition apply to multiplication as well. For example, $[2] *_5 [4] = [2 \cdot 4] = [8] = [3]$, and this is also written $2 *_5 4 = 3$. Table 8–2 defines multiplication for Z_5.

TABLE 8–2

$*_5$	0	1	2	3	4
0	0	0	0	0	0
1	0	1	2	3	4
2	0	2	4	1	3
3	0	3	1	4	2
4	0	4	3	2	1

Using the terminology of Section 5.3, we can say that Z_5 is an algebra with two operations. More generally, $(Z_d, +_d, *_d)$ is an algebra for any positive integer d. Is $(Z_d, +_d, *_d, 0, 1)$ an integral domain? (Note that here, 0 stands for $[0]$ and 1 stands for $[1]$.)

It is an exercise to prove that the commutative, associative, and distributive laws hold in Z_d. It is also easy to verify that $x +_d 0 = x$ and $x *_d 1 = x$ for all x in Z_d. For the existence of the additive inverse of every element in Z_d, we use the equivalence class notation. Note that $[x] +_d [-x] = [0]$, where $-x$ is the additive inverse of x in $\mathcal{Z}$. So, Axioms 1–5 hold for the algebra Z_d.

In order to see whether Z_d is an integral domain, we only need check Axiom 6: If $x \cdot y = 0$, then $x = 0$ or $y = 0$.

Example 4

Show that Z_5 is an integral domain.

Solution Assume $[x] *_5 [y] = [0]$. Then $xy \equiv_5 0$, and hence, xy is divisible by 5. So $5|x$ or $5|y$, by Theorem 3 of Section 8.4.

Example 5

Show that Z_4 is not an integral domain.

Solution $[2] *_4 [2] = [0]$, but $[2] \neq [0]$.

Theorem 3 Z_d is an integral domain if and only if d is a prime.

Example 6

Find a homomorphism from $(\mathcal{Z}, +, \cdot, 0, 1)$ to $(Z_3, +_3, *_3, 0, 1)$.

Solution Define $h : \mathcal{Z} \rightarrow Z_3$ by $h(x) = [x]$; for example, $h(20) = [20] = [2]$. It is easy to see that h is an onto function. Furthermore, h preserves both binary operations. We verify this in the case of addition by $h(x + y) = [x + y] = [x] +_3 [y]$.

Also, $h(0) = [0] = 0$ and $h(1) = [1] = 1$.

EXERCISE SET 8.5

1. Use the division algorithm to find the remainder r such that:
   (a) $2{,}043 \equiv_3 r$ and $0 \leq r < 3$.
   (b) $759 \equiv_8 r$ and $0 \leq r < 8$.

2. Find the integer mod 7 for each of the following:
   a. 738   b. 93

3. Find the integer mod 6 for each of the following:
   a. 731   b. 9,304

4. Find the integer mod d for 517 when:
   a. $d = 3$ b. $d = 6$

5. Find the integer mod d for 843 when:
   a. $d = 7$ b. $d = 8$

6. Prove Theorem 2.

7. Without using Theorem 2 or any equivalent statement, prove that if $x \equiv_d y$, then $w \cdot x \equiv_d w \cdot y$ for any integer w.

8. Prove that each of the following laws holds in the algebra Z_d.
   (a) Commutative law
   (b) Associative law
   (c) Distributive law

9. Find a proof or a counterexample for the following conjecture: If $ab \equiv_d 0$, then $a \equiv_d 0$ or $b \equiv_d 0$.

10. Find a proof or a counterexample for the following conjecture: If p is a prime and $ab \equiv_p 0$, then $a \equiv_p 0$ or $b \equiv_p 0$.

11. Prove Theorem 3.

12. Prove: If p is a prime and $a^2 \equiv_p b^2$, then $a \equiv_p \pm b$.

13. Prove: If $ac \equiv_n bc$ and c and n have no common divisors other than ± 1, then $a \equiv_n b$.

14. Prove: If m is a factor of n and $a \equiv_n b$, then $a \equiv_m b$.

15. Find a homomorphism from $(\mathcal{Z}, +, \cdot, 0, 1)$ to $(Z_5, +_5, *_5, 0, 1)$.

16. Find a homomorphism from $(\mathcal{Z}, +, \cdot, 0, 1)$ to $(Z_d, +_d, *_d, 0, 1)$.

KEY CONCEPTS

Integral domain	Prime factorization theorem
Divisibility in $\mathcal{Z}$	Least common multiple

Division algorithm	Congruent modulo d
Greatest common divisor	Z_d
Euclidean algorithm	Addition mod d $(+_d)$
Prime	Multiplication mod d $(*_d)$

REVIEW EXERCISES

1. Use mathematical induction to prove that $5|(8^n - 3^n)$ for $n = 0, 1, 2, \ldots$.

2. *Prove:* $4^n \equiv 1 \pmod 3$ for $n = 0, 1, 2, \ldots$.

3. Let $\gcd(a, b) = 1$. Prove that if $a|c$ and $b|c$, then $ab|c$.

4. *Prove:* $6|(n^3 - n)$ for $n = 0, 1, 2, \ldots$.

5. *Prove:* $n^5 \equiv n \pmod{30}$ for $n = 0, 1, 2, \ldots$. (*Hint:* Use Exercises 3 and 4.)

6. *Prove:* If a positive integer m has a factor greater than $\sqrt{m}$, then m also has a positive factor greater than 1 and less than $\sqrt{m}$.

7. Find a proof or a counterexample for the following conjecture: If $\gcd(a, b) = d$ and ad is a factor of cb, then a is a factor of c.

8. *Prove:* If p is a prime and p is a factor of a^n, then p is a factor of a, for $n = 1, 2, \ldots$.

9. *Prove:* $\gcd(ac, bc) = c \cdot \gcd(a, b)$.

10. *Prove:* $\operatorname{lcm}(ac, bc) = c \cdot \operatorname{lcm}(a, b)$.

11. Define $f : \mathcal{N} \times \mathcal{N} \to \mathcal{N}$ by $f(m, n) = 2^m 3^n$. Prove that f is a 1–1 function.

12. Suppose p is a prime. Prove that either $\gcd(a, p) = 1$ or $\gcd(a, p) = p$.

Both the gcd and the lcm can be considered binary operations on $\mathcal{Z}$. In Exercises 13–15, let $*$ and Θ be defined by $a * b = \gcd(a, b)$ and $a \Theta b = \operatorname{lcm}(a, b)$.

13. Prove that $*$ and Θ are commutative and associative.

14.   **(a)** Find an integer e such that $a \Theta e = a$ for all a.
     (b) Either find an integer f such that $a * f = a$ for all a, or explain why no such element exists.

15. Prove the distributive law $a \Theta (b * c) = (a \Theta b) * (a \Theta c)$.

SUPPLEMENTARY EXERCISES

1. Suppose $a > 0$ is not a perfect nth power (that is, $a \neq b^n$ for any $b > 0$ and any natural number n). Prove that $\sqrt[n]{a}$ is irrational. (*Hint:* If a is not a perfect nth power, then some prime in the prime factorization of a does not occur a multiple of n times.)

2. *Prove:* If a and b are arbitrary integers, then the equation $ax + by = c$ has a solution in integers if and only if $g|c$, where $g = \gcd(|a|, |b|)$. (*Hint:* If a and b are positive, then this is the corollary to Theorem 2 in Section 8.3. So assume, for example, that $a < 0$. Then $-a > 0$. If the equation $-ax + by = c$ has a solution, then so does $ax + by = c$.)

A **Diophantine equation** is a polynomial equation with integer coefficients and one or more unknowns. Any solution to a Diophantine equation must consist of integers. The term "Diophantine" comes from Diophantus (ca. 250 A. D.), a Greek mathematician who is given credit for introducing abbreviations into algebraic equations. Exercises 3 and 4 are concerned with a certain Diophantine equation.

3. We show in this exercise how to find the general solution in integers, if *one exists*, of the Diophantine equation $ax + by = c$, where a, b, and c are arbitrary integers. Supply all the missing details.

   Let $d = \gcd(|a|, |b|)$. Then, by Exercise 2 above, $ax + by = c$ has a solution if and only if $d|c$.

   First prove that if $b|ac$, then $b|dc$. (See Exercise 19 in Section 8.3.) Let $a = de$ and $b = df$ and assume $ax + by = c$ has a solution, say x_0, y_0. Then $ax_0 + by_0 = c$. Assume also that x, y is a solution. Then $a(x - x_0) + b(y - y_0) = 0$. Hence, $a(x - x_0) = b(y_0 - y)$, so that $b|a(x - x_0)$. Therefore, $b|d(x - x_0)$, and it follows that $x = x_0 + fk$ for some k. Substituting into $a(x - x_0) = b(y_0 - y)$, we obtain $y = y_0 - ek$. Verify that $x = x_0 + fk$, $y = y_0 - ek$ is a solution for any integer k.

   So the general solution is $x = x_0 + fk$, $y = y_0 - ek$, where k is an integer.

4. Find the general solution for the Diophantine equation $4x - 6y = 8$.

In Exercises 5–14, we will construct $\mathcal{Z}$ from $\mathcal{N}$. Imagine that we do not have the system of integers, but we do have the system of natural numbers. We wish to construct (in other words, represent as sets) a system of numbers which contains, for example, -3. We may try to represent -3 by the ordered pair $(2, 5)$ since $2 - 5 = -3$ in $\mathcal{Z}$. However, this attempt is not adequate, since $(4, 7)$ will also represent -3. Nonetheless, the representations $(2, 5)$ and $(4, 7)$ are somehow related. Their relationship is expressed in $\mathcal{N}$ by $2 + 7 = 5 + 4$. (We would like to say $2 - 5 = 4 - 7$, but we cannot do this in $\mathcal{N}$.) In what follows, we see that this relationship is actually an equivalence relation on $\mathcal{N} \times \mathcal{N}$.

In Exercise 15 of the review exercises for Chapter 5, we defined a relation T on $\mathcal{N} \times \mathcal{N}$ by $(m, n)T(p, q)$ if and only if $m + q = n + p$. Recall (or prove) that T is an equivalence relation on $\mathcal{N} \times \mathcal{N}$. We write the equivalence class of (m, n) as $[n, m]$. Let $\mathcal{Z} = (\mathcal{N} \times \mathcal{N})/T$.

5. (a) Define $+_{\mathcal{Z}}$ by $[m, n] +_{\mathcal{Z}} [p, q] = [m+p, n+q]$. Show that $+_{\mathcal{Z}}$ is a binary operation on $\mathcal{Z}$.
   (b) Define $\cdot_{\mathcal{Z}}$ by $[m, n] \cdot_{\mathcal{Z}} [p, q] = [mp + nq, mq + np]$. Show that $\cdot_{\mathcal{Z}}$ is a binary operation on $\mathcal{Z}$.

To prove that we have constructed $\mathcal{Z}$ from the natural numbers, we must prove that $\mathcal{Z} = (\mathcal{N} \times \mathcal{N})/T$ satisfies Axioms 1–10 for a well-ordered integral domain. In Exercises 6–14 we verify these axioms.

6. Show that $+_{\mathcal{Z}}$ and $\cdot_{\mathcal{Z}}$ satisfy Axioms 1–3.

7. Show that $+_{\mathcal{Z}}$ and $\cdot_{\mathcal{Z}}$ satisfy Axiom 4. (*Hint:* Consider $[0, 0]$ and $[1, 0]$.)

8. (a) Show that $+_{\mathcal{Z}}$ and $\cdot_{\mathcal{Z}}$ satisfy Axiom 5.
   (b) Show that $-[n, m] = [m, n]$.

9. Show that $+_{\mathcal{Z}}$ and $\cdot_{\mathcal{Z}}$ satisfy Axiom 6.

10. Define a function $f : \mathcal{N} \to \mathcal{Z}$ by $f(n) = [n, 0]$. Show that:
   (1) f is 1–1.
   (2) $f(n + m) = f(n) +_z f(m)$.
   (3) $f(nm) = f(n) \cdot_z f(m)$.
   (Properties (1), (2), and (3) show that $f[\mathcal{N}] \subseteq \mathcal{Z}$ is isomorphic to $\mathcal{N}$; in other words, there is a "copy" of $\mathcal{N}$ included in $\mathcal{Z}$. When we identify the element n of $\mathcal{N}$ with the element $[n, 0]$ of $\mathcal{Z}$—for example, 0 is identified with $[0, 0]$—we then have, essentially, $\mathcal{N} \subseteq \mathcal{Z}$.)

11. Verify that Axioms 7–9 hold; that is, prove that the set $f\big[\mathcal{N}\big]$ is closed under $\cdot_z$ and $+_z$, and for all $n, m \in \mathcal{N}$, exactly one of the following holds: $[m, n] \in f\big[\mathcal{N} - \{0\}\big]$ or $-[m, n] \in f\big[\mathcal{N} - \{0\}\big]$ or $[m, n] = f(0)$.

12. Define an order relation $\leq_z$ on $\mathcal{Z}$ by $[m, n] \leq_z [p, q]$ if $[m, n] +_z [k, 0] = [p, q]$ for some k. Show that $\leq_z$ is a linear order.

13. Let $\leq$ be the linear order on $\mathcal{N}$. Refer to Exercises 10 and 12. Prove that $m \leq n$ if and only if $f(m) \leq_z f(n)$.

14. Prove the well-ordering principle. That is, show that every nonempty subset of $f[\mathcal{N}]$ has a least element.

9

Limits and the Real Numbers

The concepts of a function and a limit are fundamental to the development of calculus and analysis. Indeed, the language of calculus is largely a language of functions, and the definitions of the derivative and the integral are based on the limit concept.

The history of calculus dates back to the fourth century B. C., with the work of Eudoxus (ca. 370 B. C.) and Archimedes (287–212 B. C.). Many of the major discoveries were made in the latter part of the 17th century by Newton (1642–1727) and Leibniz (1646–1716). However, it was not until the 18th century that the function concept was precisely defined and not until the early 19th century that Cauchy (1789–1857) gave the modern definition of a limit.

We have already discussed and applied functions extensively. In this chapter, we turn our attention to limits. There are several important types of limits for real-valued functions. We will restrict ourselves to two of these.

9.1 LIMITS OF THE FORM $\lim_{x \to \infty} f(x)$

Recall that a real-valued function is a function $f : D \to \mathcal{R}$. In this chapter, we restrict our attention to real-valued functions for which D is a subset of $\mathcal{R}$.

We can intuitively define $\lim_{x \to \infty} f(x) = L$ to mean that as x gets large, $f(x)$ gets close to L. The problem with this definition is that "gets close to" and "gets large" are vague. If we are to prove theorems involving limits, we should have a precise definition of these phrases.

Accordingly, we interpret the above definition as follows: $\lim_{x \to \infty} f(x) = L$ means that given a distance ϵ, $f(x)$ is within ϵ of L whenever x is large enough. We say "x is large enough" when $x > n$ for some n that depends on the given ϵ. More formally, we have the following definition:

Definition. $\displaystyle\lim_{x \to \infty} f(x) = L$ if, for every positive distance ϵ, there is an $n \in \mathcal{N}$ such that $f(x)$ is within distance ϵ of L whenever $x > n$.

It is helpful to consider the geometric interpretation of this definition. The ϵ (usually small) specifies a band about the value L of width 2ϵ. After the ϵ-band is given, n must be found so that to the right of n the function values are all in the ϵ-band (see Figure 9–1).

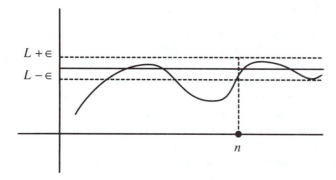

Figure 9–1

In the figure, the given n works, since $f(x)$ is within ϵ of L for all $x > n$. Geometrically, "$f(x)$ is within ϵ of L" means that the graph of f is between the two horizontal lines determined by ϵ. For the ϵ in Figure 9–2, the given n does not work, since at the indicated x, $f(x)$ is not within ϵ of L.

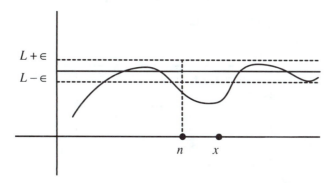

Figure 9–2

The following facts about absolute values will be useful in what follows:

1. $|xy| = |x||y|$
2. $\left|\dfrac{x}{y}\right| = \dfrac{|x|}{|y|}$
3. $|x + y| \le |x| + |y|$

Also, note that w is within distance ϵ of b if and only if $b - \epsilon < w < b + \epsilon$. This is equivalent to $-\epsilon < w - b < \epsilon$, which is equivalent to $|w - b| < \epsilon$. Hence, "$|w - b| < \epsilon$" means that w is within distance ϵ of b.

For limits of the type $\lim_{x \to \infty} f(x)$, we will assume that for all $n \in \mathcal{N}$, there exists x such that $x > n$ and x is in the domain of f. This is equivalent to assuming that $(n, \infty) \cap \text{dom}(f) \neq \emptyset$ for all $n \in \mathcal{N}$.

We are now ready to give a precise definition of $\lim_{x \to \infty} f(x) = L$.

Definition. $\lim_{x \to \infty} f(x) = L$ if for every positive number ϵ, there is an $n \in \mathcal{N}$ such that for all x in the domain of f, $x > n$ implies $|f(x) - L| < \epsilon$.

The limit, if it exists, is unique. We prove this later.

Suppose we have an n such that $x > n$ implies $|f(x) - L| < \epsilon$ for all x in the domain of f. If $n_1 > n$, then $x > n_1$ also implies $|f(x) - L| < \epsilon$ for all x in the domain of f. This means that we may always choose n as large as we wish.

If $x > n$ it may happen that x is not in the domain of f. However, since we are assuming that $(n, \infty) \cap \text{dom}(f) \neq \emptyset$ for all n, there exists $x > n$ such that $f(x)$ is defined. When using the limit definition, we assume that x is in $(n, \infty) \cap \text{dom}(f)$ whenever we write $|f(x) - L| < \epsilon$.

Before considering several examples, we recall a property of $\mathcal{R}$, first stated in Chapter 3.

> *The ϵ-property for $\mathcal{R}$.* Given any positive real number ϵ, there exists a positive integer m such that $1/m < \epsilon$.

We will prove the ϵ-property in Section 9.4. It is equivalent to the following statement.

> *The ϵ-property (restated).* Given any positive real number w, there exists a positive integer m such that $m > w$.

Example 1

Show that $\lim_{x \to \infty} \dfrac{1}{x} = 0$.

Solution Let $\epsilon > 0$ be an arbitrary real number. We need to find a positive integer n such that $|1/x - 0| < \epsilon$ if $x > n$. But $|1/x - 0| = |1/x| = 1/x$ since $x > 1$. So we need to find an n such that $x > n$ implies $1/x < \epsilon$. We solve $1/x < \epsilon$ for x to obtain $x > 1/\epsilon$. By the ϵ-property, there is a positive integer n such that $n > 1/\epsilon$.

We verify that $x > n$ implies $|1/x - 0| < \epsilon$. Assume $x > n$. Then $x > 1/\epsilon$, since $n > 1/\epsilon$. Hence, $1/x < \epsilon$, and we have $|1/x - 0| < \epsilon$. Therefore, we have found a positive integer n such that $x > n$ implies $|1/x - 0| < \epsilon$.

The strategy used in the solution to Example 1 is worth noting. We solved an inequality $(1/x < \epsilon)$ for x in order to find how big $(x > 1/\epsilon)\, x$ must be. Then we chose $n\, (n > 1/\epsilon)$ so that all $x > n$ are big enough. This strategy is also applied in Example 2.

Example 2

Show that $\lim\limits_{x \to \infty} x/(x-1) = 1$.

Solution Let $\epsilon > 0$. We need to find an n such that $|x/(x-1) - 1| < \epsilon$ if $x > n$. First we note that $|x/(x-1) - 1| = |1/(x-1)| = 1/(x-1)$ since $x > 1$. So we want $1/(x-1) < \epsilon$. Solving for x, we see that this is equivalent to $x > 1/\epsilon + 1$. Thus, we choose n such that $n > 1/\epsilon + 1$.

If $x > n$, then $x > 1/\epsilon + 1$. It then follows that $1/(x-1) < \epsilon$. Hence, $|x/(x-1) - 1| < \epsilon$.

The solutions in Examples 1 and 2 begin with "let $\epsilon > 0$." We may always assume that ϵ is smaller than any given positive real number. Suppose that we start with $\epsilon > 0$, and for some reason we would like ϵ less than a positive real number w. If the given ϵ is already less than w, then there is no problem. If the given ϵ is greater than or equal to w, then we choose ϵ_1 such that $0 < \epsilon_1 < w$. Note that $0 < \epsilon_1 < w \le \epsilon$, and hence, $\epsilon_1 < \epsilon$. Then we find an n such that $x > n$ implies $|f(x) - L| < \epsilon_1$. It then follows that $x > n$ implies $|f(x) - L| < \epsilon$.

Example 3

Show that

$$\lim_{x \to \infty} \frac{5x^2 - 6}{x^2 + 1} = 5$$

Solution Let $\epsilon > 0$. Now,

$$\left| \frac{5x^2 - 6}{x^2 + 1} - 5 \right| = \left| \frac{-11}{x^2 + 1} \right| = \frac{11}{x^2 + 1}$$

We want

$$\frac{11}{x^2 + 1} < \epsilon$$

So we solve for x^2 to obtain $x^2 > 11/\epsilon - 1$. At this point, we would like to solve for x, but $11/\epsilon - 1$ may be less than zero. We would like $11/\epsilon - 1 > 0$, that is, $\epsilon \le 11$. By the remarks preceding this example, we may assume that $\epsilon \le 11$. With this assumption, we then obtain $x > \sqrt{11/\epsilon - 1}$. So we choose $n > \sqrt{11/\epsilon - 1}$. It is now a routine matter to verify that $x > n$ implies

$$\left| \frac{5x^2 - 6}{x^2 + 1} - 5 \right| < \epsilon$$

Example 4

If $f(x) = b$ for all x, then $\lim\limits_{x \to \infty} f(x) = b$.

Solution Let $\epsilon > 0$. We have $|f(x) - b| = 0$. Hence, $|f(x) - b| < \epsilon$, no matter how we choose n. Therefore, $x > n$ implies $|f(x) - b| < \epsilon$ for any n.

Exercise Set 9.1

EXERCISE SET 9.1

1. Show that

$$\lim_{x \to \infty} \frac{5x - 1}{x + 2} = 5$$

2. Show that

$$\lim_{x \to \infty} \frac{2x + 1}{3x + 2} = \frac{2}{3}$$

3. Show that

$$\lim_{x \to \infty} \frac{\pi x^2 - 3}{x^2 + 2} = \pi$$

4. Show that

$$\lim_{x \to \infty} \frac{2x^2 - 3}{3x^2 + 2} = \frac{2}{3}$$

5. Let

$$f(x) = \begin{cases} 0 \text{ if } x \in \mathcal{N} \\ 1/x \text{ otherwise} \end{cases}$$

Prove that $\lim\limits_{x \to \infty} f(x) = 0$.

6. Let

$$f(x) = \begin{cases} 0 \text{ if } x \notin \mathcal{N} \\ \pi/x \text{ otherwise} \end{cases}$$

Prove that $\lim\limits_{x \to \infty} f(x) = 0$.

7. (a) Write the definition of $\lim\limits_{x \to \infty} f(x) = L$ using quantifiers.

    (b) Write the definition of $\lim\limits_{x \to \infty} f(x) \neq L$ using quantifiers.

In Exercises 8–14, make a conjecture about $\lim\limits_{x \to \infty} f(x)$, and prove your conjecture.

8. $f(x) = \begin{cases} x \text{ if } x \in \mathcal{N} \\ 1/x \text{ otherwise} \end{cases}$

9. $f(x) = 3 + 1/x$

10. $f(x) = 5/x$

11. $f(x) = 1/x^2$

12. $f(x) = 10^9 + 3/x$

13. $f(x) = (\pi x - 3)/(7x + 2)$

14. $f(x) = (\sin x)/x$

9.2 THEOREMS ON LIMITS

Following are some important theorems on limits. The domains of all functions in this section include elements from (n, ∞) for all $n \in \mathcal{N}$.

For real numbers x and y, $\mathbf{max}(x, y)$ is the maximum of x and y.

Theorem 1 (Uniqueness of Limits) If $\lim\limits_{x \to \infty} f(x) = L$ and $\lim\limits_{x \to \infty} f(x) = K$, then $L = K$.

Proof. Suppose $L \neq K$. Then

$$\frac{|L - K|}{2} > 0$$

Now, by the definition of a limit, there is an n_1 such that $x > n_1$ implies

$$|f(x) - L| < \frac{|L - K|}{2}$$

Similarly, there is an n_2 such that $x > n_2$ implies

$$|f(x) - K| < \frac{|L - K|}{2}$$

Let $x > \max(n_1, n_2)$. Then

$$|L - K| = |(f(x) - K) - (f(x) - L)| \leq |(f(x) - K)| + |(f(x) - L)| <$$

$$\frac{|L - K|}{2} + \frac{|L - K|}{2} = |L - K|$$

Hence, $|L - K| < |L - K|$, a contradiction. It follows that $L = K$. $\square$

Theorem 2 $\lim\limits_{x \to \infty} f(x) = L$ if and only if $\lim\limits_{x \to \infty} (f(x) - L) = 0$.
Hint for proof: Write out the definitions of $\lim\limits_{x \to \infty} f(x) = L$ and $\lim\limits_{x \to \infty} (f(x) - L) = 0$. $\square$

Suppose $\lim\limits_{x \to \infty} f(x) = L$. By definition, for *any* positive number α, there is a positive integer m such that $x > m$ implies $|f(x) - L| < \alpha$. From this, it follows that if ϵ is an arbitrary positive number and k is any positive constant, then there is a positive integer n such that $x > n$ implies $|f(x) - L| < k\epsilon$. For example, if $\epsilon > 0$ is arbitrary, there exists an n such that $x > n$ implies $|f(x) - L| < \epsilon/2$. We use this result in the proof of Theorem 3 following and many times thereafter.

Theorem 3 (Sum Theorem) If $\lim\limits_{x \to \infty} f(x) = L$ and $\lim\limits_{x \to \infty} g(x) = K$, then $\lim\limits_{x \to \infty} (f(x) + g(x)) = L + K$.

Proof. Let $\lim\limits_{x \to \infty} f(x) = L$ and $\lim\limits_{x \to \infty} g(x) = K$. Let $\epsilon > 0$ be arbitrary. Then there is an n_1 such that $x > n_1$ implies $|f(x) - L| < \epsilon/2$, and there is an n_2 such that $x > n_2$ implies $|g(x) - K| < \epsilon/2$. Choose $n = \max(n_1, n_2)$.

Then $x > n$ implies $|f(x) - L| < \epsilon/2$ and $|g(x) - K| < \epsilon/2$. Therefore, $x > n$ implies $|(f(x) + g(x)) - (L + K)| = |(f(x) - L) + (g(x) - K)| \leq |f(x) - L| + |g(x) - K| < \epsilon/2 + \epsilon/2 = \epsilon$. $\square$

Definition. Let c be an arbitrary real number. If there exists a real number q such that $|f(x)| < q$ for all $x \in \text{dom}(f) \cap [c, \infty)$, we say f is **bounded on** $[c, \infty)$.

Let f be bounded on $[c, \infty)$. Then the images $f(u)$ stay within distance q of the x-axis for all $u \geq c$.

Theorem 4 If $\lim_{x \to \infty} f(x) = L$, then there exists a real number c in the domain of f such that f is bounded on $[c, \infty)$.

Proof. Assume $\lim_{x \to \infty} f(x) = L$. Then, for every positive number ϵ, there exists a positive integer n such that $x > n$ implies $|f(x) - L| < \epsilon$. Let $\epsilon = 1$. Then $|f(x) - L| < 1$ if $x > n$. Hence, $L - 1 < f(x) < L + 1$ if $x > n$. Let $q = \max(|L - 1|, |L + 1|)$. Then $|f(x)| < q$ for all $x > n + 1$. If $c = n + 1$, then f is bounded on $[c, \infty)$. $\square$

The converse of Theorem 4 is false. Consider the function $f(x) = \cos x$. Then, since $|\cos x| \leq 1$ for all x, f is bounded on $[c, \infty)$ for any c. However, in Example 1 following, we see that $\lim_{x \to \infty} \cos x$ does not exist. (In general, when there is no real number L such that $\lim_{x \to \infty} f(x) = L$, we say that **the limit,** $\lim_{x \to \infty} f(x)$, **does not exist.**)

Example 1

$\lim_{x \to \infty} \cos x$ does not exist.

Proof We give a proof by contradiction. Suppose there is a real number L such that $\lim_{x \to \infty} \cos x = L$. Let $\epsilon = 1$. Then there exists an n such that $x > n$ implies $|\cos x - L| < 1$. Hence, $-1 + \cos x < L < 1 + \cos x$ for $x > n$. Now, there exists a positive integer m such that $2m\pi > n$. So $-1 + \cos 2m\pi < L$. Hence, $0 < L$ since $\cos 2m\pi = 1$. Similarly, $L < 1 + \cos(2m + 1)\pi$. Hence, $L < 0$ since $\cos(2m + 1)\pi = -1$. Therefore, $L < 0$ and $L > 0$, a contradiction, and it follows that $\lim_{x \to \infty} \cos x$ does not exist.

Example 2

If $f(x) = x$, then $\lim_{x \to \infty} f(x)$ does not exist.

Proof We give a proof by contradiction. Suppose there is a real number L such that $\lim_{x \to \infty} f(x) = L$. It is an exercise to show that f is not bounded on $[c, \infty)$ for any number c (see Exercise 4). By Theorem 4, we then have a contradiction, and it follows that $\lim_{x \to \infty} f(x)$ does not exist.

Theorem 5 If $\lim_{x \to \infty} g(x) = 0$ and f is bounded on $[c, \infty)$ for some number c, then $\lim_{x \to \infty} f(x)g(x) = 0$

Proof. Let $\lim_{x \to \infty} g(x) = 0$, and let f be bounded on $[c, \infty)$. Assume $\epsilon > 0$. Then there exists a positive integer n_1 such that $x > n_1$ implies $|f(x)| < q$ for some positive number q. Also, there exists a positive integer n_2 such that $x > n_2$ implies $|g(x)| < \epsilon/q$.

Choose $n = \max(n_1, n_2)$. Let $x > n$. Then $|g(x)| < \epsilon/q$. Hence, $x > n$ implies $|f(x)g(x) - 0| = |f(x)||g(x)| < q(\epsilon/q) = \epsilon$. $\square$

The proof of Theorem 6 following makes use of the algebraic identity $ab - cd = ab - ad + ad - cd = a(b - d) + d(a - c)$.

Theorem 6 (Product Theorem) If $\lim_{x \to \infty} f(x) = L$ and $\lim_{x \to \infty} g(x) = K$, then $\lim_{x \to \infty} (f(x) \cdot g(x)) = L \cdot K$.

Proof. Let $\lim_{x \to \infty} f(x) = L$ and $\lim_{x \to \infty} g(x) = K$. Then $\lim_{x \to \infty} (f(x) - L) = 0$ and $\lim_{x \to \infty} (g(x) - K) = 0$. We only need prove $\lim_{x \to \infty} (f(x)g(x) - LK) = 0$. First we write $f(x)g(x) - LK$ in a more useful form, namely, $f(x)g(x) - LK = f(x)g(x) - f(x)K + f(x)K - LK = f(x)(g(x) - K) + K(f(x) - L)$. Now, by Theorem 4, f is bounded on $[c, \infty)$ for some c. Also, clearly, the constant K is bounded on $[c, \infty)$.

Hence, $\lim_{x \to \infty} (f(x)(g(x) - K)) = 0$ and $\lim_{x \to \infty} K(f(x) - L) = 0$, by Theorem 5. It then follows from the sum theorem that $\lim_{x \to \infty} (f(x)g(x) - LK) = 0$. $\square$

Theorem 7 If $\lim_{x \to \infty} g(x) = K$ and $K \neq 0$, then there exists an n such that

$$|g(x)| > \frac{|K|}{2}$$

for all x such that $x > n$.
Hint for proof: There is an n such that $x > n$ implies

$$|g(x) - K| < \frac{|K|}{2}$$

Now use $|K| = |K - g(x) + g(x)| \leq |K - g(x)| + |g(x)|$. $\square$

Theorem 8 (Reciprocal Theorem) If $\lim_{x \to \infty} g(x) = K$ and $K \neq 0$, then $\lim_{x \to \infty} (1/g(x)) = 1/K$.

Proof. We only need prove that $\lim_{x \to \infty} (1/g(x) - 1/K) = 0$. By Theorem 7, there exists an n_1 such that

$$\frac{1}{|g(x)|} < \frac{2}{|K|}$$

for all x such that $x > n_1$. Hence,

$$\left| \frac{1}{g(x)} - \frac{1}{K} \right| = \left| \frac{K - g(x)}{Kg(x)} \right| < \left(\frac{2}{K^2} \right) |g(x) - K|$$

for all x such that $x > n_1$. Now, let $\epsilon > 0$ be arbitrary, and choose $n_2 > 0$ such that $x > n_2$ implies $|g(x) - K| < \epsilon(K^2/2)$. Let $n = \max(n_1, n_2)$. Then $x > n$ implies $|1/g(x) - 1/K| < (2/K^2)|g(x) - K| < (2/K^2)\epsilon(K^2/2) = \epsilon$. $\square$

Corollary. (Quotient Theorem) If $\lim_{x \to \infty} f(x) = L$ and $\lim_{x \to \infty} g(x) = K$ and $K \neq 0$, then $\lim_{x \to \infty} (f(x)/g(x)) = L/K$.

Theorem 9 (Squeezing Theorem) If $\lim_{x \to \infty} f(x) = L = \lim_{x \to \infty} g(x)$ and $f(x) < h(x) < g(x)$ for all x, then $\lim_{x \to \infty} h(x) = L$.

Hint for proof: Let $\epsilon > 0$ be arbitrary. There is an n_1 such that $x > n_1$ implies $-\epsilon + L < f(x)$, and there is an n_2 such that $x > n_2$ implies $g(x) < \epsilon + L$. $\quad\square$

EXERCISE SET 9.2

1. Prove Theorem 2.
2. Prove that $\lim_{x \to \infty} x^2$ does not exist.
3. Prove that $\lim_{x \to \infty} \sin x$ does not exist.
4. Let $f(x) = x$. Prove that f is not bounded on $[c, \infty)$ for any number c.
5. Let

$$f(x) = \begin{cases} 2 & \text{if } x \in \mathcal{N} \\ 1/x & \text{otherwise} \end{cases}$$

   Prove that $\lim_{x \to \infty} f(x)$ does not exist.

6. Let

$$f(x) = \begin{cases} -3 & \text{if } x \in \mathcal{N} \\ 1/x^2 & \text{otherwise} \end{cases}$$

   Prove that $\lim_{x \to \infty} f(x)$ does not exist.

7. Prove that $\lim_{x \to \infty} \frac{\sin x}{x} = 0$.
8. Prove Theorem 7.
9. Prove the quotient theorem.
10. Prove the squeezing theorem.

    In Exercises 11–18, assume $\lim_{x \to \infty} f(x) = L$.

11. Prove that $\lim_{x \to \infty} k f(x) = kL$.
12. Let $k > 0$. Prove that $\lim_{x \to \infty} f(kx) = L$.
13. Prove that $\lim_{x \to \infty} f(x + k) = L$.
14. Prove that if $L > 0$, then there exists an n such that $f(x) > 0$ for all $x > n$.
15. Let $g(x) = 1/f(x)$. Prove that if $L \neq 0$, then there exists a real number c such that g is bounded on $[c, \infty)$.
16. Prove that $\lim_{x \to \infty} (k f(x) + b) = kL + b$.
17. Prove that $\lim_{x \to \infty} f(x)^n = L^n$ for $n = 0, 1, 2, \ldots$.
18. Let $L > 0$. Prove that $\lim_{x \to \infty} \sqrt{f(x)} = \sqrt{L}$.

19. Suppose $b \neq 0$.

    (a) Explain why the quotient theorem cannot be used directly to prove that

$$\lim_{x \to \infty} \frac{ax + c}{bx + d} = \frac{a}{b}$$

    (b) Prove that

$$\lim_{x \to \infty} \frac{ax + c}{bx + d} = \frac{a}{b}$$

        (*Hint for Part b:* Divide the numerator and the denominator by x, and then apply the appropriate theorems on limits.)

20. Let $b \neq 0$. Prove that

$$\lim_{x \to \infty} \frac{ax^2 + c}{bx^2 + d} = \frac{a}{b}$$

    In Exercises 21–26, you may assume that $\lim_{x \to \infty} (1 + 1/x)^x = e$ and $p, q \in \mathbb{Z}$.

21. Prove that $\lim_{x \to \infty} (1 + 1/x)^{px} = e^p$ for $p \geq 0$. (*Hint:* Use Exercise 17.)

22. Prove that $\lim_{x \to \infty} (1 - 1/x)^x = e^{-1}$. (*Hint:*

$$1 - \frac{1}{x} = \frac{1}{1 + \frac{1}{x-1}}$$

So

$$\left(1 - \frac{1}{x}\right)^x = \frac{1}{(1 + 1/(x-1))^x} = \frac{1}{(1 + 1/(x-1))^{x-1}(1 + 1/(x-1))}$$

Now use Exercise 13 and appropriate theorems on limits.)

23. Prove that $\lim_{x \to \infty} (1 + 1/x)^{qx} = e^q$, where $q < 0$. (*Hint:* $-q > 0$ and $(1 + 1/x)^{qx} = 1/(1 + 1/x)^{-qx}$.)

24. Prove that $\lim_{x \to \infty} (1 + p/x)^x = e^p$, where $p > 0$. (*Hint:* Note that $p/x = 1/[(1/p)x]$.)

25. Prove that $\lim_{x \to \infty} (1 + p/x)^x = e^p$, where $p < 0$.

26. Prove that $\lim_{x \to \infty} (1 + p/x)^{qx} = e^{pq}$.

27. Look up the definition of the limit of a sequence in your calculus text or elsewhere. Compare that definition with the definition of $\lim_{x \to \infty} f(x) = L$, and discuss the similarities and differences between the definitions.

9.3 LIMITS OF THE FORM $\lim_{x \to a} f(x)$

In the previous two sections, we have investigated the type of limit where x "gets large" or x "approaches infinity." We now study a limit in which x "approaches a real number."

    Intuitively, $\lim_{x \to a} f(x) = L$ means that as x gets close to a, $f(x)$ gets close to L. As before, this intuitive notion can be made precise. Let a and L be real numbers, where a is not necessarily in the domain of the function f.

Definition. $\lim_{x \to a} f(x) = L$ if for every positive real number ϵ, there exists a positive real number δ such that for all x in the domain of f, if $0 < |x - a| < \delta$, then $|f(x) - L| < \epsilon$.

Figure 9–3 provides an example of a function f such that $\lim_{x \to a} f(x) \neq L$. Note the break in the graph at (a, L).

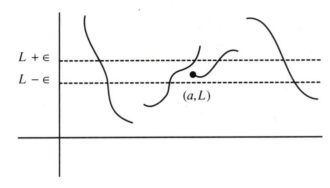

Figure 9–3

For the definition of $\lim_{x \to \infty} f(x) = L$ in Section 9.1, we assumed that for all n, there exists an x such that $x > n$ and x is in the domain of f. Similarly, for the definition of $\lim_{x \to a} f(x) = L$, we assume that for all $\delta > 0$ there is an x such that $0 < |x - a| < \delta$ and x is in the domain of f.

For real numbers x and y, $\mathbf{min}(x, y)$ is the minimum of x and y.

Example 1

Show that $\lim_{x \to 2} (x^2 + 1) = 5$.

Solution Let $\epsilon > 0$ be arbitrary. We need to find a δ such that $0 < |x - 2| < \delta$ implies $|(x^2 + 1) - 5| = |x^2 - 4| < \epsilon$. First we note that $x^2 - 4 = (x - 2)(x + 2)$. Now if we assume $\delta \leq 1$, then $|x - 2| < \delta$ implies $-1 < x - 2 < 1$, which in turn implies $3 < x + 2 < 5$. So $|x - 2| < \delta$ implies $|x^2 - 4| = |x + 2||x - 2| < 5|x - 2|$.

Now choose $\delta = \min(1, \epsilon/5)$. Then it is easy to verify that $0 < |x - 2| < \delta$ implies $|x^2 - 4| < 5|x - 2| < 5\delta < \epsilon$.

Example 2

Let a, b, and k be any real numbers. Show that $\lim_{x \to a} (kx + b) = ka + b$.

Solution Let ϵ be an arbitrary positive number. First we note that $|f(x) - L| = |(kx + b) - (ka + b)| = |k||x - a|$. There are two cases.

Case 1. Assume $k = 0$. Then $|f(x) - L| < \epsilon$ for all x. Hence, no matter how δ is chosen, $0 < |x - a| < \delta$ implies $|f(x) - L| < \epsilon$, for all x.

Case 2. Assume $k \neq 0$. Choose $\delta = \epsilon/|k|$. Then $0 < |x - a| < \delta$ implies $|f(x) - L| = |k||x - a| < |k|\delta = \epsilon$, for all x.

There are two special consequences of Example 2: (1) When $k = 0$, we see that $\lim_{x \to a} b = b$; this result is analogous to Example 4 in Section 9.1. (2) When $b = 0$ and $k = 1$, we obtain $\lim_{x \to a} x = a$; compare this result to Example 2 in Section 9.2.

Example 3

Show that $\lim_{x \to 2} \sqrt{x} = \sqrt{2}$.

Solution Let $\epsilon > 0$. We wish to find a $\delta > 0$ such that $0 < |x - 2| < \delta$ implies $|\sqrt{x} - \sqrt{2}| < \epsilon$. By rationalizing, we obtain

$$\left|\sqrt{x} - \sqrt{2}\right| = \frac{|x - 2|}{\sqrt{x} + \sqrt{2}}$$

Now, $\sqrt{x} + \sqrt{2} > \sqrt{2}$. Hence,

$$\frac{1}{\sqrt{x} + \sqrt{2}} < \frac{1}{\sqrt{2}}$$

and

$$\frac{|x - 2|}{\sqrt{x} + \sqrt{2}} < \frac{|x - 2|}{\sqrt{2}}$$

We chose $\delta = \epsilon\sqrt{2}$. Assume $0 < |x - 2| < \delta$. Then

$$\left|\sqrt{x} - \sqrt{2}\right| = \frac{|x - 2|}{\sqrt{x} + \sqrt{z}} < \frac{|x - 2|}{\sqrt{2}} < \frac{\delta}{\sqrt{2}} = \frac{\epsilon}{\sqrt{2}}\sqrt{2} = \epsilon$$

It is possible to state and prove theorems analogous to Theorems 1 through 9 in Section 9.2. We prove one of these as an example.

Theorem 1   (Uniqueness of Limits) If $\lim_{x \to a} f(x) = L$ and $\lim_{x \to a} f(x) = K$, then $L = K$.

Proof. Suppose $L \neq K$. Then

$$\frac{|L - K|}{2} > 0$$

By the definition of a limit, there is a δ_1 such that $0 < |x - a| < \delta_1$ implies

$$|f(x) - L| < \frac{|L - K|}{2}$$

Similarly, there is a δ_2 such that $0 < |x - a| < \delta_2$ implies

$$|f(x) - K| < \frac{|L - K|}{2}$$

Let $0 < |x - a| < \min(\delta_1, \delta_2)$. Then

$$|L - K| = |(f(x) - K) - (f(x) - L)| < |(f(x) - K)| + |(f(x) - L)|$$

$$< \frac{|L - K|}{2} + \frac{|L - K|}{2} = |L - K|$$

But this is a contradiction, and it follows that $L = K$. $\qquad\square$

For convenience's sake we state the remaining theorems. Exercise Set 9.3 requests a proof of each.

Theorem 2 $\lim_{x \to a} f(x) = L$ if and only if $\lim_{x \to a} (f(x) - L) = 0$.

Theorem 3   (Sum Theorem) If $\lim_{x \to a} f(x) = L$ and $\lim_{x \to a} g(x) = K$, then $\lim_{x \to a} (f(x) + g(x)) = L + K$.

For Theorems 4 and 5, a definition analogous to the definition of "f is bounded on $[c, \infty)$" is needed. Accordingly, let a be an arbitrary real number, and suppose that for all $\delta > 0$, there is an x such that $0 < |x - a| < \delta$ and x is in the domain of f.

Definition. f is **bounded at** a if there exist real numbers δ and q such that $|f(x)| < q$ for all $x \in \text{dom}(f)$ such that $0 < |x - a| < \delta$.

Theorem 4   If $\lim_{x \to a} f(x) = L$, then f is bounded at a.

Theorem 5   If $\lim_{x \to a} g(x) = 0$ and f is bounded at a, then $\lim_{x \to a} f(x)g(x) = 0$.

Theorem 6   (Product Theorem) If $\lim_{x \to a} f(x) = L$ and $\lim_{x \to a} g(x) = K$, then $\lim_{x \to a} (f(x) \cdot g(x)) = L \cdot K$.

Theorem 7   If $\lim_{x \to a} g(x) = K$ and $K \neq 0$, then there exists a $\delta > 0$ such that

$$|g(x)| > \frac{|K|}{2}$$

for all x such that $0 < |x - a| < \delta$.

Theorem 8   (Reciprocal Theorem) If $\lim_{x \to a} g(x) = K$ and $K \neq 0$, then $\lim_{x \to a} (1/g(x)) = 1/K$.

Corollary. (Quotient Theorem) If $\lim_{x \to \infty} f(x) = L$ and $\lim_{x \to a} g(x) = K$ and $K \neq 0$, then $\lim_{x \to a} (f(x)/g(x)) = L/K$.

Theorem 9 (Squeezing Theorem) If $\lim_{x \to a} f(x) = L = \lim_{x \to a} g(x)$ and $f(x) < h(x) < g(x)$ for all x, then $\lim_{x \to a} h(x) = L$.

In the definition of $\lim_{x \to \infty} f(x)$, the condition $0 < |x - a| < \delta$ implies that $x \neq a$. Hence, if $\lim_{x \to a} f(x) = L$ and $g(x) = f(x)$ for all $x \neq a$, then $\lim_{x \to a} g(x) = L$.

Example 4

Let

$$g(x) = \begin{cases} x^2 + 1 & \text{if } x \neq 2 \\ 7 & \text{if } x = 2 \end{cases}$$

Show that $\lim_{x \to 2} g(x) = 5$.

Solution By the remarks preceding this example, $\lim_{x \to 2} g(x) = \lim_{x \to 2} (x^2 + 1)$. Hence, $\lim_{x \to 2} g(x) = 5$, by Example 1.

Note that in Example 4 the result would be the same even if the function g were not defined at $x = 2$.

Example 2

Show that

$$\lim_{x \to 1} \frac{x^3 - 1}{x^2 - 1} = \frac{3}{2}$$

Solution We use some of the preceding theorems. First, note that

$$\frac{x^3 - 1}{x^2 - 1} = \frac{x^2 + x + 1}{x + 1}$$

for $x \neq 1$. Hence, we need only show that

$$\lim_{x \to 1} \frac{x^2 + x + 1}{x + 1} = \frac{3}{2}$$

By Example 2, $\lim_{x \to 1} x = 1$ and $\lim_{x \to 1} 1 = 1$. By the product theorem, $\lim_{x \to 1} x^2 = 1$. By the sum theorem, $\lim_{x \to 1} (x + 1) = 2$ and $\lim_{x \to 1} (x^2 + x + 1) = 3$. The conclusion now follows from the quotient theorem.

EXERCISE SET 9.3

1. Write the definition of $\lim_{x \to a} f(x) = L$ using quantifiers.

2. Write the definition of $\lim_{x \to a} f(x) \neq L$ using quantifiers.

3. Prove that $\lim_{x \to a} x^2 = a^2$ for any a. Do not use the product theorem in the proof.

4. Prove Theorem 2.

5. Prove Theorem 3.

6. Prove Theorem 4.

7. Prove Theorem 5.

8. Prove Theorem 6.

9. Prove Theorem 7.

10. Prove Theorem 8.

11. Prove the quotient theorem.

12. Prove the squeezing theorem.

13. Prove that $\lim_{x \to a} \sqrt{x} = \sqrt{a}$ for any nonnegative a.

14. *Prove:* If $\lim_{x \to a} f(x) = L$ and $L > 0$, then $\lim_{x \to a} \sqrt{f(x)} = \sqrt{L}$.

15. Let $\lim_{x \to a} f(x) = L$. Prove that $\lim_{x \to a} f(x)^n = L^n$ for $n = 0, 1, 2, \ldots$.

16. *Prove:* If $\lim_{x \to 0} f(x) = L$ and k is any constant, then $\lim_{x \to 0} f(kx) = L$.

For Exercises 17–20 you may assume that $\lim_{x \to 0}(1 + x)^{1/x} = e$. In each of these exercises, guess the limit and prove your conjecture.

17. $\lim_{x \to 0}(1 + x)^{2/x}$

18. $\lim_{x \to 0}(1 + x)^{1/2x}$

19. $\lim_{x \to 0}(1 + 2x)^{1/x}$

20. $\lim_{x \to 0}(1 + ax)^{1/bx}$, where $ab \neq 0$

21. Prove that $\lim_{x \to 0} 1/x$ does not exist.

22. Prove that $\lim_{x \to 3} 1/(x - 3)$ does not exist.

23. Prove that $\lim_{x \to 0} x \sin(1/x) = 0$.

24. Prove that $\lim_{x \to 0} \sin(1/x)$ does not exist.

In Exercises 25, 26, 29, and 30 you may assume that $\lim_{x \to 0} \sin x = 0$, $\lim_{x \to 0} \cos x = 1$, and $\cos \tau < (\sin \tau)/\tau < 1$ for $0 < |\tau| < 1$.

25. Prove that $\lim_{x \to 0} \sin x/x = 1$.

26. Prove that $\lim_{x \to 0}(1 - \cos x)/x = 0$.

In Exercises 27–30, guess the limit and prove your conjecture.

27. $\lim_{x \to 1} g(x)$ where

$$g(x) = \begin{cases} x^3 - 1 & \text{if } x \neq 1 \\ 7 & \text{if } x = 1 \end{cases}$$

28. $\lim_{x \to 1} \frac{(x^4 - 1)}{(x^3 - 1)}$

29. $\lim_{x \to 0}(\sin \pi x)/x$

30. $\lim_{x \to 0}(1 - \cos x)/x^2$

We have already seen how important the ϵ-property is. For example, we used it to prove that $\lim_{x \to \infty} 1/x = 0$. In what follows, we will introduce the **completeness property** of the real number system. It will be used to prove the ϵ-property.

> *The Completeness Property.* If A is a nonempty subset of $\mathcal{R}$ that is bounded above, then there exists a $v \epsilon \mathcal{R}$ such that v is a least upper bound of A.

In other words, every nonempty subset of the real numbers that is bounded above has a least upper bound. The completeness property is sometimes called the **least upper bound property.** It guarantees that the real number line, unlike the rational numbers, has no gaps.

Theorem 1 (The ϵ-property) For every $\epsilon > 0$, there exists a positive integer m such that $1/m < \epsilon$.

Proof. Let $\epsilon > 0$ be given, and suppose there is no positive integer m such that $1/m < \epsilon$. Then $1/n \geq \epsilon$ for all n, and hence, $n\epsilon \leq 1$ for all integers n. Let $A = \{n\epsilon : n$ is an integer$\}$. By assumption, A is bounded above by 1. Also, by the completeness property, there is a least upper bound v of A. Since v is the *least* upper bound of A, $v - \epsilon$ is not an upper bound of A. Hence, there exists a positive integer m such that $m\epsilon > v - \epsilon$. So $(m + 1)\epsilon > v$. But $(m + 1)\epsilon$ is an element of A. Hence, we have a contradiction to the fact that v is an upper bound of A. Therefore, there exists a positive integer m such that $1/m < \epsilon$. $\qquad\square$

Recall that the ϵ-property is equivalent to the property that for any real number w, there is a positive integer m such that $m > w$. Thus, the ϵ-property says that the positive integers are not bounded above in the real numbers.

The following corollary, called the **Archimedean Property,** is equivalent to the ϵ-property.

Corollary. (The Archimedean Property) For any positive real numbers a and b, there exists a positive integer q such that $a < qb$.

The Archimedean property is named for Archimedes. It was originally considered to be an axiom concerning magnitudes and can be stated informally as follows: Given two magnitudes of the same kind, we can find a multiple of the smaller that exceeds the larger.

Rationals and Irrationals

Recall that a rational number is any number of the form m/n, where m and n are integers and $n \neq 0$. From Exercise 2, if r and s are rational numbers, then so are $r + s$ and $r \cdot s$.

Theorem 2   There is a rational number in any interval (a, b) of $\mathcal{R}$.

Proof. We use the ϵ-property for $\mathcal{R}$. Assume a and b are any two real numbers such that $a < b$.

Case 1. Assume $0 < a$. By the ϵ-property, there exists a positive integer m such that $1/m < b - a$. We will find a positive integer p such that p/m is in the interval (a, b) (see Figure 9–4). By the Archimedean property, there exists a positive integer n such that $n/m > a$. Now, choose the least positive integer p such that $p/m \geq a$. We show that p/m is in the interval (a, b). Suppose not. Then $p/m > b$. But then, $(p - 1)/m = p/m - 1/m > b - (b - a) = a$, since $1/m < b - a$. But this contradicts the fact that p is the least positive integer n such that $n/m > a$.

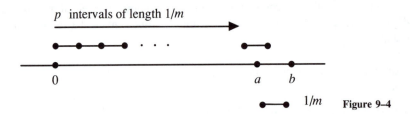

p intervals of length 1/m

0   a   b

1/m   **Figure 9–4**

Case 2. Assume $a \leq 0 < b$. Then there exists a positive integer m such that $1/m < b$. Clearly, $1/m$ is in the interval (a, b).

Case 3. Assume $a < b \leq 0$. Then $0 \leq -b < -a$. Now, by Cases 1 and 2, there is a rational number q in the interval $(-b, -a)$. Therefore, $-q$ is in the interval (a, b). □

Before proving the next theorem, recall that $\sqrt{2}$ is irrational. Also, the product of an irrational number with a rational number is an irrational number (see Exercise 4).

Theorem 3   There is an irrational number in any interval (a, b) of $\mathcal{R}$.

Proof. Let a and b be any two real numbers such that $a < b$. Then $a/\sqrt{2} < b/\sqrt{2}$. By Theorem 2, there is a rational number q in the interval $(a/\sqrt{2}, b/\sqrt{2})$. Hence, $q\sqrt{2}$ is in the interval (a, b). Now, since $q\sqrt{2}$ is the product of an irrational number and a rational number, it is irrational. Therefore, $q\sqrt{2}$ is an irrational number in (a, b). □

Example 1

Let

$$f(x) = \begin{cases} x & \text{if } x \text{ is rational} \\ 1 & \text{if } x \text{ is irrational} \end{cases}$$

a. Prove that $\lim_{x \to 1} f(x) = 1$.

b. Prove that $\lim_{x \to a} f(x)$ does not exist for $a \neq 1$.

Solution

a. Let $\epsilon > 0$ be arbitrary. Now, $|f(x) - 1| = |x - 1|$ if x is rational and $|f(x) - 1| = 0$ if x is irrational. In either case, $0 < |x - 1| < \delta$ implies $|f(x) - 1| < \delta$. So choose $\delta = \epsilon$.

b. Assume $a \neq 1$, and suppose $\lim_{x \to a} f(x) = L$. We will derive a contradiction.

Case 1. Assume $L \neq 1$. Let $\epsilon = |L - 1|/2$. Then there exists a $\delta > 0$ such that $0 < |x-a| < \delta$ implies $|f(x)-L| < |L-1|/2$. But no matter how small δ is chosen, there is an irrational number z such that $0 < |z-a| < \delta$. Hence, $f(z) = 1$ and $|f(z)-L| < |L-1|/2$, a contradiction.

Case 2. Assume $L = 1$. Let $\epsilon = |a - 1|/2$. Then there exists a $\delta > 0$ such that $0 < |x - a| < \delta$ implies $|f(x) - 1| < |a - 1|/2$.
If $a > 1$, choose a rational number q in $(a, a+\delta)$. Then $|q-1| < (a-1)/2$, since $f(q) = q$. Hence, $q < 1+(a-1)/2 = 1/2+a/2 < a$. But $q < a$ is a contradiction to q is in $(a, a+\delta)$. If $a < 1$, choose a rational number q in $(a - \delta, a)$. Then $|q - 1| < (a - 1)/2$. Hence, $q > 1 - (a - 1)/2 = 3/2 - a/2 > 3a/2 - a/2 = a$, a contradiction.

EXERCISE SET 9.4

1.   (a) Prove the Archimedean property.
     (b) Use the Archimedean property to prove the ϵ-property.

2. Let r and s be rational numbers.
     (a) Prove that $r + s$ is a rational number.
     (b) Prove that $r \cdot s$ is a rational number.

3. Prove that the sum of an irrational number and a rational number is an irrational number.

4. Prove that the product of an irrational number and a rational number is an irrational number.

5. Prove that every interval (a, b) contains infinitely many rational numbers.

6. Prove that every interval (a, b) contains infinitely many irrational numbers.

7. Let

$$f(x) = \begin{cases} 1 \text{ if } x \text{ rational} \\ -1 \text{ if } x \text{ is irrational} \end{cases}$$

Prove that $\lim_{x \to a} f(x)$ does not exist for any real number a.

8. Let

$$f(x) = \begin{cases} 1 \text{ if } x \text{ is rational} \\ x - 1 \text{ if } x \text{ is irrational} \end{cases}$$

     (a) Prove that $\lim_{x \to 2} = 1$.
     (b) Prove that $\lim_{x \to a} f(x)$ does not exist for $a \neq 2$.

9. Let $B = \{x : x^2 < 2\}$.
     (a) Prove that B is bounded above.
     (b) Find the least upper bound of B.

10. Let $B = \{x : x^3 < 8\}$.
     (a) Prove that B is bounded above.
     (b) Find the least upper bound of B.

11. Let p/q be a rational number such that $0 < p/q < 1$. Prove that there exists a positive integer n such that

$$\frac{1}{n+1} \leq \frac{p}{q} < \frac{1}{n}$$

12. This exercise shows that the set of rational numbers Q does not satisfy the completeness property: If A is a nonempty subset of Q that is bounded above, then there exists a $v \in Q$ such that v is a least upper bound of A.

Let $A = \{x : x \in Q : \text{and } x^2 < 2\}$.

    (a) Prove that A is bounded above by a rational number.

    (b) Prove that no rational number can be a least upper bound of A.

13. *Prove:* If A is a nonempty subset of $\mathcal{R}$ that is bounded below, then there exists a $v \in \mathcal{R}$ such that v is a greatest lower bound of A.

14. In this exercise we outline the steps showing that there is a positive real number u such that $u^2 = 2$. In other words, we are verifying the existence of $\sqrt{2}$. Verify each step. Let $B = \{r : r \in Q : \text{and } r^2 < 2\}$.

    (1) B has a least upper bound. Call it u.

    (2) $1 < u$.

    (3) $u^2 \geq 2$. For if $u^2 < 2$, there exists a positive integer m such that $1/m < (2 - u^2)/3u$. Note that $1/m^2 < u/m$, and obtain $(u + 1/m)^2 < 2$, yielding a contradiction.

    (4) $u^2 \leq 2$. For if $u^2 > 2$, there exist a positive integer m such that $1/m < (u^2 - 2)/2u$. Hence, $2 < (u - 1/m)^2$, a contradiction.

15. Prove the existence of $\sqrt{3}$ by steps similar to those used in Exercise 14.

16. In this exercise we prove that every positive real number has a positive nth root. Let v be a positive number and n be a positive integer. Prove that there is a positive number u such that $u^n = v$.

KEY CONCEPTS

The ϵ-property for $\mathcal{R}$

Definition of $\lim_{x \to \infty} f(x)$

Bounded on $[c, \infty)$

Theorems involving limits
    of the type $\lim_{x \to \infty} f(x)$

Definition of $\lim_{x \to a} f(x)$

Bounded at a

Theorems involving limits
    of the type $\lim_{x \to a} f(x)$

Completeness property for $\mathcal{R}$

Proof of the ϵ-property for $\mathcal{R}$

Archimedean property for $\mathcal{R}$

Rational numbers and irrational numbers

REVIEW EXERCISES

1. Show that $\lim_{x \to \infty} \frac{3x^3 - 2x + 1}{5x^3 + 2} = \frac{3}{5}$.

2. Let

$$f(x) = \begin{cases} 1 & \text{if } x \in \mathcal{N} \\ (x+1)/x & \text{otherwise} \end{cases}$$

Find $\lim_{x \to \infty} f(x)$, and prove that your answer is correct.

3. Let

$$f(x) = \begin{cases} 1 \text{ if } x \text{ is an even integer} \\ -1 \text{ if } x \text{ is an odd integer} \\ 0 \text{ otherwise} \end{cases}$$

Prove that $\lim_{x \to \infty} f(x)$ does not exist.

4. Prove that $\lim_{x \to \infty} \frac{1}{x^2} \sin \frac{x^3}{x+1} = 0$

5. Assume $\lim_{x \to a} f(x) = L$ and $L > 0$. Prove that if $\lim_{x \to a} f(x)^{1/q}$ exists, then $\lim_{x \to a} f(x)^{1/q} = L^{1/q}$ for $q = 1, 2, \ldots$. (*Hint:* First show that there exists a $\delta > 0$ such that $f(x) \geq 0$ for all $x, 0 < |x - a| < \delta$. Then let $g(x) = f(x)^{1/q}$. Hence, $g(x)^q = f(x)$. Now use Exercise 15 in Section 9.3.)

6. Assume $\lim_{x \to a} f(x) = L$ and $L > 0$. Prove that if $\lim_{x \to a} f(x)^{1/q}$ exists, then $\lim_{x \to a} f(x)^{p/q} = L^{p/q}$ for $p = 0, 1, 2, \ldots$ and $q = 1, 2, 3, \ldots$. (*Hint:* Use Exercise 15 in Section 9.3 and Exercise 5 above.)

7. Is the sum of two irrationals irrational? Is the product of two irrationals irrational? Explain.

8. Let

$$f(x) = \begin{cases} x \text{ if } x \text{ is rational} \\ -x \text{ if } x \text{ is irrational} \end{cases}$$

For which values of a does $\lim_{x \to a} f(x)$ exist? Prove that your answer is correct.

9. *Prove:* If $\lim_{x \to 0} f(x) = L$, then $\lim_{x \to \infty} f(1/x) = L$.

10. Find an example to show that the converse of Exercise 9 is false.

11. Let $A = \{1 - 1/(n + 1) : n \epsilon \mathcal{N}\}$.
   (a) Find lub(A).
   (b) Prove that your answer in part a is correct.

12. Let a and b be real numbers such that $0 < a < b$. Prove that there are natural numbers k and n such that $a < k/2^n < b$.

SUPPLEMENTARY EXERCISES

   A function f is **asymptotic to the straight line** $y = kx + b$ if $\lim_{x \to \infty} (f(x) - (kx + b)) = 0$. If $k = 0$, then f is asymptotic to the horizontal line $y = b$, and we say that f has a **horizontal asymptote.** In particular, if $k = 0$ and $b = 0$, the horizontal asymptote is the x-axis.

   In Exercises 1–4, a function and a straight line are given. Prove that the function is asymptotic to the straight line.

   1. $f(x) = 2/(x + 5)$; the x-axis.
   2. $f(x) = (2 - \pi x)/(3x + 1)$; $y = -\pi/3$.
   3. $f(x) = (x^2 + 2\pi x + 1)/(x + \pi)$; $y = x + \pi$.
   4. $f(x) = \sqrt{x^2 + 6x}$; $y = x + 3$.

For the functions given in Exercises 5–10, guess the asymptote and prove your conjecture. Assume that a, b, and c are arbitrary real numbers.

5. $f(x) = (2 + ax)/(bx + 1)$
6. $f(x) = (x^2 + 2bx + c)/(x + b)$
7. $f(x) = (x^2 + 2bx + c)/(x + a)$
8. $f(x) = \sqrt{x^2 + 2bx}$
9. $f(x) = (1 + 1/x)^x$
10. $f(x) = x \cdot \sin(1/x)$
11. Let $f : \mathcal{R} \to \mathcal{R}$ be an increasing function.
    (a) Prove that $\{f(x) : x < a\}$ has a least upper bound, say, u.
    (b) Let $\lim_{x \to a} f(x) = L$. Prove that $L = u$.
12. Let $f : \mathcal{R} \to \mathcal{R}$ be a decreasing function.
    (a) Prove that $\{f(x) : x > a\}$ has a least upper bound, say, u.
    (b) Let $\lim_{x \to a} f(x) = L$. Prove that $L = u$.
13. *Prove:* $\lim_{x \to \infty} f(x) = L$ if and only if for every $\epsilon > 0$, there is an $n \epsilon \mathcal{N}$ such that $f[(n, \infty)] \subseteq (L - \epsilon, L + \epsilon)$.
14. *Prove:* $\lim_{x \to \infty} f(x) = L$ if and only if for every $\epsilon > 0$, there is an $n \epsilon \mathcal{N}$ such that $(n, \infty) \subseteq f^{-1}[(L - \epsilon, L + \epsilon)]$.

The following information pertains to Exercises 15 and 16. For $\delta > 0$, let $D(a, \delta) = (a - \delta, a + \delta) - \{a\}$. The set $D(a, \delta)$ is called a **deleted neighborhood** of a.

15. *Prove:* $\lim_{x \to a} f(x) = L$ if and only if for every $\epsilon > 0$, there is a $\delta > 0$ such that $f[D(a, \delta)] \subseteq (L - \epsilon, L + \epsilon)$.
16. *Prove:* $\lim_{x \to a} f(x) = L$ if and only if for every $\epsilon > 0$, there is a $\delta > 0$ such that $D(a, \delta) \subseteq f^{-1}[(L - \epsilon, L + \epsilon)]$.

Appendix A

References

1. Apostol, T. M. *Introduction to Analytic Number Theory.* New York: Springer-Verlag, 1976.
2. Boole, G. E. *The Laws of Thought.* New York: Dover, 1951.
3. Enderton, H. *Elements of Set Theory.* New York: Academic Press, 1977.
4. Enderton, H. *Mathematical Introduction to Logic.* New York: Academic Press, 1972.
5. Gardner, M. *Aha! Gotcha.* New York: W. H. Freeman, 1982.
6. Halmos, P. *Naive Set Theory.* New York: Springer-Verlag, 1974.
7. Henkin, L. *On Mathematical Induction.* American Mathematical Monthly, 67(1960) pp. 323-337.
8. Lay, S. *Analysis: An Introduction to Proof.* 2nd ed. Englewood Cliffs, N.J: Prentice Hall, 1990.
9. Niven, I., and Zuckerman, H. S. *Introduction to the Theory of Numbers.* New York: John Wiley and Sons, 1980.
10. Pinter, C.A. *A Book of Abstract Algebra.* New York: McGraw-Hill, 1982.
11. Solow, D. *How to Read and Do Proofs.* New York: John Wiley and Sons, 1982.
12. Stein, S. *Calculus and Analytic Geometry.* 4th ed. New York: McGraw-Hill, 1987.
13. Wilder, R. L. *Introduction to the Foundations of Mathematics.* 2nd ed. New York: John Wiley and Sons, 1965.

Appendix B

Solutions

Section 1.1

1. *Goal:* Prove that $x = y$.

   Hypothesis: x and y are nonnegative numbers, $\sqrt{xy} = (x + y)/2$.

   Definitions necessary: None is used.

   Previously proven facts or laws of logic: Facts about algebra.

3. *Goal:* Prove that the area of a triangle is $c^2/4$.

   Hypothesis: a and b are the lengths of the sides, and c is the length of the hypotenuse, of a right triangle that is also isosceles.

   Definitions necessary: Isosceles triangle.

   Previously proven facts or laws of logic: The Pythagorean theorem, the formula for the area of a triangle, facts about algebra.

   Proof: Since the triangle is isosceles, $a = b$, and hence, $(a - b)^2 = 0$. Therefore, $a^2 - 2ab + b^2 = 0$. Adding $2ab$ to both sides yields $a^2 + b^2 = 2ab$. The Pythagorean theorem says that $a^2 + b^2 = c^2$ for a right triangle with hypotenuse of length c, and hence, $2ab = c^2$. Therefore, $ab/2 = c^2/4$. The area of the given triangle is $ab/2$. So the area of the triangle is $c^2/4$.

5. *Goal:* Prove that $x + y$ is even.

   Hypothesis: x and y are odd integers.

   Definitions necessary: Even and odd integer.

   Previously proven facts or laws of logic: Facts about algebra.

7. *Goal:* Prove that $x + 1/x \geq 2$.

Hypothesis: x is a positive real number.

Definitions necessary: None is used.

Previously proven facts or laws of logic: Facts about algebra.

Proof: Since $(x-1)^2 \geq 0$ we have $x^2 - 2x + 1 \geq 0$. Therefore, $x^2 + 1 \geq 2x$. Dividing by x yields $x + 1/x \geq 2$.

9. *Goal:* Prove that $\sin x \leq x$ for all nonnegative real numbers x.

Hypothesis: x is a nonnegative real number.

Definitions necessary: Nondecreasing function.

Previously proven facts or laws of logic: The theorem of calculus that states that a function with a nonnegative derivative is nondecreasing.

Section 1.2

1. 1. p, q, and r are statements by S1. 2. $\neg q$ is a statement by S6. 3. $p \Rightarrow \neg q$ is a statement by S4 and Steps 1 and 2. 4. $(p \Rightarrow \neg q) \wedge r$ is a statement by S2 and Steps 1 and 3. 5. $(p \Rightarrow \neg q) \wedge r \Rightarrow q$ is a statement by S4 and Steps 1 and 4. The main connective is the $\Rightarrow$ on the right.

3. 1. p, q, and r are statements by S1. 2. $p \Rightarrow q$ is a statement by S4 and Step 1. 3. $p \wedge r$ is a statement by S2 and Step 1. 4. $q \wedge r$ is a statement by S2 and Step 1. 5. $(p \wedge r) \Rightarrow (q \wedge r)$ is a statement by S4 and Steps 3 and 4. 6. $(p \Rightarrow q) \Rightarrow [(p \wedge r) \Rightarrow (q \wedge r)]$ is a statement by S4 and Steps 2 and 5. The main connective is the middle $\Rightarrow$.

5. 1. p, q, r, and s are statements by S1. 2. $p \Rightarrow q$ is a statement by S4 and Step 1. 3. $r \Rightarrow s$ is a statement by S4 and Step 1. 4. $(p \Rightarrow q) \wedge (r \Rightarrow s)$ is a statement by S2 and Steps 2 and 3. 5. $p \vee r$ is a statement by S3 and Step 1. 6. $q \vee s$ is a statement by S3 and Step 1. 7. $(p \vee r) \Rightarrow (q \vee s)$ is a statement by S4 and Steps 5 and 6. 8. $[(p \Rightarrow q) \wedge (r \Rightarrow s)] \Rightarrow [(p \vee r) \Rightarrow (q \vee s)]$ is a statement by S4 and Steps 4 and 7. The main connective is the third $\Rightarrow$ from the left.

7. **(a)**

P	Q	$P \wedge Q \Rightarrow P$	
T	T	T	T
T	F	F	T
F	T	F	T
F	F	F	T

(b)

P	Q	$P \Rightarrow P \vee Q$	
T	T	T	T
T	F	T	T
F	T	T	T
F	F	T	F

9. (a)

P	Q	$P \wedge (\neg P \vee Q) \Rightarrow Q$			
T	T	T	F	T	T
T	F	F	F	F	T
F	T	F	T	T	T
F	F	F	T	T	T

(b)

P	Q	$\neg P \wedge (P \Rightarrow Q) \Rightarrow Q$			
T	T	F	F	T	T
T	F	F	F	F	T
F	T	T	T	T	T
F	F	T	T	T	F

11.

P	Q	R	$(P \Rightarrow Q) \Rightarrow [(P \vee R) \Rightarrow (Q \vee R)]$				
T	T	T	T	T	T	T	T
T	T	F	T	T	T	T	T
T	F	T	F	T	T	T	T
T	F	F	F	T	T	F	F
F	T	T	T	T	T	T	T
F	T	F	T	T	F	T	T
F	F	T	T	T	T	T	T
F	F	F	T	T	F	T	F

13.

P	Q	R	$(P \vee Q \Rightarrow R) \Rightarrow (P \Rightarrow R) \wedge (Q \Rightarrow R)$					
T	T	T	T	T	T	T	T	T
T	T	F	T	F	T	F	F	F
T	F	T	T	T	T	T	T	T
T	F	F	T	F	T	F	F	T
F	T	T	T	T	T	T	T	T
F	T	F	T	F	T	T	F	F
F	F	T	F	T	T	T	T	T
F	F	F	F	T	T	T	T	T

15.

P	Q	R	$(P \wedge Q \Rightarrow R) \Rightarrow (P \Rightarrow R) \vee (Q \Rightarrow R)$					
T	T	T	T	T	T	T	T	T
T	T	F	T	F	T	F	F	F
T	F	T	F	T	T	T	T	T
T	F	F	F	T	T	F	T	T
F	T	T	F	T	T	T	T	T
F	T	F	F	T	T	T	T	F
F	F	T	F	T	T	T	T	T
F	F	F	F	T	T	T	T	T

17.

P	Q	R	¬ (P ∧ ¬ Q) ∨ (R ⇒ Q)				
T	T	T	T	F F	T	T
T	T	F	T	F F	T	T
T	F	T	F	T T	F	F
T	F	F	F	T T	T	T
F	T	T	T	F F	T	T
F	T	F	T	F F	T	T
F	F	T	T	F T	T	F
F	F	F	T	F T	T	T

19. **(a)**

P	P ⊻ P
T	F
F	F

(b)

P	P ⊻ ¬ P	
T	T	F
F	T	T

21. **(a)**

P	Q	R	P ⊻ (Q ∧ R)	
T	T	T	F	T
T	T	F	T	F
T	F	T	T	F
T	F	F	T	F
F	T	T	T	T
F	T	F	F	F
F	F	T	F	F
F	F	F	F	F

(b)

P	Q	R	(P ⊻ Q) ∧ (P ⊻ R)		
T	T	T	F	F	F
T	T	F	F	F	T
T	F	T	T	F	F
T	F	F	T	T	T
F	T	T	T	T	T
F	T	F	T	F	F
F	F	T	F	F	T
F	F	F	F	F	F

23. $(\neg P \wedge Q) \vee (P \wedge \neg Q)$

25. No truth value has been assigned to p. A truth value must be T or F, but not both. If p is true, thèn p is false, and if p is false, then p is true. So we did not assign the truth value T to p.

Section 1.3

1. **(a)**

P Q	$\neg (P \lor Q)$		$\neg P \land \neg Q$		
T T	F	T	F	F	F
T F	F	T	F	F	T
F T	F	T	T	F	F
F F	T	F	T	T	T

(b)

P Q	$\neg (P \land Q)$		$\neg P \lor \neg Q$		
T T	F	T	F	F	F
T F	T	F	F	T	T
F T	T	F	T	T	F
F F	T	F	T	T	T

3. **(a)**

P Q R	$P \lor (Q \land R)$		$(P \lor Q) \land (P \lor R)$		
T T T	T	T	T	T	T
T T F	T	F	T	T	T
T F T	T	F	T	T	T
T F F	T	F	T	T	T
F T T	T	T	T	T	T
F T F	F	F	T	F	F
F F T	F	F	F	F	T
F F F	F	F	F	F	F

(b)

P	$P \lor \neg P$		$\mathcal{I}$
T	T	F	T
F	T	T	T

(c)

P	$P \land \neg P$		$\mathcal{O}$
T	F	F	F
F	F	T	F

5. **(a)**

P Q	$P \land Q \Rightarrow P$		$\mathcal{I}$
T T	T	T	T
T F	F	T	T
F T	F	T	T
F F	F	T	T

(b)

P	Q	R	$P \vee Q \Rightarrow R$		$(P \Rightarrow R) \wedge (Q \Rightarrow R)$		
T	T	T	T	T	T	T	T
T	T	F	T	F	F	F	F
T	F	T	T	T	T	T	T
T	F	F	T	F	F	F	T
F	T	T	T	T	T	T	T
F	T	F	T	F	T	F	F
F	F	T	F	T	T	T	T
F	F	F	F	T	T	T	T

7. (a)

p	q	$(p \vee \neg q) \vee (q \Rightarrow p)$			
T	T	T F	T	T	
T	F	T T	T	T	
F	T	F F	F	F	⟵ *counterexample*: pF, qT
F	F	T T	T	T	

(b)

p	q	r	$(p \wedge q) \vee r \Rightarrow (p \vee q) \wedge r$				
T	T	T	T	T	T	T	T
T	T	F	T	T	F	T	F	⟵ *counterexample*: pT, qT, rF
T	F	T	F	T	T	T	T
T	F	F	F	F	T	T	F
F	T	T	F	T	T	T	T
F	T	F	F	F	T	T	F
F	F	T	F	T	F	F	F	⟵ *counterexample*: pF, qF, rT
F	F	F	F	F	T	F	F

9.

p	q	$\neg (p \Rightarrow q)$		(a) $p \Rightarrow \neg q$		(b) $\neg p \Rightarrow \neg q$			(c) $\neg p \Rightarrow q$	
T	T	F	T	F	F	F	T	F	F	T
T	F	T	F	T	T	F	T	T	F	T
F	T	F	T	T	F	T	F	F	T	T
F	F	F	T	T	T	T	T	T	T	F

None is equivalent to $\neg(p \Rightarrow q)$.

11.

p	q	$q \Rightarrow p$	(a) $p \Rightarrow \neg q$		(b) $\neg p \Rightarrow \neg q$			(c) $\neg p \Rightarrow q$	
T	T	T	F	F	F	T	F	F	T
T	F	T	T	T	F	T	T	F	T
F	T	F	T	F	T	F	F	T	T
F	F	T	T	T	T	T	T	T	F

Only $\neg p \Rightarrow \neg q$ is equivalent to $q \Rightarrow p$.

13. (a) $P \Rightarrow Q \vee R \equiv \neg P \vee (Q \vee R)$
  (b) $P \wedge Q \Rightarrow R \equiv (\neg P \vee \neg Q) \vee R$

15. (a) $\neg(P \Rightarrow Q \vee R) \equiv \neg(\neg P \vee (Q \vee R))$

(b) $\neg(P \wedge Q \Rightarrow R) \equiv \neg((\neg P \vee \neg Q) \vee R)$

17. A counterexample to $\neg P$ would put an F on some line in the main column of the truth table for $\neg P$. It would then put a T on the same line of the truth table for P. However, the other lines of the truth table for P may still contain an F, and hence, P may not be a tautology.

Section 1.4

1.
1. P	(Pr)
2. $P \Rightarrow P \vee Q$	(Pr IV)
3. $P \vee Q$	(1, 2, MP)
4. $P \vee Q \Rightarrow S \wedge T$	(Pr)
5. $S \wedge T$	(3, 4, MP)

3.
1. P	(Hyp)
2. $P \Rightarrow \neg Q$	(Pr)
3. $\neg Q$	(1, 2, MP)
4. $S \vee Q$	(Pr)
5. $\neg Q \wedge (S \vee Q)$	(3, 4, Adj)
6. $\neg Q \wedge (S \vee Q) \Rightarrow S$	(Pr II)
7. S	(5, 6, MP)
8. $P \Rightarrow S$	(1, 7, DPI)

5.
1. R	(Hyp)
2. $R \Rightarrow \neg N$	(Pr)
3. $\neg N$	(1, 2, MP)
4. $P \Rightarrow N$	(Pr)
5. $(P \Rightarrow N) \Leftrightarrow (\neg N \Rightarrow \neg P)$	(Pr I)
6. $\neg N \Rightarrow \neg P$	(4, 5, Sub)
7. $\neg P$	(3, 6, MP)
8. $\neg P \Rightarrow \neg S$	(Pr)
9. $\neg S$	(7, 8, MP)
10. $R \Rightarrow \neg S$	(1, 9, DPI)

7.
1. R	(Hyp)
2. $P \Rightarrow Q$	(Pr)
3. $(P \Rightarrow Q) \Leftrightarrow (\neg Q \Rightarrow \neg P)$	(Pr I)
4. $\neg Q \Rightarrow \neg P$	(2, 3, Sub)
5. $\neg Q$	(Pr)
6. $\neg P$	(4, 5, MP)
7. $\neg P \wedge R$	(6, 1, Adj)
8. $\neg P \wedge R \Rightarrow S$	(Pr)
9. S	(7, 8, MP)
10. $R \Rightarrow S$	(1, 9, DPI)

9.

1. $\neg S$	(Neg of concl)
2. $R \Rightarrow S$	(Pr)
3. $(R \Rightarrow S) \Leftrightarrow (\neg S \Rightarrow \neg R)$	(Pr I)
4. $\neg S \Rightarrow \neg R$	(2, 3, Sub)
5. $\neg R$	(1, 4, MP)
6. $P \vee R$	(Pr)
7. $\neg R \wedge (P \vee R)$	(5, 6, Adj)
8. $\neg R \wedge (P \vee R) \Rightarrow P$	(PR II)
9. P	(7, 8, MP)
10. $P \Rightarrow S$	(PR)
11. S	(9, 10, MP)
12. $\neg S \wedge S$	(1, 11, Adj)
13. S	(1, 12, Contra)

11.　**(a)**

1. r	(Hyp)
2. $(p \Rightarrow q) \Leftrightarrow (\neg q \Rightarrow \neg p)$	(Pr)
3. $p \Rightarrow q$	(Pr)
4. $\neg q \Rightarrow \neg p$	(2, 3, Sub)
5. $\neg q$	(Pr)
6. $\neg p$	(4, 5, MP)
7. $\neg p \wedge r$	(1, 6, Adj)
8. $\neg p \wedge r \Rightarrow s$	(Pr)
9. s	(7, 8, MP)
10. $r \Rightarrow s$	(1, 9, DPI)

　(b) A counterexample is pF, qF, rF, sT.

13.　**(a)** A counterexample is dF, jT, nF, pT, rF.

　(b)

1. n	(Hyp)
2. $n \Rightarrow \neg j$	(Pr)
3. $\neg j$	(1, 2, MP)
4. $\neg j \Rightarrow \neg d$	(Pr)
5. $\neg d$	(3, 4, MP)
6. $\neg d \Rightarrow p$	(Pr)
7. p	(5, 6, MP)
8. $n \Rightarrow p$	(1, 7, DPI)

15. Let S be any statement.

1.　　$\neg S$	(Neg of concl)
2.　　.	(Steps 2 through n are steps in the proof of $P \wedge \neg P$.)
3.　　.	
⋮	
n.　　$P \wedge \neg P$	(We are given that $P \wedge \neg P$ can be derived from the premises.)
$n+1$. S	(1, n, Contra)

Section 1.5

1. 1. $P \wedge (Q \wedge R) \Leftrightarrow \neg\neg[P \wedge (Q \wedge R)]$ (T_1)
    2. $\Leftrightarrow \neg[\neg P \vee \neg(Q \wedge R)]$ $(T_{11}b)$
    3. $\Leftrightarrow \neg[\neg P \vee (\neg Q \vee \neg R)]$ $(T_{11}b)$
    4. $\Leftrightarrow \neg[(\neg P \vee \neg Q) \vee \neg R)]$ (T_7a)
    5. $\Leftrightarrow \neg(\neg P \vee \neg Q) \wedge \neg\neg R$ $(T_{11}a)$
    6. $\Leftrightarrow (\neg\neg P \wedge \neg\neg Q) \wedge \neg\neg R$ $(T_{11}a)$
    7. $\Leftrightarrow (P \wedge Q) \wedge R$ (T_1)
    8. $P \wedge (Q \wedge R) \Leftrightarrow (P \wedge Q) \wedge R$ $(1, 2, \ldots, 8)$

3. 1. $P \Rightarrow Q \Leftrightarrow \neg P \vee Q$ (T_{10})
    2. $\Leftrightarrow \neg P \vee \neg\neg Q$ (T_1)
    3. $\Leftrightarrow \neg\neg Q \vee \neg P$ (T_6a)
    4. $\Leftrightarrow \neg Q \Rightarrow \neg P$ (T_{10})
    5. $P \Rightarrow Q \Leftrightarrow \neg Q \Rightarrow \neg P$ $(1, 2, 3, 4)$

5. (a) 1. $\neg P$ (Hyp)
    2. $\neg P \Rightarrow \neg P \vee Q$ (T_5)
    3. $\neg P \vee Q$ $(1, 2, \text{MP})$
    4. $(P \Rightarrow Q) \Leftrightarrow (\neg P \vee Q)$ (T_{10})
    5. $P \Rightarrow Q$ $(3, 4, \text{Sub})$
    6. $\neg P \Rightarrow (P \Rightarrow Q)$ $(1, 5, \text{DPI})$

   (b) 1. $Q \Rightarrow (P \Rightarrow Q) \Leftrightarrow Q \Rightarrow (\neg Q \Rightarrow \neg P)$ (T_{13})
    2. $\Leftrightarrow \neg\neg Q \Rightarrow (\neg Q \Rightarrow \neg P)$ (T_1)
    3. $Q \Rightarrow (P \Rightarrow Q) \Leftrightarrow \neg\neg Q \Rightarrow (\neg Q \Rightarrow \neg P)$ $(1, 2)$
    4. $\neg\neg Q \Rightarrow (\neg Q \Rightarrow \neg P)$ (Exercise 5a)
    5. $Q \Rightarrow (P \Rightarrow Q)$ $(3, 4, \text{Sub})$

7. 1. $P \Rightarrow (R \Rightarrow Q) \Leftrightarrow \neg P \vee (R \Rightarrow Q)$ (T_{10})
    2. $\Leftrightarrow \neg P \vee (\neg R \vee Q)$ (T_{10})
    3. $\Leftrightarrow (\neg P \vee \neg R) \vee Q$ (T_7a)
    4. $\Leftrightarrow \neg(P \wedge R) \vee Q$ $(T_{11}b)$
    5. $\Leftrightarrow P \wedge R \Rightarrow Q$ (T_{10})
    6. $P \Rightarrow (R \Rightarrow Q) \Leftrightarrow P \wedge R \Rightarrow Q$ $(1, 2, 3, 4, 5)$

9. 1. $\neg Q \wedge (\neg Q \Rightarrow \neg P) \Rightarrow \neg P$ (T_9)
    2. $\neg Q \wedge (\neg Q \Rightarrow \neg P) \Rightarrow \neg P \Leftrightarrow (\neg Q \Rightarrow \neg P) \wedge \neg Q \Rightarrow \neg P$ (T_6b)
    3. $\Leftrightarrow (P \Rightarrow Q) \wedge \neg Q \Rightarrow \neg P$ (T_{13})
    4. $\neg Q \wedge (\neg Q \Rightarrow \neg P) \Rightarrow \neg P \Leftrightarrow (P \Rightarrow Q) \wedge \neg Q \Rightarrow \neg P$ $(2, 3)$
    5. $(P \Rightarrow Q) \wedge \neg Q \Rightarrow \neg P$ $(1, 4, \text{Sub})$

11. 1. $(P \wedge Q) \vee (P \wedge \neg Q) \Leftrightarrow P \wedge (Q \vee \neg Q)$ (T_8b)
    2. $\Leftrightarrow (Q \vee \neg Q) \wedge P$ (T_6b)
    3. $\Leftrightarrow P$ (Example 6)
    4. $(P \wedge Q) \vee (P \wedge \neg Q) \Leftrightarrow P$ $(1, 2, 3)$

13. **(a)**

P	Q	$P \wedge (\neg P \vee Q) \Rightarrow Q$			
T	T	T	F	T	T
T	F	F	F	F	T
F	T	F	T	T	T
F	F	F	T	T	T

(b) 1. $P \wedge (\neg P \vee Q) \Rightarrow Q \Leftrightarrow P \wedge (P \Rightarrow Q) \Rightarrow Q$ (T_{10})
2. $P \wedge (P \Rightarrow Q) \Rightarrow Q$ (T_9)
3. $P \wedge (\neg P \vee Q) \Rightarrow Q$ (1, 2, Sub)

Chapter 1 Review

1. *Goal:* Prove that x is even.

Hypothesis: x^3 is even.

Definitions necessary: Even and odd integer.

Previously proven facts or laws of logic: Facts about algebra and the contrapositive law of logic.

Proof: Let x be an odd integer. Then $x = 2n + 1$ for some integer n. $x^3 = (2n + 1)^3 = 8n^3 + 12n^2 + 6n + 1 = 2(4n^3 + 6n^2 + 3n) + 1$. Hence, x^3 is odd. So we have proved that if x is odd, then x^3 is odd. But this means the same as if x^3 is not odd, then x is not odd. Since not odd is the same as even, it follows that if x^3 is even, then x is even.

3. *Goal:* Prove that $x = y = z$.

Hypothesis: $y^2 + z^2 = 2x(y + z - x)$.

Definitions necessary: Square of a real number.

Previously proven facts or laws of logic: The square of a real number is nonnegative; if the square of a real number is zero, then the real number is zero.

5. $(p \Rightarrow q) \Leftrightarrow ((\neg r \vee p) \Rightarrow q)$

    1. p, q, and r are statements.   (S1)
    2. $p \Rightarrow q$ is a statement.   (1, S4)
    3. $\neg r$ is a statement.   (1, S6)
    4. $\neg r \vee p$ is a statement.   (1, 3, S3)
    5. $(\neg r \vee p) \Rightarrow q$ is a statement.   (1, 4, S4)
    6. $(p \Rightarrow q) \Leftrightarrow ((\neg r \vee p) \Rightarrow q)$ is a statement.   (2, 5, S5)

7. **(a)**   1. $\neg c$ (Neg of concl)
        2. $a \Rightarrow c$ (Pr)
        3. $(a \Rightarrow c) \Leftrightarrow (\neg c \Rightarrow \neg a)$ (Pr)
        4. $\neg c \Rightarrow \neg a$ (2, 3, Sub)
        5. $\neg a$ (1, 4, MP)
        6. $\neg a \Rightarrow \neg b$ (Pr)
        7. $\neg b$ (5, 6, MP)
        8. b (Pr)
        9. $b \wedge \neg b$ (7, 8, Adj)
        10. c (1, 9, Contra)

    (b) A counterexample is pF, qT, rF, sF.

9.
1. $(P \Leftrightarrow Q) \Leftrightarrow (P \Rightarrow Q) \wedge (Q \Rightarrow P)$ (T_{16})
2. $\Leftrightarrow (Q \Rightarrow P) \wedge (P \Rightarrow Q)$ (T_6b)
3. $\Leftrightarrow (Q \Leftrightarrow P)$ (T_{16})
4. $(P \Leftrightarrow Q) \Leftrightarrow (Q \Leftrightarrow P)$ (1, 2, 3)

Section 2.1

1. Assume $x > 1$. Since $1 > 0$, $x > 0$. Multiplying $x > 1$ by x, we obtain $x^2 > x$. Hence, $x^2 > 1$.

3. Assume $0 < x$ and $x < y$. Then $0 < y$. Multiply $x < y$ by x to obtain $x^2 < xy$. Multiply $x < y$ by y to obtain $xy < y^2$. Therefore, $x^2 < y^2$.

5. Assume $x \leq y$ and $y \leq z$. There are four cases.

Case 1. $x < y$ and $y < z$. Then $x < z$, and it follows that $x \leq z$.

Case 2. $x = y$ and $y < z$. Then $x < z$, and hence, $x \leq z$. The other two cases are similar.

7. Assume $z > 0$ and $x \leq y$.

Case 1. $x < y$. Then $xz < yz$, and hence, $xz \leq yz$.

Case 2. $x = y$. Then $xz = yz$, from which it follows that $xz \leq yz$.

9. Assume $x \leq y$ and $u \leq v$.

Case 1. $x < y$ and $u < v$. By Example 1, $x + u < y + v$. Therefore, $x + u \leq y + v$.

Case 2. $x < y$ and $u = v$. Then $x + u < y + u$. Hence, $x + u < y + v$. It now follows that $x + u \leq y + v$. The two remaining cases are similar to those just proven.

11. ($\Rightarrow$) Assume that $|x| \leq b$. By Example 6 and Exercise 10, $-|x| \leq x \leq |x|$. Therefore, $-b \leq -|x| \leq x \leq |x| \leq b$, and hence, $-b \leq x \leq b$.

($\Leftarrow$) Assume that $-b \leq x \leq b$. Then $-x \leq b$ and $x \leq b$. If $x \geq 0$, then $|x| = x \leq b$, and if $x \leq 0$, then $|x| = -x \leq b$.

13. By Exercise 12, $-(|a| + |b|) \leq a + b \leq |a| + |b|$. By Exercise 11, $|a + b| \leq |a| + |b|$.

15. Assume that x is even. Then $x = 2k$ for some k. It follows that $x^2 = (2k)^2 = 2(2k^2)$, which is even.

17. Assume that x and y are even. Then $x = 2k$ and $y = 2m$, for some k and m. Hence, $x + y = 2k + 2m = 2(k + m)$, which is even.

Section 2.2

1. Assume that $x > 0$ and $1/x \leq 0$. Then $x(1/x) \leq x \cdot 0$. Hence, $1 \leq 0$, which is a contradiction. Therefore, $1/x > 0$.

3. Assume that $x < y$ and $x^{1/3} \geq y^{1/3}$. By Exercise 2, $(x^{1/3})^3 \geq (y^{1/3})^3$, and hence, $x \geq y$, a contradiction.

5. Assume that $m \cdot n = 100$ and $m < 10$ and $n \leq 10$. If $n = 10$, then $m \cdot n < 100$, and this contradicts $m \cdot n = 100$. If $n < 10$, then by Practice Problem 1 in Section 2.1, $m \cdot n < 100$, a contradiction.

7. Assume that $x > 0$ and $y > 0$. Then, by Exercise 1, $1/y > 0$. Hence, $x(1/y) > x \cdot 0$; that is, $x/y > 0$.

9. Assume that $x \geq 0$, $y < 0$, and $x/y > 0$. Then $y(x/y) < y \cdot 0$; that is, $x < 0$, a contradiction.

11. We prove the contrapositive. Assume the negation of the conclusion, that is, x is even or y is even. In either case, xy is even.

13. We prove the contrapositive. Assume x is even and y is even. Then $x + y$ is even.

15. ($\Rightarrow$) Prove the contrapositive. Assume that x is odd and y is odd. Then $x = 2k + 1$ and $y = 2m+1$ for some integers k and m. Hence, $xy = (2k+1)(2m+1) = 4km+2m+2k+1 = 2(2km + m + k) + 1$, which is odd.

($\Leftarrow$) Use Exercise 11.

Section 2.3

1. *Basis step:* $1 = 2 - 1/2^0$. *Induction hypothesis:* $1 + \frac{1}{2} + \frac{1}{4} + \ldots + 1/2^k = 2 - 1/2^k$. *Induction step:* $1 + \frac{1}{2} + \frac{1}{4} + \ldots + 1/2^k + 1/2^{k+1} = 2 - 1/2^k + 1/2^{k+1} = 2 - 1/2^{k+1}$.

3. *Basis step:*

$$1 = \frac{1 - r^{0+1}}{1 - r}$$

Induction hypothesis:

$$1 + r + r^2 + \ldots + r^k = \frac{1 - r^{k+1}}{1 - r}$$

Induction step:

$$1 + r + r^2 + \ldots + r^k + r^{k+1} = \frac{1 - r^{k+1}}{1 - r} + r^{k+1} = \frac{1 - r^{k+1} + r^{k+1}(1 - r)}{1 - r} = \frac{1 - r^{k+2}}{1 - r}$$

5. *Basis step:* $1 \leq 2 \cdot 1/(1 + 1)$. *Induction hypothesis:* $1/1^2 + 1/2^2 + \ldots + 1/k^2 \leq 2k/(k + 1)$. Hence, $1/1^2 + 1/2^2 + \ldots + 1/k^2 + 1/(k+1)^2 \leq 2k/(k+1) + 1/(k+1)^2 = (2k^2 + 2k + 1)/(k+1)^2$. Now verify that $(2k^2 + 2k + 1)/(k + 1)^2 < 2(k + 1)/(k + 2)$ by the backwards/forwards method.

7. *Basis step:* $\sum_{i=1}^{1}(2i - 1)^3 = (2 \cdot 1 - 1)^3 = 1 = 1^2(2 \cdot 1^2 - 1)$. *Induction hypothesis:* $\sum_{i=1}^{k}(2i - 1)^3 = k^2(2k^2 - 1)$. Hence, $\sum_{i=1}^{k+1}(2i - 1)^3 = \sum_{i=1}^{k}(2i - 1)^3 + (2(k+1) - 1)^3 = k^2(2k^2 - 1) + (2(k + 1) - 1)^3 = (k + 1)^2(2(k + 1)^2 - 1)$. The last equality can be verified by multiplying both sides out and comparing the results.

9. *Basis step:* We are given a line segment of unit length. *Induction hypothesis:* We can construct a line segment of length $\sqrt{k}$. Using the straightedge and compass, construct a right triangle with legs of length 1 and $\sqrt{k}$. The hypotenuse of this triangle is of length $\sqrt{1^2 + \left(\sqrt{k}\right)^2} = \sqrt{k + 1}$.

11. *Conjecture:*

$$\sum_{i=1}^{n} \frac{1}{i(i + 1)} = \frac{n}{n + 1}$$

Basis step:

$$\sum_{i=1}^{1} \frac{1}{1(1 + 1)} = \frac{1}{1(1 + 1)} = \frac{1}{1 + 1}$$

Induction hypothesis:

$$\sum_{i=1}^{k} \frac{1}{i(i + 1)} = \frac{k}{k + 1}$$

Then

$$\sum_{i=1}^{k+1} \frac{1}{i(i+1)} = \sum_{i=1}^{k} \frac{1}{i(i+1)} + \frac{1}{(k+1)(k+2)} = \frac{k}{k+1} + \frac{1}{(k+1)(k+2)} = \frac{k+1}{k+2}$$

13. $D^1(f(x)) = D(f(x))$, where $D(f(x)) = f'(x)$

   $D^{k+1}(f(x)) = D(D^k(f(x)))$ [or $D^{k+1}(f(x)) = D^k(D(f(x)))$]

15. *Basis step:* $D^1(cf(x)) = D(cf(x)) = cD(f(x))$. *Induction hypothesis:* $D^k(cf(x)) = cD^k(f(x))$. $D^{k+1}(cf(x)) = D(D^k(cf(x))) = D(cD^k(f(x))) = cD(D^k(f(x))) = cD^{k+1}(f(x))$.

17. *Basis step:* $D^1(e^{bx}) = D(e^{bx}) = be^{bx} = b^1 e^{bx}$. *Induction hypothesis:* $D^k(e^{bx}) = b^k e^{bx}$. $D^{k+1}(e^{bx}) = D(D^k(e^{bx})) = D(b^k e^{bx}) = b^k D(e^{bx}) = b^k be^{bx} = b^{k+1} e^{bx}$.

19. In the induction hypothesis, we assume that $7^j = 1$ for all $0 \le j \le k$, for an arbitrary k, including $k = 0$. In the induction step, we state that $7^{k+1} = (7^k \cdot 7^k)/7^{k-1}$. If $k = 0$, then the denominator 7^{k-1} is 7^{-1}, and the induction hypothesis says nothing about 7^{-1}.

Section 2.4

1. $U = \mathcal{N}$.   (a) $\emptyset$ (b) $\{0, 1, 2\}$ (c) $\mathcal{N}$ (d) $\emptyset$

   $U = \mathcal{R}$.   (a) $\{\frac{3}{5}\}$ (b) $\{x : x < \frac{7}{3}\}$ (c) $\mathcal{R}$ (d) $\{\pi/2 + 2n\pi : n \in \mathcal{Z}\}$

3. (a) F   (b) T   (c) T   (d) T

5. (a) $\exists x(x^2 + 2x - 3 \ne 0)$
   (b) $\forall x(x^2 + 2x - 3 \ne 0)$
   (c) $\exists x(\sin(x + \pi/2) \ne \cos x)$
   (d) $\forall x(\sin(x + \pi/2) \ne \cos x)$

7. (a) $\exists x \forall m(x^2 \ge m)$, false.
   (b) $\forall m \exists x(x^2 \ge m)$, true.

9. (a) $\exists z(\forall \epsilon > 0)(0 \le z < \epsilon \wedge z \ne 0)$, false.
   (b) $\forall m \exists x(x/(|x| + 1) \ge m)$, false.

11. (a) w is a lower bound of B if $(\forall x \in B)(w \le x)$.
    (b) w is a least element of B if $w \in B \wedge (\forall x \in B)(w \le x)$.
    (c) $\forall x[0 < x \le 5 \Rightarrow \exists y(0 < y \le 5 \wedge y < x)]$
    (d) Let x and y be least elements of B. Then $x \le y$ since x is a least element of B, and $y \le x$ since y is a least element. Therefore, $x = y$.

13. $p(1) \wedge (\forall k \in \mathcal{N})[p(k) \Rightarrow p(k + 1)] \Rightarrow (\forall n \in \mathcal{N})p(n)$

15. Let U be any universe. Assume that $\exists x(p(x) \vee q(x))$ is true in U. Then there is an $a \in U$ such that $p(a)$ or $q(a)$. *Case 1.* $p(a)$. Then $\exists x p(x)$ is true in U. *Case 2.* $q(a)$. Then $\exists x q(x)$ is true in U. In either case, $\exists x p(x) \vee \exists x q(x)$ is true in U. Conversely, assume that $\exists x p(x) \vee \exists x q(x)$ is true in U. *Case 1.* $\exists x p(x)$ is true in U. Then $p(a)$ for some $a \in U$. Hence, $p(a)$ or $q(a)$. Therefore, $\exists x(p(x) \vee q(x))$ is true in U. *Case 2.* $\exists x q(x)$ is true in U. Similar to case 1.

Section 2.5

1. Let $x = 2$. Verification: $2^2 - 5 \cdot 2 + 6 = 0$

3. **(a)** 9 divides 486 since $486 = 9 \cdot 54$.
   (b) Assume that 9 divides a and 9 divides b. Then $a = 9k$ and $b = 9m$, for some k and m. So $a + b = 9k + 9m = 9(k + m)$. Therefore, 9 divides $a + b$.

5. Assume that d divides a. Then $a = dq$ for some q. Therefore, $ab = d(qb)$, and hence, d divides ab.

7. Let $\delta = 0.0401$. Assume that $4 < x < 4 + \delta$. Then $4 < x < 4 + 4(0.01) + 0.0001$. Hence, $4 < x < (2 + 0.01)^2$, and it follows that $2 < \sqrt{x} < 2.01$. Note that $\delta = 0.0001$ also works.

9. Let $\delta = 0.0001$. Assume that $d^2 < x < d^2 + \delta$. Then $d^2 < x < d^2 + 0.0001$, and hence, $d^2 < x < (d + 0.01)^2$. Therefore, $d < x < d + 0.01$.

11. Let $\delta = \epsilon^2$. Assume that $9 < x < 9 + \delta$. Then $9 < x < 9 + 6\epsilon + \epsilon^2$. Hence, $9 < x < (3 + \epsilon)^2$.

13. *False:* $(-1)^{1/3} = -1$.

15. *False:* $-2 < 1$ and $(-2) \cdot 1 \neq 0$, but $1/1 > 1/(-2)$.

17. *False:* 6 divides $3 \cdot 4$, but 6 does not divide 3 and 6 does not divide 4.

Chapter 2 Proofs to Evaluate

1. The conjecture is correct. It is incorrect to assume the conclusion $1/x < 0$. Assume the negation of the conclusion, that is, $1/x \geq 0$. Then $x^2(1/x) \geq x^2 \cdot 0$. Hence, $x \geq 0$, proving the contrapositive.

3. The conjecture is correct. However, x is odd or y is odd is not the negation of the conclusion. The negation of the conclusion is x is odd and y is odd. Therefore, $x = 2k+1$ and $y = 2m+1$ for some k and m. Hence, $xy = (2k + 1)(2m + 1) = 2(2km + m + k) + 1$, which is odd.

5. False for $n = 1$. The proof lacks a basis step.

Chapter 2 Review

1. If $y \geq \sqrt{z}$, then, since $x > \sqrt{z}$, we have $xy > z$, contradicting $xy < z$.

3. **(a)** Assume $x < y$. Multiplying by x, we obtain $x^2 < xy$, and multiplying by y, we obtain $xy < y^2$. Therefore, $x^2 < xy < y^2$, from which it follows that $x < \sqrt{xy} < y$.
   (b) As in part a, we obtain $x^2 < xy < y^2$. Therefore, $x^2 + xy < 2xy < y^2 + xy$. Hence, $x(x + y) < 2xy < y(x + y)$, and it follows that $x < 2xy/(x + y) < y$.

5. **(a)** $0 \leq (x - y)^2$. Therefore, $0 \leq x^2 - 2xy + y^2$. Adding $4xy$ yields $4xy \leq x^2 + 2xy + y^2$; that is, $4xy \leq (x + y)^2$. Taking square roots, we obtain $2\sqrt{xy} \leq x + y$. Therefore, $\sqrt{xy} \leq (x + y)/2$.
   (b) By part a, $2\sqrt{xy} \leq x + y$. Therefore, $2\sqrt{xy}/(x + y) \leq 1$. Multiplying by $\sqrt{xy}$, we obtain $2xy/(x + y) \leq \sqrt{xy}$.

7. *Basis step:*

$$\sum_{i=1}^{1} \frac{1}{(i + 1)^2 - 1} = \frac{1}{(1 + 1)^2 - 1} = \frac{1}{3} = \frac{3}{4} - \frac{1}{2(1 + 1)} - \frac{1}{2(1 + 2)}$$

Induction hypothesis:

$$\sum_{i=1}^{k} \frac{1}{(i + 1)^2 - 1} = \frac{3}{4} - \frac{1}{2(k + 1)} - \frac{1}{2(k + 2)}$$

Induction step:

$$\sum_{i=1}^{k+1} \frac{1}{(i+1)^2 - 1} = \sum_{i=1}^{k} \frac{1}{(i+1)^2 - 1} + \frac{1}{(k+1+1)^2 - 1}$$

$$= \frac{3}{4} - \frac{1}{2(k+1)} - \frac{1}{2(k+2)} + \frac{1}{(k+2)^2 - 1}$$

$$= \frac{3}{4} - \frac{1}{2(k+2)} - \frac{1}{2(k+1)} + \frac{1}{(k+2)^2 - 1}$$

$$= \frac{3}{4} - \frac{1}{2(k+2)} - \frac{1}{2(k+1)} + \frac{1}{(k+1)(k+3)}$$

$$= \frac{3}{4} - \frac{1}{2(k+2)} - \frac{2 - (k+3)}{2(k+1)(k+3)}$$

$$= \frac{3}{4} - \frac{1}{2(k+2)} + \frac{k+1}{2(k+1)(k+3)}$$

$$= \frac{3}{4} - \frac{1}{2(k+2)} + \frac{1}{2(k+3)}$$

9. *Basis step:* $(x+1)^1 = x + 1 \geq 1 + 1 \cdot x.$

Induction hypothesis: $(1+x)^k \geq 1 + k \cdot x.$ Hence, $(1+x)^{k+1} = (1+x)(1+x)^k \geq (1+x)(1+kx) = 1 + (k+1)x + kx^2 \geq 1 + (k+1)x.$

11. *Basis step:* $(\cos x + i \cdot \sin x)^1 = \cos 1 \cdot x + i \cdot \sin 1 \cdot x.$

Induction hypothesis: $(\cos x + i \cdot \sin x)^k = \cos kx + i \cdot \sin kx.$ $(\cos x + i \cdot \sin x)^{k+1} = (\cos x + i \cdot \sin x)^k (\cos x + i \cdot \sin x) = (\cos kx + i \cdot \sin kx)(\cos x + i \cdot \sin x) = \cos kx \cos x - \sin kx \sin x + i(\sin kx \cos x + \cos kx \sin x) = \cos(kx + x) + i \sin(kx + x) = \cos((k+1)x) + i \sin((k+1)x)$

13. *Basis step:* $(2 \cdot 1)! < 2^{2 \cdot 1}(1!)^2.$

Induction hypothesis: $(2k)! < 2^{2k}(k!)^2.$ First we show that $(2k+2)(2k+1) < 4(k+1)^2.$ This statement is equivalent to $4k^2 + 6k + 2 < 4k^2 + 8k + 4$, which in turn is equivalent to $0 < 2k + 2.$ This last statement is clearly true for $k \geq 0.$ Now, $(2(k+1))! = (2k+2)! = (2k+2)(2k+1)(2k)! \leq (2k+2)(2k+1)2^{2k}(k!)^2 < 4(k+1)^2 2^{2k}(k!)^2 = 4 \cdot 2^{2k}(k+1)^2(k!)^2 = 2^{2(k+1)}((k+1)!)^2.$

15. (a) $R_1 = 2$, $R_2 = 4$, $R_3 = 7$, $R_4 = 11$

(b) If there are k such lines in the plane, any additional line drawn that satisfies all the conditions would have to cross all k lines. The new line divides the region on each side of the lines it crosses into two regions. There are $k + 1$ regions involved, and hence, $R_{k+1} = R_k + (k+1).$

(c) $R_n = 1 + \frac{n(n+1)}{2}$

(d) *Basis step:*

$$R_0 = 1 + \frac{0(0+1)}{2}$$

Induction hypothesis:

$$R_k = 1 + \frac{k(k+1)}{2}$$

Then

$$R_{k+1} = R_k + (k+1) = 1 + \frac{k(k+1)}{2} + (k+1)$$

$$= 1 + \frac{k(k+1) + 2(k+1)}{2} = 1 + \frac{(k+1)(k+2)}{2}$$

17. Assume that $\exists x(p(x) \land q(x))$ is true in a universe U. Then $p(a)$ and $q(a)$ for some $a \in U$. Hence, $p(a)$ for some $a \in U$, from which it follows that $\exists x p(x)$ is true in U. Similarly, $\exists x q(x)$ is true in U. Therefore, $\exists x p(x) \land \exists x q(x)$ is true in U.

19. Let the universe be $\mathcal{N}$. Let $p(x)$ be "x is even" and $q(x)$ be "x is odd." Both $\exists x p(x)$ and $\exists x q(x)$ are true in $\mathcal{N}$. However, $\exists x(p(x) \land q(x))$ is not true in $\mathcal{N}$.

Section 3.1

1. (a) True   **(b)** False   **(c)** True   **(d)** False   **(e)** False   **(f)** True

3. $\{-5, 5\}$

5. (a) $\{N, O\}$ **(b)** $\{M, A, D\}$ **(c)** $\{M, A, T, H, E, I, C, S\}$ **(d)** $\{R, A, C, E\}$

7. (a) Yes   **(b)** No   **(c)** No   **(d)** Yes

9. (a) $P(\{a\}) = \{\emptyset, \{a\}\}$ **(b)** $P(\{a, b, c, d\}) = \{\emptyset, \{a\}, \{b\}, \{c\}, \{d\}, \{a, b\}, \{a, c\},$ $\{a, d\}, \{b, c\}, \{b, d\}, \{c, d\}, \{a, b, c\}, \{a, c, d\}, \{b, c, d\}, \{a, b, d\}, \{a, b, c, d\}\}$

11. (a) $\{a, b, c, d, e, f\}$ **(b)** $\{d\}$ **(c)** $\{d, e, f\}$ **(d)** $\emptyset$ **(e)** No

13. (a) $\{3, -2\}$ **(b)** $\emptyset$ **(c)** $\emptyset$ **(d)** $\{3, -2\}$

15. (a) True   **(b)** False   **(c)** True   **(d)** True   **(e)** False   **(f)** False   **(g)** False

17. $C - D = C$, $D - C = D$, $C \neq D$

19. (a) $C' = \{7, 8, 9\}$ **(b)** $\emptyset$ **(c)** C **(d)** C' **(e)** U

21. (a) $\{\emptyset\}$ **(b)** $\emptyset$ **(c)** $\{\emptyset\}$ **(d)** $\{\{\emptyset\}\}$ **(e)** $\{\emptyset, \{\emptyset\}\}$

23. If the barber shaves himself, then he is one of those that he does not shave. On the other hand, if he does not shave himself, then he is one of those that he does shave. So there cannot be such a barber.

Section 3.2

1.   (a)

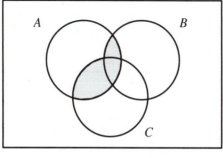

$$A \cap (B \cup C)$$

(b)

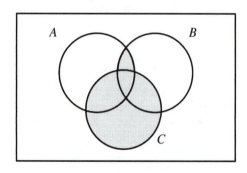

$(A \cap B) \cup C$

3. **(a)**

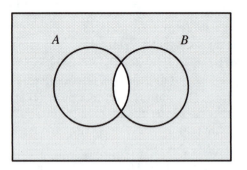

$(A \cap B)'$

(b)

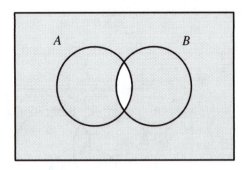

$A' \cup B'$

5. *False:* Let $U = \{1\}$, $A = \emptyset$, $B = \emptyset$, and $C = \{1\}$.

7. True.

9. *False:* Let $A = \{1, 2\}$, $B = \{2, 3\}$, and $D = \{2, 4\}$.

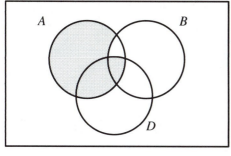

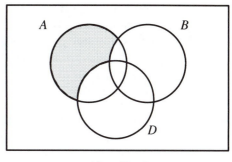

$$A-(B-D)$$

$$(A - B)-D$$

11. True.

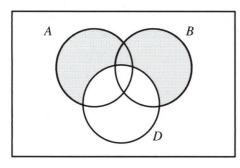

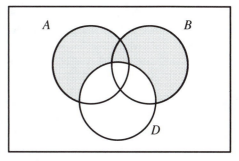

$$(A \cup B)-D$$

$$(A - D) \cup (B-D)$$

13.   (a) True

     (b) True

     (c) True

15. *No:* Let $A = D = \emptyset$ and $B = C = \{1\}$.

17.   (a) $E = D' \cap (A \cup B)$

     (b) $F = (A' \cup D) \cup B$

Section 3.3

1.   (a) Let $x \in B$. Then $x \in A$ or $x \in B$. Therefore, $x \in A \cup B$.

     (b) Let $x \in (A')'$. Then $x \notin A'$, and hence, $x \in A$. Reversing the steps yields $A \subseteq (A')'$.

3.   (a) Let $x \in A \cap U$. Then $x \in A$ and $x \in U$. Therefore, $x \in A$. Conversely, let $x \in A$. We always have $x \in U$. Therefore, $x \in A \cap U$.

     (b) Suppose that $A \cap A' \neq \emptyset$. Then there is an $x \in A \cap A'$. It follows that $x \in A$ and $x \notin A$, a contradiction. Therefore, $A \cap A' = \emptyset$.

5. Assume that $A \subseteq B$. Let $x \in D - B$. Then $x \in D$ and $x \notin B$. Since $x \notin B$ and $A \subseteq B$, it follows that $x \notin A$. Therefore, $x \in D$ and $x \notin A$. Hence, $x \in D - A$.

7. ($\Rightarrow$) Assume that $A \not\subseteq B$. Then there is an $x \in A$ such that $x \notin B$. Hence, $x \in B'$. Therefore, $x \in A \cap B'$ and $A \cap B' \neq \emptyset$.

($\Leftarrow$) Assume that $A \cap B' \neq \emptyset$. Then there is an $x \in A \cap B'$. Hence, $x \in A$ and $x \in B'$. Therefore, $x \in A$ and $x \notin B$. It follows that $A \not\subseteq B$.

9. ($\Rightarrow$) Assume that $A \subseteq B$. We show that $A \cup B \subseteq B$. Let $x \in A \cup B$.

   Case 1. $x \in A$. Then by the assumption that $A \subseteq B$, $x \in B$.

   Case 2. $x \in B$. Therefore, $x \in B$. To show that $B \subseteq A \cup B$, see Practice Problem 3. ($\Leftarrow$) Assume that $A \cup B = B$. Let $x \in A$. Then $x \in A$ or $x \in B$, and hence, $x \in A \cup B$. Therefore, $x \in B$.

11. (a) Let $x \in A - (A - B)$. Then $x \in A$ and $x \notin A - B$. So by the definition of $A - B$, $x \notin A$ or $x \in B$. Since $x \in A$, we cannot have $x \notin A$. Therefore, $x \in B$. Hence, $x \in A \cap B$. For the other set inclusion, let $x \in A \cap B$. Then $x \in A$ and $x \in B$. Since $x \in B$, $x \notin A - B$, and hence, $x \in A - (A - B)$.

    (b) $A - (A - B) = A \cap (A - B)' = A \cap (A \cap B')' = A \cap (A' \cup B'') = A \cap (A' \cup B) = (A \cap A') \cup (A \cap B) = \emptyset \cup (A \cap B) = A \cap B$

13. (a) Let $x \in (A \cap B) \cup (C - A) \cup (B \cap C)$.

    Case 1. $x \in A \cap B$. Then $x \in (A \cap B) \cup (C - A)$. *Case 2.* $x \in C - A$. Then $x \in (A \cap B) \cup (C - A)$. *Case 3.* $x \in B \cap C$. Then $x \in B$ and $x \in C$. If $x \in A$, then $x \in A \cap B$. If $x \notin A$, then $x \in C - A$. In either case, $x \in (A \cap B) \cup (C - A)$.

    Conversely, let $x \in (A \cap B) \cup (C - A)$. *Case 1.* $x \in A \cap B$. *Case 2.* $x \in C - A$. In either case, $x \in (A \cap B) \cup (C - A) \cup (B \cap C)$.

    (b) $x \in (A \cap B) \cup (C - A) \cup (B \cap C) \Leftrightarrow x \in A \cap B$ or $x \in C - A$ or $x \in B \cap C$
    $\Leftrightarrow (x \in A$ and $x \in B)$ or $(x \in C$ and $x \notin A)$ or $(x \in B$ and $x \in C)$
    $\Leftrightarrow (x \in A$ and $x \in B)$ or $(x \in C$ and $x \notin A)^*$
    $\Leftrightarrow x \in (A \cap B)$ or $x \in (C - A)$
    $\Leftrightarrow x \in (A \cap B) \cup (C - A)$ *

    (c) First note that $X \cup (X \cap Y) = X$. Then $((A \cap B) \cup (C - A)) \cup (B \cap C)$
    $= (A \cap B) \cup (A' \cap C) \cup [(A \cup A') \cap (B \cap C)]$
    $= (A \cap B) \cup (A' \cap C) \cup [(A \cap B \cap C) \cup (A' \cap B \cap C)]$
    $= [(A \cap B) \cup (A \cap B \cap C)] \cup [(A' \cap C) \cup (A' \cap C \cap B)]$
    $= (A \cap B) \cup (A' \cap C) = (A \cap B) \cup (C - A)$.

15. Assume that $C \cup B = U$ and $A \cap C = \emptyset$. Let $x \in A$. If $x \in C$, then $x \in A \cap C$, which is a contradiction since $A \cap C = \emptyset$. Therefore, $x \notin C$. However, $x \in U$, and hence, $x \in C \cup B$. Since $x \notin C$, we must have $x \in B$.

17. Assume that $B \cap E = \emptyset$, $C \cup B = U$, and $A \cap C = \emptyset$. Also, let $x \in A \cap E$. Then $x \in A$ and $x \in E$. Since $B \cap E = \emptyset$ and $x \in E$, it follows that $x \notin B$. Also, since $A \cap C = \emptyset$ and $x \in A$, it follows that $x \notin C$. Therefore, $x \notin C \cup B$. Hence, $x \notin U$, which is impossible. Therefore, we cannot have $x \in A \cap E$, and it follows that $A \cap E = \emptyset$.

Section 3.4

1. (a) $\{5\}$ (b) $\{1, 2, 3, 4, 5, 6\}$

3. (a) $\{0, 1\}$ (b) $\mathcal{N}$

5. (a) $\emptyset$, since $A_1 = (0, 0) = \emptyset$ (b) $(-1, 1)$

7. (a) $\emptyset$ (b) $\mathcal{Z}$

* This step follows from the tautology

$$(P \wedge Q) \vee (R \wedge \neg P) \vee (Q \wedge R) \Leftrightarrow (P \wedge Q) \vee (R \wedge \neg P)$$

9. *Basis step:* $A \cap (B_1) = A \cap B_1$.

Induction hypothesis: $A \cap (B_1 \cup B_2 \cup \ldots \cup B_k) = (A \cap B_1) \cup (A \cap B_2) \cup \ldots \cup (A \cap B_k)$

Induction step: $A \cap (B_1 \cup B_2 \cup \ldots \cup B_k \cup B_{k+1}) = A \cap [(B_1 \cup B_2 \cup \ldots \cup B_k) \cup B_{k+1}]$
$= [A \cap (B_1 \cup B_2 \cup \ldots \cup B_k)] \cup (A \cap B_{k+1}) = (A \cap B_1) \cup (A \cap B_2) \cup \ldots \cup (A \cap B_k) \cup (A \cap B_{k+1})$

11. *Basis step:* $D - (A_1) = D - A_1$.

Induction hypothesis: $D - (A_1 \cup A_2 \cup \ldots \cup A_k) = (D - A_1) \cap (D - A_2) \cap \ldots \cap (D - A_k)$

Induction step: $D - (A_1 \cup A_2 \cup \ldots \cup A_k \cup A_{k+1}) = D - [(A_1 \cup A_2 \cup \ldots \cup A_k) \cup A_{k+1}]$
$= [D - (A_1 \cup A_2 \cup \ldots \cup A_k)] \cap (D - A_{k+1}) = (D - A_1) \cap (D - A_2) \cap \ldots \cap (D - A_k) \cap (D - A_{k+1})$

13. Let $x \in D - \cap \Omega$. Then $x \in D$ and $x \notin \cap \Omega$. Hence, $x \notin X$, for some $X \in \Omega$. Therefore, $x \in D - X$ for some $X \in \Omega$. Consequently, $x \in \cup \{D - A : A \in \Omega\}$. Conversely, let $x \in \cup \{D - A : A \in \Omega\}$. Then $x \in D - X$ for some $X \in \Omega$. Therefore, $x \in D$ and $x \notin X$ for some $X \in \Omega$. Hence, $x \notin \cap \Omega$, and it follows that $x \in D - \cap \Omega$.

15. Assume that $A \subseteq D$ for all $A \in \Upsilon$. Let $x \in \cup \Upsilon$. Then $x \in A$ for some $A \in \Upsilon$. Since $A \in \Upsilon$, $A \subseteq D$. Therefore, $x \in D$.

17. Assume $\Upsilon \neq \emptyset$. Let $A \in \Upsilon$. To show that $\cap \Upsilon \subseteq \cup \Upsilon$, let $x \in \cap \Upsilon$. Then $x \in X$ for all $X \in \Upsilon$. In particular, $x \in A$. Therefore, $x \in \cup \Upsilon$.

19. Assume that $\Omega \subseteq \Upsilon$. Let $x \in \cap \Upsilon$. Therefore, $x \in X$ for all $X \in \Upsilon$. To show that $x \in \cap \Omega$, let $Y \in \Omega$. Since $\Omega \subseteq \Upsilon$, $Y \in \Upsilon$. Therefore, $x \in Y$. Hence, $x \in \cap \Omega$.

21. True. Assume that $A \cup B = A \cap B$. Let $x \in A$. Then $x \in A \cup B$, and hence, $x \in A \cap B$. Therefore, $x \in B$. So we have $A \subseteq B$. Similarly, $B \subseteq A$.

23. False: Let $\Omega = \{\{1\}\}$ and $\Upsilon = \{\{1\}, \{1, 2\}\}$.

Chapter 3 Proofs to Evaluate

1. The conjecture is incorrect. To show that $A \subseteq D$, start with $u \in A$, not with $u \in A \cap B$. A counterexample is $A = \{1\}$, $B = \emptyset$, and $D = \{2\}$.

3. The conjecture is correct. The proof is valid.

5. The conjecture is incorrect. It is not correct to cancel C from both sides of $A \cap C = B \cap C$. By doing so, the conjecture is being used to prove the conjecture. A counterexample is $A = \{1\}, B = \{2\}$, and $C = \emptyset$.

Chapter 3 Review

1.   (a) True
    (b) False
    (c) True
    (d) False
    (e) False

3. *False:* Let $A = \emptyset$, $B = \{1\}$, and $D = \emptyset$.

5. *False:* Let $A = \{1\}$, $B = \{\{1\}\}$, and $D = \{\{\{1\}\}\}$.

7. True. By Exercise 11 of Section 3.3, $B - (B - A) = B \cap A$. Also, $A \subseteq B$ implies $B \cap A = A$.

9. True. Let $x \in (A - D) \cup (B - D)$.

Case 1. $x \in A - D$. Then $x \in A$ and $x \notin D$. Hence, $x \in A \cup B$. Therefore, $x \in (A \cup B) - D$.

Case 2. $x \in B - D$. Similar to case 1.

11. True. If $x \in (A - B) \cap D$, then $x \notin B$. If $x \in (A - D) \cap B$, then $x \in B$. So we cannot have $x \in [(A - B) \cap D] \cap [(A - D) \cap B]$.

13. ($\Rightarrow$) Assume that $D \cup (A \cap B) = D \cap (A \cup B)$. Let $x \in D$. Then $x \in D \cup (A \cap B)$, and hence, $x \in D \cap (A \cup B)$. Therefore, $x \in A \cup B$. So we have $D \subseteq A \cup B$. Now let $x \in A \cap B$. Then $x \in D \cup (A \cap B)$. Hence, $x \in D \cap (A \cup B)$. So $x \in D$.

($\Leftarrow$) Assume that $D \subseteq A \cup B$ and $A \cap B \subseteq D$. Then $D \cap (A \cup B) = D$ and $D \cup (A \cap B) = D$. Hence, $D \cup (A \cap B) = D = D \cap (A \cup B)$.

15.   (a) $A \triangle B = (A - B) \cup (B - A) = (B - A) \cup (A - B) = B \triangle A$
      (b) $A \triangle \emptyset = (A - \emptyset) \cup (\emptyset - A) = A \cup \emptyset = A$
      (c) $A \triangle A = (A - A) \cup (A - A) = \emptyset \cup \emptyset = \emptyset$

17. $A \cap (B \triangle D) = A \cap ((B - D) \cup (D - B)) = [A \cap (B - D)] \cup [A \cap (D - B)] = [A \cap (B \cap D')] \cup [A \cap (D \cap B')]$

   $(A \cap B) \triangle (A \cap D) = [(A \cap B) - (A \cap D)] \cup [(A \cap D) - (A \cap B)]$
   $= [(A \cap B) \cap (A \cap D)'] \cup [(A \cap D) \cap (A \cap B)']$
   $= [(A \cap B) \cap (A' \cup D')] \cup [(A \cap D) \cap (A' \cup B')]$
   $= [(A \cap B) \cap A'] \cup [(A \cap B) \cap D'] \cup [(A \cap D) \cap A'] \cup [(A \cap D) \cap B']$
   $= \emptyset \cup [A \cap (B \cap D')] \cup \emptyset \cup [A \cap (D \cap B')]$

19. The conjecture is correct. Assume that $\cup \Omega = \cap \Omega$. Suppose that $A, B \in \Omega$ and that $A \neq B$.

   Case 1. There is an $a \in A$ such that $a \notin B$. Then $a \in \cup \Omega$. Therefore, $a \in \cap \Omega$, and hence, $a \in B$, contradicting $a \notin B$.

   Case 2. There is an $a \in B$ such that $a \notin A$. This case also leads to a contradiction. Therefore, $A = B$. Hence, Ω contains at most one member. We now show that $\Omega \neq \emptyset$. Assume that $\Omega = \emptyset$. Then $\cap \Omega = U$ (verify this; see Exercise 16 of Section 3.4), where U is the universe and $\cup \Omega = \emptyset$ (verify this). Hence, $U = \emptyset$, contrary to one of the assumptions about the universal set.

Section 4.1

1.   (a) $\{(x, y) : 1 \leq x \leq 5 \text{ and } y = 2\}$

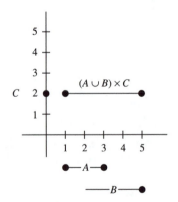

   (b) $\{(x, y) : (1 \leq x \leq 3 \text{ and } y = 2) \text{ or } (2 < x \leq 5 \text{ and } y = 2)\}$

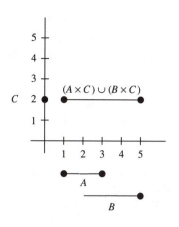

3.   (a) $\{(x,y) : (1 \leq x \leq 3 \text{ or } x = 5) \text{ and } (4 \leq y \leq 6)\}$

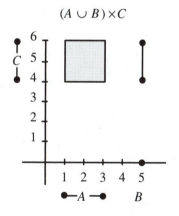

(b) $\{(x,y) : (1 \leq x \leq 3 \text{ and } 4 \leq y \leq 6) \text{ or } (x = 5 \text{ and } 4 \leq y \leq 6)\}$

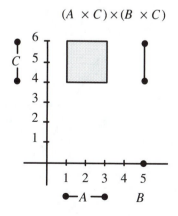

5. ($\Rightarrow$) Assume that $(a, b) = (c, d)$. Then $\{\{a\}, \{a, b\}\} = \{\{c\}, \{c, d\}\}$. Hence, $\cap\{\{a\}, \{a, b\}\} = \cap\{\{c\}, \{c, d\}\}$; that is, $\{a\} = \{c\}$. Therefore, $a = c$. Also, $\cup\{\{a\}, \{a, b\}\} = \cup\{\{c\}, \{c, d\}\}$; that is, $\{a, b\} = \{c, d\}$. Therefore, $\{a, b\} = \{a, d\}$. Hence, $\{a, b\} - \{a\} = \{a, d\} - \{a\}$. Therefore, $\{b\} = \{d\}$, and hence, $b = d$.

($\Leftarrow$) The proof is routine.

7. Let $(x, y) \in A \times (B \cup C)$. Then $x \in A$ and $y \in B \cup C$.

Case 1. $y \in B$. Then $(x, y) \in A \times B$, and hence, $(x, y) \in (A \times B) \cup (A \times C)$.

Case 2. $y \in C$. Then $(x, y) \in (A \times C)$, and hence, $(x, y) \in (A \times B) \cup (A \times C)$. The other set inclusion is similar.

9. True. Assume that $A \times B \neq \emptyset$. Let $(a, b) \in A \times B$. Then $a \in A$ and $b \in B$.

($\Rightarrow$) Assume that $A \times B = B \times A$. Let $x \in A$. Then $(x, b) \in A \times B$, and hence, $(x, b) \in B \times A$. It follows that $x \in B$. Hence, $A \subseteq B$. Similarly, $B \subseteq A$.

($\Leftarrow$) Assume that $A = B$. Then $A \times B = A \times A = B \times A$.

11. *False:* Let $B = E = D = \emptyset$ and $A = \{1\}$.

13. True. Let $(x, y) \in (A \times C) \cap (B \times D)$. Therefore, $(x, y) \in A \times C$ and $(x, y) \in B \times D$. Hence, $x \in A, y \in C, x \in B$, and $y \in D$. Therefore, $x \in A \cap B$ and $y \in C \cap D$, and it follows that $(x, y) \in (A \cap B) \times (C \cap D)$.

15. Prove $A \times (B - C) = (A \times B) - (A \times C)$: Let $(x, y) \in A \times (B - C)$. Then $x \in A$, $y \in B$, and $y \notin C$. It follows that $(x, y) \in A \times B$ and $(x, y) \notin A \times C$. Hence, $(x, y) \in (A \times B) - (A \times C)$. For the other set inclusion, let $(x, y) \in (A \times B) - (A \times C)$. Then $x \in A, y \in B$, and $(x, y) \notin A \times C$. Hence, $x \notin A$ or $y \notin C$. But $x \in A$; so $y \notin C$. Therefore, $(x, y) \in A \times (B - C)$.

17. Let $(x, y) \in A \times \cap \Omega$. Then $x \in A$ and $y \in B$ for all $B \in \Omega$. Hence, $(x, y) \in A \times B$ for all $B \in \Omega$, and it follows that $(x, y) \in \cap\{A \times B : B \in \Omega\}$. Conversely, let $(x, y) \in \cap\{A \times B : B \in \Omega\}$. Then $(x, y) \in A \times B$ for all $B \in \Omega$. Hence, $x \in A$ and for all $B \in \Omega, y \in B$. Therefore, $y \in \cap\Omega$, and hence, $(x, y) \in A \times \cap\Omega$.

Section 4.2

1. R is reflexive on $\{1, 2, 3\}$, is not symmetric, is antisymmetric, and is transitive.

3. **(a)** $R = \{(1, 4), (1, 9), (1, 21), (1, 25), (2, 4), (3, 9), (3, 21)\}$

   (b) R is reflexive on $A \cup B$, is not symmetric, is antisymmetric, and is transitive.

5. **(a)**

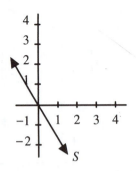

   (b) S is not reflexive since $(1, 1) \notin S$. S is not symmetric since $(1, -2) \in S$ but

$(-2, 1) \notin S$. S is antisymmetric: Assume $(x, y) \in S$ and $(y, x) \in S$. Then $y = -2x$ and $x = -2y$. Therefore, $y = -2(-2y)$, and hence, $y = 0$. Therefore, $x = 0$, and hence, $x = y$. S is not transitive since $(1, -2) \in S, (-2, 4) \in S$, but $(1, 4) \notin S$.

7. (a)

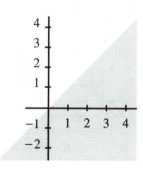

(b) S is not reflexive since $\neg 1S1$. S is not symmetric since $1S0$ but $\neg 0S1$. S is vacuously antisymmetric. S is transitive: Assume that xSy and ySz. Then $y < x$ and $z < y$. Therefore, $z < x$, and hence, xSz.

9. (a) $R = \{(1, 1), (2, 2), (3, 3), (4, 4), (5, 5), (2, 4), (4, 2), (3, 5), (5, 3), (1, 3), (3, 1), (1, 5), (5, 1)\}$

(b) R is reflexive, is symmetric, is not antisymmetric, and is transitive.

11. (a) Yes
    (b) Yes
    (c) Yes

13. R is reflexive since $x = 1 \cdot x$. R is not symmetric since $2R1$ but $\neg 1R2$. R is not antisymmetric since $1R(-1)$ and $(-1)R1$ but $1 \neq -1$. R is transitive: Assume that xRy and yRz. Then $x = ky$ for some $k \in \mathcal{Z}$ and $y = mz$ for some $m \in \mathcal{Z}$. Therefore, $x = kmz$, and hence, xRz.

15. R is not reflexive since $\neg(1, 1)R(1, 1)$. R is not symmetric since $(1, 1)R(2, 2)$, but $\neg(2, 2)R(1, 1)$. R is not antisymmetric since $(1, 2)R(2, 1)$ and $(2, 1)R(1, 2)$, but $(1, 2) \neq (2, 1)$. R is not transitive since $(3, 2)R(1, 3)$ and $(1, 3)R(2, 1)$, but $\neg(3, 2)R(2, 1)$.

17. R is reflexive since $x = x^1$. R is not symmetric since $4R2$ but $\neg 2R4$. R is antisymmetric: Assume that xRy and yRx. Then $x = y^k$ and $y = x^m$ for some k and m in $\mathcal{N}$. Therefore, $x = x^{mk}$, and hence, $mk = 1$. Since k and m are in $\mathcal{N}$, $m = k = 1$, and therefore, $x = y$. R is transitive: Assume that xRy and yRz. Then $x = y^k$ and $y = z^m$. Hence, $x = z^{mk}$. Therefore, xRz.

19. (a) $x = 2 + y$
    (b) $y = 4 + x$

21. ($\Rightarrow$) Assume that R is reflexive. Let $(x, y) \in 1_A$. Then $x = y$. Since R is reflexive, $(x, y) \in R$. Hence, $1_A \subseteq R$.

    ($\Leftarrow$) Assume that $1_A \subseteq R$. Let $x \in A$. Then $(x, x) \in 1_A$. Hence, $(x, x) \in R$, and R is reflexive.

23. Assume that $R \subseteq S$. Let $(x, y) \in R^{-1}$. Then $(y, x) \in R$. Therefore, $(y, x) \in S$, and hence, $(x, y) \in S^{-1}$.

25. False. Let $R = \{(1,2),(5,4)\}$ and $T = \{(0,1),(4,3)\}$. Then $(R \circ T)^{-1} = \{(2,0)\}$ and $R^{-1} \circ T^{-1} = \{(3,5)\}$. So $(R \circ T)^{-1} \subseteq R^{-1} \circ T^{-1}$ and $R^{-1} \circ T^{-1} \subseteq (R \circ T)^{-1}$ are both false.

27. $\text{dom}(R) \cap \text{dom}(S) \subseteq \text{dom}(R \cap S)$ is false. Let $R = \{(1,2)\}, S = \{(1,3)\}$. Prove that $\text{dom}(R \cap S) \subseteq \text{dom}(R) \cap \text{dom}(S)$: Let $x \in \text{dom}(R \cap S)$. Then for some y, $(x,y) \in R \cap S$. Therefore, $(x,y) \in R$ and $(x,y) \in S$. Hence, $x \in \text{dom}(R)$ and $x \in \text{dom}(S)$. So, $x \in \text{dom}(R) \cap \text{dom}(S)$.

29. ($\Rightarrow$) Assume that R is symmetric. Let $xR^{-1}y$. Then yRx. Since R is symmetric, xRy. Therefore, $R^{-1} \subseteq R$. Reverse the steps to show that $R \subseteq R^{-1}$.
($\Leftarrow$) Assume that $R^{-1} = R$. Let xRy. Then $xR^{-1}y$, and hence, yRx.

31. ($\Rightarrow$) Assume that R is symmetric and transitive. By Exercise 29, $R = R^{-1}$, and by Practice Problem 8, we only need to prove that $R \subseteq R \circ R$. Assume that xRy. Since R is symmetric, yRx. Therefore, xRx, since R is transitive. So, we have xRx and xRy. Hence, $xR \circ Ry$.
($\Leftarrow$) Assume that $R^{-1} \circ R = R$. To show that R is symmetric, we assume xRy. Then $xR^{-1} \circ Ry$, and hence, xRz and $zR^{-1}y$ for some z. Therefore, yRz and $zR^{-1}x$, from which it follows that $yR^{-1} \circ Rx$ and hence, yRx. By Exercise 29, $R^{-1} = R$, and hence, $R \circ R = R$. By Practice Problem 8, R is transitive.

Section 4.3

1. f is a function from A to B.

3. (a) $\text{dom}(f) = \{x : x \in \mathcal{R} \text{ and } x \neq -3\}$
(b) $\text{dom}(f) = \{x : x \in \mathcal{R} \text{ and } x \geq -5\}$

5.

$$s(n) = \begin{cases} n & \text{if } n \text{ is even} \\ 1/n & \text{if } n \text{ is odd} \end{cases}$$

7. (a) $t_1 = 1$
(b) $t_5 = 15$
(c) $t_6 = 21$, $t_7 = 28$

9. (a) $s_2 = 4$, $s_3 = 9$, $s_5 = 25$
(b) $s_n = n^2$.
Basis step: $s_1 = 1 = 1^2$.
Induction hypothesis: $s_k = k^2$. Then $s_{k+1} = (2(k+1)-1)+s_k = (2k+1)+k^2 = (k+1)^2$.

11. (a) $\lfloor 2.3 \rfloor = 2$
(b) $\lfloor 3.0 \rfloor = 3$
(c) $\lfloor -1.3 \rfloor = -2$

13. (a) Yes, since $x^2 - 2x + 2 \in \mathcal{N}$ whenever $x \in \mathcal{N}$.
(b) No, since $1^2 - 1 \cdot 2 \notin \mathcal{N}$.

15. Assume $f : D \to C$ and $g : D \to C$.
($\Rightarrow$) Assume that $f = g$. By Theorem 1, $f(x) = g(x)$ for all $x \in \text{dom}(f) = D$.
($\Leftarrow$) Assume that $f(x) = g(x)$ for all $x \in D$. Since $f : D \to C$ and $g : D \to C$, $\text{dom}(f) = D = \text{dom}(g)$. Therefore, by Theorem 1, $f = g$.

17. ($\Rightarrow$) Assume that $f \cup g$ is a function. Let $x \in \text{dom}(f) \cap \text{dom}(g)$. Then $(x, f(x)) \in f$ and $(x, g(x)) \in g$. Hence, $(x, f(x)) \in f \cup g$ and $(x, g(x)) \in f \cup g$. Since $f \cup g$ is a function, $f(x) = g(x)$.

($\Leftarrow$) Assume that $f(x) = g(x)$ for all $x \in \text{dom}(f) \cap \text{dom}(g)$. Let $(x, y) \in f \cup g$ and $(x, z) \in f \cup g$.

Case 1. $(x, y) \in f$ and $(x, z) \in f$. Then $y = z$ since f is a function.

Case 2. $(x, y) \in g$ and $(x, z) \in g$. Similar to Case 1.

Case 3. $(x, y) \in f$ and $(x, z) \in g$. Then $x \in \text{dom}(f) \cap \text{dom}(g)$, $y = f(x)$, and $z = g(x)$. Therefore, $y = f(x) = g(x) = z$.

Case 4. $(x, y) \in g$ and $(x, z) \in f$. Similar to case 3.

Section 4.4

1. f is a function, but f is not 1–1 since $f(a) = f(b)$.

3. **(a)**

$$f \circ g = \begin{pmatrix} 1 & 2 & 3 & 4 \\ 2 & 4 & 3 & 1 \end{pmatrix}, g \circ f = \begin{pmatrix} 1 & 2 & 3 & 4 \\ 3 & 1 & 2 & 4 \end{pmatrix}$$

(b)

$$f^{-1} = \begin{pmatrix} 1 & 2 & 3 & 4 \\ 2 & 1 & 4 & 3 \end{pmatrix}$$

$f \circ f^{-1}(1) = f(f^{-1}(1)) = f(2) = 1$, $f \circ f^{-1}(2) = f(f^{-1}(2)) = f(1) = 2$, $f \circ f^{-1}(3) = f(f^{-1}(3)) = f(4) = 3$, $f \circ f^{-1}(4) = f(f^{-1}(4)) = f(3) = 4$. A similar computation holds for $f^{-1} \circ f$.

(c)

$$g^{-1} = \begin{pmatrix} 1 & 2 & 3 & 4 \\ 1 & 4 & 2 & 3 \end{pmatrix}$$

$g^{-1} \circ g(1) = g^{-1}(g(1)) = g^{-1}(1) = 1$, $g^{-1} \circ g(2) = g^{-1}(g(2)) = g^{-1}(3) = 2$, $g^{-1} \circ g(3) = g^{-1}(g(3)) = g^{-1}(4) = 3$, $g^{-1} \circ g(4) = g^{-1}(g(4)) = g^{-1}(2) = 4$. A similar computation holds for $g \circ g^{-1}$.

(d)

$$(f \circ g)^{-1} = \begin{pmatrix} 1 & 2 & 3 & 4 \\ 4 & 1 & 3 & 2 \end{pmatrix} = g^{-1} \circ f^{-1}$$

5. $v = 2 - 5u$, $u = (2 - v)/5$, $f^{-1}(x) = (2 - x)/5$, $f \circ f^{-1}(x) = f(f^{-1}(x)) = f((2 - x)/5) = 2 - 5((2 - x)/5) = x$, $f^{-1} \circ f(x) = f^{-1}(f(x)) = f^{-1}(2 - 5x) = (2 - (2 - 5x))/5 = x$

7. $f \circ g$ is well defined; $(f \circ g)(n) = f(g(n)) = f(2 - n) = 3(2 - n) - 2 = 4 - 3n$; $g \circ f$ is not well defined because the codomain of f is not a subset of $\text{dom}(g)$.

9. $f \circ g$ is well defined; $(f \circ g)(n) = f(g(n)) = f(2^n - 1) = \log_2((2^n - 1) + 1) = n$; $g \circ f$ is not well defined.

11. $f \circ g$ is not well defined; $g \circ f$ is not well defined.

13. f is not 1–1 because $f(1) = 4 = f(-3)$; f is not onto because there is no x such that $f(x) = -1$.

15. f is not 1–1 because $f(\{2\}) = 2 = f(\{1, 3\})$; f is not onto because there is no $X \in A$ such that $f(X) = \sqrt{2}$. Use the fact that $\sqrt{2}$ is irrational to show this.

17. f is 1–1: Let $f(n) = f(m)$. Then the largest element in $f(n)$ is the same as the largest element in $f(m)$. But the largest element in $f(n)$ is n, and the largest element in $f(m)$ is m. Therefore, $n = m$. f is not onto: There is no $n \in \mathcal{N} - \{0\}$ such that $f(n) = \{4\}$, since $1 \in f(n)$ for all $n \in \mathcal{N} - \{0\}$.

19. (a) $(\forall x \in D)(\forall y \in D)(f(x) = f(y) \Rightarrow x = y)$
   (b) $(\forall y \in C)(\exists x \in D)(f(x) = y)$

21. We think of f as a relation. Then f^{-1} is a relation. We show that f^{-1} is a function. Let $(x, y) \in f^{-1}$ and $(x, z) \in f^{-1}$. Then $(y, x) \in f$ and $(z, x) \in f$. Therefore, $f(y) = x = f(z)$. Since f is 1–1, $y = z$. Now, note that $\mathrm{dom}(f) = \mathrm{ran}(f^{-1})$ and $\mathrm{dom}(f^{-1}) = \mathrm{ran}(f)$. We show that $\mathrm{dom}(f^{-1}) = C$. Let $y \in \mathrm{dom}(f^{-1})$. Then $y \in \mathrm{ran}(f) \subseteq C$. Conversely, let $y \in C$. Then, since f is onto, $f(x) = y$ for some $x \in D$. Therefore, xfy; that is, $yf^{-1}x$. Hence, $y \in \mathrm{dom}(f^{-1})$. To verify that $\mathrm{ran}(f^{-1}) \subseteq D$, let $x \in \mathrm{ran}(f^{-1})$. Then $x \in \mathrm{dom}(f) = D$.

Observe that for all $x \in D$, $f(x) = y$ if and only if $x = f^{-1}(y)$. Let $y \in C$. Then $y = f(x)$ for some $x \in D$. Therefore, $x = f^{-1}(y)$, and hence, $y = f(f^{-1}(y))$; that is, $f \circ f^{-1}(y) = y = 1_C(y)$. Now, let $x \in D$ and $y = f(x)$. Then $x = f^{-1}(y)$, and hence, $f^{-1}(f(x)) = x$. So $f^{-1} \circ f(x) = x = 1_D(x)$.

23. Assume that $f^{-1}(u) = f^{-1}(v)$. Then $u = 1_C(u) = f \circ f^{-1}(u) = f(f^{-1}(u)) = f(f^{-1}(v)) = f \circ f^{-1}(v) = 1_C(v) = v$. Therefore, f^{-1} is 1–1. To show that f^{-1} is onto, assume that $x \in D$. Let $y = f(x)$. Then $f^{-1}(y) = f^{-1}(f(x)) = x$.

25. (a) The conjecture is true. Assume $g(u) = g(v)$. Then $f \circ g(u) = f(g(u)) = f(g(v)) = f \circ g(v)$. Hence, $u = v$, since $f \circ g$ is 1–1.
   (b) The conjecture is false. Let $D = \{a\} = C$, $A = \{1, 2\}$, $g(a) = 1$, $f(1) = a$, and $f(2) = a$.

Section 4.5

1. $f[A] = \{2x + 1 : 0 < x < 1\} = \{t : 0 < (t - 1)/2 < 1\} = \{t : 1 < t < 3\} = (1, 3), f^{-1}[B] = \{x : -1 \le 2x + 1 \le 1\} = \{x : -1 \le x \le 0\} = [-1, 0]$

3. $f[A] = \{\sin x : 0 \le x \le \pi\} = [0, 1], f^{-1}[B] = \{x : 0 \le \sin x \le 2\} = \{x : \text{for some even integer } k, k\pi \le x \le (k + 1)\pi\} = \cup\{[k\pi, (k + 1)\pi] : k \text{ is even}\}$.

5. Assume that f is 1–1. By Theorem 3, we have $f[A \cap B] \subseteq f[A] \cap f[B]$. Let $x \in f[A] \cap f[B]$. Then $x = f(y)$ for some $y \in A$, and $x = f(z)$ for some $z \in B$. Hence, $f(y) = f(z)$. Since f is 1–1, $y = z$, and we have $y \in A \cap B$. Therefore, $x \in f[A \cap B]$.

7. Assume that f is 1–1. Let $x \in f[A - B]$. Then $x = f(y)$ for some $y \in A - B$. Hence, $y \in A$ and $y \notin B$. Since $y \in A$, $x \in f[A]$. Assume that $x \in f[B]$. Then $x = f(z)$, for some $z \in B$. Since f is 1–1 and $f(y) = f(z)$, $y = z$. Therefore, $y \in B$, contradicting $y \notin B$. Therefore, $x \notin f[B]$. Hence, $x \in f[A] - f[B]$. Conversely, assume that $x \in f[A] - f[B]$. Then $x \in f[A]$ and $x \notin f[B]$. Hence, $x = f(y)$ for some $y \in A$. But if $y \in B$, then $x = f(y) \in f[B]$, contradicting $x \notin f[B]$. Therefore, $y \notin B$. Hence, $y \in A - B$ and $x \in f[A - B]$.

9. The conjecture is true. Let $x \in f[\cup \Omega]$. Then $x = f(y)$ for some $y \in \cup \Omega$. Hence, $y \in A$ for some $A \in \Omega$. So $x \in f[A]$ for some $A \in \Omega$, and it follows that $x \in \cup\{f[A] : A \in \Omega\}$. Reverse the steps to obtain the other set inclusion.

11. The conjecture is true. Let $x \in f^{-1}[\cup \Upsilon]$. Then $f(x) \in \cup \Upsilon$. Hence, $f(x) \in A$ for some $A \in \Upsilon$. Therefore, $x \in f^{-1}[A]$ for some $A \in \Upsilon$. So $x \in \cup\{f^{-1}[A] : A \in \Upsilon\}$. Reverse the steps to obtain the other set inclusion.

13. Assume that f is 1–1. Let $x \in f[\cap \Omega]$. Then $x = f(y)$ for some $y \in \cap \Omega$. Hence, $y \in A$ for all $A \in \Omega$. So $x \in f[A]$ for all $A \in \Omega$, and hence, $x \in \cap\{f[A] : A \in \Omega\}$. Now let $x \in \cap\{f[A] : A \in \Omega\}$. Then $x \in f[A]$ for all $A \in \Omega$. Since $\Omega \neq \emptyset$, let $B \in \Omega$. Then $x \in f[B]$, and hence, $x = f(y)$ for some $y \in B$. We show that $y \in \cap \Omega$. Let $A \in \Omega$. Then,

$x \in f[A]$, and hence, $x = f(z)$ for some $z \in A$. Therefore, $f(z) = f(y)$. Since f is 1–1, $z = y$. Therefore, $y \in A$. So $y \in A$ for all $A \in \Omega$, and hence, $y \in \cap\Omega$. From this, it follows that $x \in f[\cap\Omega]$.

15. Let $z \in f^{-1}[B]$. By the definition for relation, $yf^{-1}z$ for some $y \in B$. Therefore, zfy, and hence, $y = f(z)$. Hence, $f(z) \in B$ and $z \in \{x : f(x) \in B\}$. For the other set inclusion, let $z \in \{x : f(x) \in B\}$. Then $f(z) \in B$. Let $y = f(z)$. Then zfy and $yf^{-1}z$. Therefore, $z \in f^{-1}[B]$.

17. $R[A] \cap R[B] \subseteq R[A \cap B]$ is false. Let $R = \{(1, 2), (2, 2)\}$, $A = \{1\}$, and $B = \{2\}$. Prove that $R[A \cap B] \subseteq R[A] \cap R[B]$: Let $y \in R[A \cap B]$. Then xRy for some $x \in A \cap B$. Since $x \in A$, $y \in R[A]$, and since $x \in B$, $y \in R[B]$. Hence, $y \in R[A] \cap R[B]$.

19. The conjecture is true. Assume $R^{-1} \circ R \subseteq 1_D$. $R[A \cap B] \subseteq R[A] \cap R[B]$ is proved as in Exercise 17. Let $y \in R[A] \cap R[B]$. Then xRy for some $x \in A$ and zRy for some $z \in B$. Hence, $yR^{-1}z$. So we have $x(R^{-1} \circ R)z$. Hence, $x1_Dz$ and $z = x$. Therefore, $x \in A \cap B$, and we have $y \in R[A \cap B]$.

21. The conjecture is true. Let $y \in R[\cup\Omega]$. Then xRy for some $x \in \cup\Omega$. Hence, $x \in A$ for some $A \in \Omega$. So, $y \in R[A]$ for some $A \in \Omega$, and hence, $y \in \cup\{R[A] : A \in \Omega\}$. Reverse the steps to obtain the other inclusion.

Chapter 4 Proofs to Evaluate

1. The conjecture is false. The steps do not reverse. A counterexample is $A = D = \emptyset$ and $B = C = \{1\}$.

3. The conjecture is false. First we must start with an $x \in A$. At this point, we do not know that there is a $y \in A$ such that $(x, y) \in R$. The argument assumes $\text{dom}(R) = A$. A counterexample is $A = \{1, 2\}$ and $R = \{(1, 1)\}$.

5. The conjecture is true and the proof is valid for the implication ($\Rightarrow$). The conjecture is false for the implication ($\Leftarrow$). In case 2, from $x \in \text{dom}(f)$, we cannot conclude that $(x, w) \in f$ and $(x, z) \in f$, which is what we need to conclude that $w = z$. (See Exercise 17 in Section 4.3.) A counterexample is $f = \{(1, 2)\}$ and $g = \{(1, 3)\}$.

Chapter 4 Review

1.　(a) $\{(2, 6), (3, 9)\}$
　　(b) R is not reflexive on $A \cup B$, is not symmetric, is antisymmetric, and is transitive.

3. R is reflexive since $x = x$; R is not symmetric since $1R0$ but $\neg 0R1$; R is antisymmetric (proof by cases); and R is transitive (proof by cases).

5. R is reflexive; R is symmetric (proof by cases); R is not antisymmetric since $0R1$ and $1R0$; R is not transitive since $1R0$ and $0R(-1)$, but $\neg 1R(-1)$.

7. The conjecture is true. Assume that R and S are symmetric. Let $(x, y) \in R \cup S$.
   Case 1. $(x, y) \in R$. Since R is symmetric, $(y, x) \in R$, and hence, $(y, x) \in R \cup S$.
   Case 2. $(x, y) \in S$. Similar to case 1.

9. The conjecture is false: Let $R = \{(1, 2)\}$ and $S = \{(2, 3)\}$.

11. Assume (i). Let $x \in A$, $y, z \in B$, xRy, and xRz. Then $yR^{-1}x$ and hence, $y(R \circ R^{-1})z$. Therefore, $y1_Bz$, and so $y = z$. Now assume (ii). Let $y(R \circ R^{-1})z$. Then $yR^{-1}x$ and xRz for some x. Hence, xRy and xRz. By (ii), $y = z$, and hence $y1_Bz$.

13. The conjecture is true. Assume $(x, y) \in (R \circ S) \circ T$. Then $(x, z) \in T$, and $(z, y) \in R \circ S$ for some z. Hence, $(z, w) \in S$ and $(w, y) \in R$ for some w. Therefore, $(x, w) \in S \circ T$

since $(x, z) \in T$ and $(z, w) \in S$. Hence, $(x, y) \in R \circ (S \circ T)$. The other set inclusion is similarly proved.

15. $D = \mathcal{N} - \{0\} = C.$ $p(s(x)) = p(x+1) = (x+1) - 1 = x,$ $s(p(x)) = s(x-1) = (x-1) + 1 = x.$

17.   (a) f is not 1–1 since $f(\{1\}) = f(\{1, 2\})$.
   (b) f is onto: Let $n \in \mathcal{N}$. Then $f(\{n\}) = n$.

Section 5.1

1.   (a) There are four equivalence classes: [0], [1], [2], and [3].
   (b) Note that $[0] = \{\ldots, -8, -4, 0, 4, 8, \ldots\}$, $[1] = \{\ldots, -7, -3, 1, 5, \ldots\}$, $[2] = \{\ldots, -2, 2, 6, \ldots\}$, and $[3] = \{\ldots, -1, 3, 7, \ldots\}$.

3.   (a) R is reflexive since x is in the same row as itself; R is symmetric since x is in the same row as y implies y is in the same row as x; R is transitive since x is in the same row as y and y is in the same row as z implies x is in the same row as z.
   (b) The equivalence classes consist of the elements in a row.

5.   (a) Note that $(x, y) \in 1_A$ if and only if $x = y$.
   (b) Since $a \in [a]$, we only need to show that $x \in [a]$ implies $x = a$. Assume $x \in [a]$. Then $(a, x) \in 1_A$, and hence, $x = a$.

7.   (a) R is reflexive since X has the same number of elements as itself; R is symmetric since X has the same number of elements as Y implies Y has the same number of elements as X; R is transitive since X has the same number of elements as Y and Y has the same number of elements as Z implies X has the same number of elements as Z.
   (b) $[\emptyset] = \{\emptyset\}$ and $[D] = \{D\}$
   (c) $[\{4\}] = \big\{\{0\}, \{1\}, \{2\}, \{3\}, \{4\}, \{5\}, \{6\}, \{7\}, \{8\}, \{9\}\big\}$
   (d) $[\{4, 7\}]$ is the set of all two-element subsets of D.
   (e) There are 11 equivalence classes.

9.   (a) T is reflexive since $\sin^2 x + \cos^2 x = 1$; T is symmetric since $\sin^2 x + \cos^2 y = 1$ implies $(1 - \cos^2 x) + (1 - \sin^2 y) = 1$, which implies $\sin^2 y + \cos^2 x = 1$; T is transitive since $\sin^2 x + \cos^2 y = 1$ and $\sin^2 y + \cos^2 z = 1$ implies $(\sin^2 x + \cos^2 y) + (\sin^2 y + \cos^2 z) = 2$, which implies $\sin^2 x + \cos^2 z = 1$.
   (b) Note that xTy if and only if $|\sin x| = |\sin y|$. $[0] = \{m\pi : m \in \mathcal{Z}\}$; $[\pi/2] = \{\pi/2 + m\pi : m \in \mathcal{Z}\}$; $[\pi] = [0]$.
   (c) $[a] = \{\pm a + m\pi : m \in \mathcal{Z}\}$

11.   (a) R is reflexive since $v \cdot u = u \cdot v$; R is symmetric since $w \cdot u = z \cdot v$ implies $v \cdot z = u \cdot w$; R is transitive since $w \cdot u = z \cdot v$ and $y \cdot z = x \cdot w$ implies $(w \cdot u)(y \cdot z) = (z \cdot v)(x \cdot w)$, which in turn implies $y \cdot u = x \cdot v$.
   (b) $[(3, 6)] = \{(1, 2), (2, 4), (3, 6), \ldots\}$; $[(6, 3)] = \{(2, 1), (4, 2), (6, 3), \ldots\}$; $[(0, 1)] = \{(0, n) : n \in \mathcal{N} \text{ and } n \neq 0\}$; $[(-7, 2)] = \{(-7n, 2n) : n \in \mathcal{N} \text{ and } n \neq 0\}$

13. We only need to show that R is reflexive. Assume $x \in A$. Then $(x, y) \in R$ for some y since $\mathrm{dom}(R) = A$. Hence, $(y, x) \in R$ since R is symmetric. Because R is transitive, it follows that $(x, x) \in R$.

15. We must show that $(x, y) \in R$ if and only if $(x, y) \in A/(A/R)$. Assume that $(x, y) \in R$. Then x and y are in the same equivalence class $[x] \in (A/R)$. Hence, $(x, y) \in A/(A/R)$. To prove the converse, assume that $(x, y) \in A/(A/R)$. Then $x \in [z]$ and $y \in [z]$, where $[z] \in A/R$ for some z. Therefore, $[x] = [y] = [z]$, and it follows that $(x, y) \in R$.

17. **(a)** $\approx$ is an equivalence relation on D since $f(x) = f(x)$, $f(x) = f(y)$ implies $f(y) = f(x)$, and $f(x) = f(y)$ and $f(y) = f(z)$ implies $f(x) = f(z)$.

(b) Assume f is onto. To show that h is 1–1, we assume $h([x]) = h([y])$. Then $f(x) = f(y)$ and $x \approx y$. Hence, $[x] = [y]$. To show that h is onto, use the fact that f is onto.

(c) $x \in [a]$ if and only if $a \approx x$ if and only if $f(a) = f(x)$ if and only if $f(x) \in \{f(a)\}$ if and only if $x \in f^{-1}[\{f(a)\}]$.

19. **(a)** $\mathcal{Z}/\approx = \{[0], [1], [2]\}$, where $[0] = \{3m : m \in \mathcal{Z}\}$, $[1] = \{3m + 1 : m \in \mathcal{Z}\}$, and $[2] = \{3m + 2 : m \in \mathcal{Z}\}$

(b) $h([0]) = 0$, $h([1]) = 1$, $h([2]) = 2$

(c) Note that $f(x) = f(y)$ if and only if $x - y$ is divisible by 3.

21. The conjecture is false. Consider $R = \{(1, 2), (2, 1), (1, 1), (2, 2), (3, 3)\}$ and $S = \{(1, 3), (3, 1), (1, 1), (2, 2), (3, 3)\}$ on $A = \{1, 2, 3\}$.

23. **(a)** We prove that $R \cap S$ is symmetric. Assume $(x, y) \in R \cap S$. Then $(x, y) \in R$ and $(x, y) \in S$. Hence, $(y, x) \in R$ and $(y, x) \in S$. Therefore, $(y, x) \in R \cap S$.

(b) $y \in [x]_{R \cap S}$ if and only if $(x, y) \in R \cap S$ if and only if $(x, y) \in R$ and $(x, y) \in S$ if and only if $y \in [x]_R$ and $y \in [x]_S$ if and only if $y \in [x]_R \cap [x]_S$

Section 5.2

1.

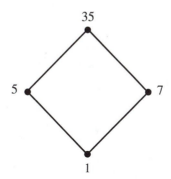

3. **(a)** e is a maximal element; a and c are minimal elements.

(b) e is a greatest element; no least elements.

(c) c is the only lower bound; e is the only upper bound.

(d) $\text{glb}(\{c, e\}) = c$

(e) $\text{lub}(\{c, e\}) = e$

5. **(a)**

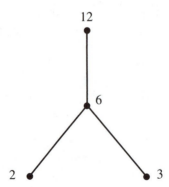

(b) 12 is a maximal element; 2 and 3 are minimal elements.

(c) 12 is a greatest element; there are no least elements.

(d) 6 and 12 are upper bounds of $\{2,3\}$; there are no lower bounds.

(e) It does not exist.

(f) $\text{lub}(\{6,3\}) = 6$

7. The proof is similar to the proof of Theorem 3.

9. **(a)** $\text{glb}(B) = -\sqrt{2}$; $\text{lub}(B) = \sqrt{2}$

(b) They do not exist.

11. $\text{glb}(B) = 2$; $\text{lub}(B)$ does not exist.

13. **(a)** Yes, verify that $\propto$ is reflexive, antisymmetric, and transitive.

(b) Not necessarily, since we can find an example of a poset with elements a, b, c, and d such that $a \preceq c$ and $d \preceq b$. Thus, neither $(a,b) \propto (c,d)$ nor $(c,d) \propto (a,b)$ holds.

15. It is easy to see that $\preceq$ is reflexive on A and transitive. We show that $\preceq$ is antisymmetric by contradiction. Assume $x \neq y$, $x \preceq y$, and $y \preceq x$. Then xRy and yRx. But since R is transitive, xRx, which contradicts the fact that R is irreflexive.

17. **(a)** Assume $[x] = [x']$ and $[y] = [y']$. Then xRx' and yRy', and so $x \preceq y$ if and only if $x' \preceq y'$.

(b) The proof uses the fact that $(A, \preceq)$ is a poset.

19. The conjecture is false: Consider $R = \{(1,2),(1,1),(2,2)\}$ and $S = \{(2,1),(1,1),(2,2)\}$ on $A = \{1,2\}$.

21. The conjecture is true. We prove that $R \cap S$ is transitive. Assume that $(x,y) \in R \cap S$ and $(y,z) \in R \cap S$. Then $(x,y) \in R$, $(x,y) \in S$, $(y,z) \in R$, and $(y,z) \in S$. Since R is transitive, $(x,z) \in R$, and since S is transitive, $(x,z) \in S$. Therefore, $(x,z) \in R \cap S$.

Section 5.3

1. **(a)** Yes

(b) No, for example $\{0,1\}^{\#}$ is not finite.

(c) No, for example $0^{\#}$ is not defined.

3. **(a)** Yes

(b) No, for example $\{0,2,4,\ldots\} * \{2,3,5,7,\ldots\}$ is not infinite.

5. First, verify that f is a bijective function. We show that f preserves the operations. $f(X \cap Y) = A - (X \cap Y) = (A - X) \cup (A - Y) = f(X) \cup f(Y)$

7.

$\oslash$	[a] [d] [e]
[a]	[a] [e] [e]
[d]	[e] [d] [e]
[e]	[e] [e] [e]

$h([a]) = q$, $h([d]) = r$, $h([e]) = p$.
It is easy to see that h is bijective.
Verify that h preserves operations.

9. We verify that φ preserves operations. $\varphi(f + g) = (f + g)(0) = f(0) + g(0) = \varphi(f) + \varphi(g)$.

11. To verify that f is a bijection, use $\log_2 x$. The function f preserves operations since $2^{x+y} = 2^x \cdot 2^y$.

13. To verify that f is a bijection, use the fact that the system of equations $x + y = a$ and $x - y = b$ has a unique solution (x, y). Also, $f((x, y) + (u, v)) = f(x + u, y + v) = (x + u + y + v, x + u - y - v) = (x + y, x - y) + (u + v, u - v) = f(x, y) + f(u, v)$.

15. (a) B is closed under multiplication, since $(m + n\sqrt{3}) \cdot (u + v\sqrt{3}) = mu + 3nv + (mv + nu)\sqrt{3}$.

    (b) Assume that $m + n\sqrt{3} = u + v\sqrt{3}$. Then $m - u = (n - v)\sqrt{3}$, and it follows that $n - v = 0$ since $\sqrt{3}$ is irrational. Hence, $n = v$ and $m = u$. Therefore, $m + n\sqrt{2} = u + v\sqrt{2}$ and $f(m + n\sqrt{3}) = f(u + v\sqrt{3})$.

    (c) The proof that f is 1–1 is similar to the proof in part b. It is routine to show that f is onto.

    (d) We show that $f((2 - \sqrt{3}) \cdot (2 + \sqrt{3})) \neq f(2 - \sqrt{3}) \cdot f(2 + \sqrt{3})$. $f((2 - \sqrt{3}) \cdot (2 + \sqrt{3})) = f(1) = 1 \neq 2 = (2 - \sqrt{2}) \cdot (2 + \sqrt{2}) = f(2 - \sqrt{3}) \cdot f(2 + \sqrt{3})$

17. Assume that $[x] = [u]$ and $[y] = [v]$. Then $[x * y] = [y] = [v] = [u * v]$. It is clear that $A/\equiv$ is closed under $\circledast$.

19. Assume that $x, y \in f[A]$. Then $x = f(a)$ and $y = f(b)$ for some $a, b \in A$. Hence, $x \# y = f(a) \# f(b) = f(a * b) \in f[A]$.

21. (a) $(B, {}^\#)$ is an algebra since ${}^\#$ is a unary operation on B.

    (b) The function f is 1–1 since $\chi_{A-X} = \chi_{A-Y}$ if and only if $X = Y$; f is onto since, for any characteristic function χ_W, there exists a set $X = A - W$ such that $f(X) = \chi_W$; f preserves the operation since $f(X') = \chi_{A-X} = \chi_X{}^\# = f(X)^\#$.

23. $g \circ f(a * b) = g(f(a * b)) = g(f(a) \# f(b)) = g(f(a)) \otimes g(f(b)) = (g \circ f)(a) \otimes (g \circ f)(b)$

Chapter 5 Proofs to Evaluate

1. The conjecture is incorrect. The flaw in the argument consists of omitting the case $x^2 + y^2 = 1$ and $z^2 + y^2 = 1$ in the transitivity part. Note that $x = 1$, $y = 0$, and $z = -1$ provides a counterexample in this case.

3. The conjecture is incorrect. One counterexample is $A = \{1, 2\}$ and $R = \{(1, 1), (2, 2), (1, 2)\}$. Then $[1] = \{1, 2\}$ and $[2] = \{2\}$. The flaw is that $z \in [y]$ implies yRz, not necessarily zRy. Also, xRy does imply $[y] \subseteq [x]$, but does not necessarily imply $[x] \subseteq [y]$. Why?

5. The conjecture is incorrect unless both $\text{lub}(X)$ and $\text{lub}(Y)$ are assumed to exist. The argument implicitly assumes existence. With the assumption of existence, the conjecture and the argument are correct.

Chapter 5 Review

1. (a) Yes. $A/R = \big\{ \{a, b\}, \{c\}, \{d\} \big\}$.

    (b) No, since R is not antisymmetric.

3. (a) No, since R is not symmetric.

(b) Yes, this is the standard "less than or equal to" ordering. 1 is a least element and a minimal element; 6 is a greatest element and a maximal element.

5. (a) No, since R is not symmetric.

(b) Yes, this is the standard "less than or equal to" ordering. There are no extremal elements.

7. Yes. We prove that R^{-1} is transitive. Assume $(x, y) \in R^{-1}$ and $(y, z) \in R^{-1}$. Then $(y, x) \in R$ and $(z, y) \in R$. Hence, $(z, x) \in R$ and $(x, z) \in R^{-1}$.

9. (a) See Exercise 5 of Section 5.1.

(b) We prove 1_A is antisymmetric. Assume $(x, y) \in 1_A$ and $(y, x) \in 1_A$. Then, by the definition of 1_A, $x = y$.

11. $R = 1_A$. Use the fact that R is a function and R is reflexive.

13. (a) We prove that T is transitive. Assume $(x, y)T(z, w)$ and $(z, w)T(u, v)$. Then $x^2 + y^2 = z^2 + w^2$ and $z^2 + w^2 = u^2 + v^2$. Therefore, $x^2 + y^2 = u^2 + v^2$, and so $(x, y)T(u, v)$.

(b) $[(0, 0)] = \{(0, 0)\}$, $[(1, 0)] = \{(x, y) : x^2 + y^2 = 1\}$

(c) $[(a, b)] = \{(x, y) : x^2 + y^2 = a^2 + b^2\}$

15. (a) This is routine.

(b) $f([(x, y)]) = x - y$. To show that f is well defined, assume $[(u, v)] = [(z, w)]$. Then $u + w = v + z$, and hence (in $\mathcal{Z}$), $u - v = z - w$. Therefore, $f([(u, v)]) = f([(z, w)])$.

(c) To show that f is 1–1, reverse the steps in part b. To show that f is onto, assume $z \in \mathcal{Z}$. If $z \geq 0$, then $f([(z, 0)]) = z$; if $z < 0$, then $f([(0, -z)]) = z$.

17. (a) Assume $[(a, b)] = [(z, w)]$ and $[(c, d)] = [(u, v)]$. Then $a + w = b + z$ and $c + v = d + u$. Hence, $a + w + c + v = b + z + d + u$. Therefore, $[(a, b)] \oplus [(c, d)] = [(a + c, b + d)] = [(z + u, w + v)] = [(z, w)] \oplus [(u, v)]$. To conclude that $[(a, b)] \odot [(c, d)] = [(ac + bd, ad + bc)] = [(zu + wv, zv + wu)] = [(z, w)] \odot [(u, v)]$, we must prove that $ac + bd + zv + wu = ad + bc + zu + wv$. To do so, multiply both sides of $a + w = b + z$ by c, multiply both sides of $b + z = a + w$ by d, multiply both sides of $c + v = d + u$ by z, multiply both sides of $d + u = c + v$ by w, and then add the resulting equations.

(b) Show that f defined by $f([(a, b)]) = a - b$ is an isomorphism.

19. Use Exercise 23 in Section 5.3 and Theorems 2 and 3 in Section 4.4.

PART II

Section 6.1

1. If there is a function $f : A \to \emptyset$, then $A = \emptyset = f$. The function f with domain $\emptyset$ and codomain $\emptyset$ is (vacuously) bijective.

3. $[N_2]$ includes all two-element subsets of U.

5. *Case 1.* Assume $A = \emptyset$. Then $A \times \{b\} = \emptyset$, and we are done by Theorem 1.

Case 2. Assume $A \neq \emptyset$. Define $f : A \times \{b\} \to A$ by $f(a, b) = a$ for all $a \in A$. Verify that f is bijective.

7. Since $A_1 \approx A_2$ and $B_1 \approx B_2$, there are bijective functions $f : A_1 \to A_2$ and $g : B_1 \to B_2$. Verify that $f \cup g : A_1 \cup B_1 \to A_2 \cup B_2$ is a well-defined bijective function.

9. Verify that $f : \mathcal{N} \to \mathcal{N} \times N_2$ defined by $f(n) = (n/2, 0)$ for n even and $f(n) = ((n-1)/2, 1)$ for n odd is bijective.

11. Since $A \approx N_k$, there is a bijection $g : A \to N_k$. Define $f : A \cup \{x\} \to N_{k+1}$ by $f(a) = g(a)$ for $a \in A$ and $f(a) = k$ if $a = x$. For an alternative proof, note that $\{k\} \approx \{x\}$ and $N_{k+1} = N_k \cup \{k\}$. Now use Theorem 4.

13. Assume $m = \text{Card}(A)$ is fixed, and use mathematical induction on $n = \text{Card}(B)$. For the basis step, $B = \emptyset$, and the result follows easily. For the induction hypothesis, assume that if $\text{Card}(B) = k$, then $A \cup B$ is a finite set and $\text{Card}(A \cup B) = m + k$. Let D be an arbitrary set such that $\text{Card}(D) = k + 1$. By the corollary to Theorem 7, $D - \{d\}$ is finite and $\text{Card}(D - \{d\}) = k$, where $d \in D$. By the induction hypothesis, $A \cup (D - \{d\})$ is finite and $\text{Card}(A \cup (D - \{d\})) = m + k$. Now use Theorem 7 to show that $A \cup D$ is finite and $\text{Card}(A \cup D) = m + k + 1$.

15. For the basis step, $B = \emptyset$, and the result follows easily. For the induction hypothesis, assume that if $\text{Card}(D) = k$, then $A \times D$ is a finite set and $\text{Card}(A \times D) = mk$. If $\text{Card}(B) = k+1$, consider $D = B - \{x\}$ for some $x \in B$. Now use the fact that $A \times B = (A \times D) \cup (A \times \{x\})$, the induction hypothesis, and Theorems 3 and 8.

17. Assume A is finite and $B \subseteq A - \{x\}$ for some $x \in A - B$. By the corollary to Theorem 7 and Theorem 10, $\text{Card}(B) \leq \text{Card}(A) - 1$. Hence, $\text{Card}(B) < \text{Card}(A)$.

19. Since $f : D \to \text{ran}(f)$ is a bijection, $\text{Card}(D) = \text{Card}(\text{ran}(f))$. By Theorem 10, $\text{Card}(\text{ran}(f)) \leq \text{Card}(C)$. Hence, $\text{Card}(D) \leq \text{Card}(C)$.

21. Suppose f is not 1–1. Then there exist $u, v \in N_k$, with $u \neq v$ and $f(u) = f(v)$. Verify that $g : N_k - \{u\} \to N_k$ defined by $g(x) = f(x)$ is onto. Now use Theorem 15 to conclude that $N_k - \{u\} = N_k$, a contradiction.

23. Assume there are bijective functions $f : A \to B$ and $g : C \to D$. Define a function $h : A \times C \to B \times D$ by $h(x, y) = (f(x), g(y))$, and verify that h is a bijection.

Section 6.2

1. If B is finite and $A \approx B$, then A is finite, by Theorem 6 of Section 6.1.

3. *Hint:* Use Theorem 10 of Section 6.1.

5. **(a)** Yes. By Theorem 5, the set of primes is countable. Hence, the set of primes is denumerable since it is infinite.

   (b) Yes. By Exercise 9 of Section 6.1, $N_2 \times \mathcal{N}$ is denumerable. By Theorem 6, $N_3 \times \mathcal{N} = (N_2 \times \mathcal{N}) \cup (\{2\} \times \mathcal{N})$ is denumerable.

7. *Case 1.* Assume C is denumerable. Use Theorem 5.

   Case 2. Assume C is finite. Use Theorem 10 of Section 6.1.

9. *Hint:* Use Theorem 5.

11. Define $g : A \to D$ by $g(x) = h(j)$, where j is the least integer such that $h(j) \in f^{-1}[\{x\}]$. Note that $f(g(x)) = x$ for all $x \in A$. If $g(u) = g(v)$, then $f(g(u)) = f(g(v))$, and hence, $u = v$.

13. By Theorem 5, $A - B$ is countable. Suppose $A - B$ is not denumerable. Then $A - B$ is finite. But then, $A \cup B = (A - B) \cup B$ is finite. By Theorem 3, however, $A \cup B$ is infinite since A is infinite, and we have a contradiction.

15. *Hint:* Use the corollary to Theorem 5.

17. *Hint:* Use mathematical induction. For the induction step, consider Exercise 5b.

19. Using the corollary to Theorem 10 of Section 6.1 and Theorem 6, prove that the union of two countable sets is countable. Then use mathematical induction on the number of sets in Ω.

21. *Hint:* Use the corollary to Theorem 5.

23. Since j_0 is the least subscript such that $d_{j_0} \in A - \{d_{i_0}d_{i_1}, d_{i_2}, \ldots\}$ and i_u is the least subscript on an element in $\{d_{i_0}, d_{i_1}, d_{i_2}, \ldots\}$ such that $j_0 < i_u$, it follows that $i_{u-1} < j_0 < i_u$. Hence, $d_{j_0} \in A - \{d_{i_0}, d_{i_1}, \ldots, d_{i_{u-1}}\}$. Recall that $d_{i_u} \in A - \{d_{i_0}, d_{i_1}, \ldots, d_{i_{u-1}}\}$ and i_u is the least subscript on the elements in $A - \{d_{i_0}, d_{i_1}, \ldots, d_{i_{u-1}}\}$. Therefore, $i_u = j_0$.

25. Let f be the function defined in Exercise 24.

    (a) Assume m is arbitrary. Then, since f is 1–1, the function $g : \mathcal{N} \to A_m$ defined by $g(x) = f(m, x)$ is bijective.

    (b) Clearly, the sets in Ω are nonempty. Now, since f is onto, $\cup\Omega = \mathcal{N}$. To show that the sets in Ω are pairwise disjoint, assume $A_k \cap A_j \neq \emptyset$. Then $f(k, x) = f(j, z)$ for some k, j, x, z. Since f is 1–1 $(k, x) = (j, z)$, and hence, $k = j$. Therefore, $A_k = A_j$.

Section 6.3

1. Assume $B = \{b_0, b_1, b_2, \ldots\}$. Let $D = \{b_1, b_2, b_3, \ldots\}$. It is easy to show that $D \approx B$.

3. Use Theorems 2 and 3.

5. Use Exercise 4 and mathematical induction on the number of elements in F.

7. Use the corollary to Theorem 3.

9. Use Exercise 8 and mathematical induction on the number of elements in F.

11. For each $x \in \mathrm{dom}(R)$, let $A_x = \{z : (x, z) \in R\}$. Let $\Omega = \{A_x : x \in \mathrm{dom}(R)\}$. Then Ω is a collection of nonempty sets. Use the axiom of choice to obtain a function $h : \Omega \to \cup\Omega$. The desired function is $f = \mathrm{ran}(h)$.

13. Assume $f : D \to C$. $(\Leftarrow)$ Assume there exists a function $g : C \to D$ such that $f \circ g = 1_C$. Let $u \in C$. Then $f(g(u)) = u$. Hence, f is onto. $(\Rightarrow)$ Assume f is onto. Let $\Omega = \{f^{-1}[\{w\}] : w \in C\}$. Since f is onto, Ω is a collection of nonempty sets. By the axiom of choice, we obtain a choice function h for Ω. Define $g : C \to D$ by $g(w) = h(f^{-1}[\{w\}])$. Use the fact that $f(g(u)) = u$ for all $u \in C$ to prove that g is 1–1.

15. *Case 1.* Assume Υ is a finite collection of countable sets. Then $\cup\Upsilon$ is countable, by Exercise 19 of Section 6.2.

    Case 2. Assume $\Upsilon = \{B_m : m \in \mathcal{N}\}$, where each B_m is countable. First note that if B is countable and A is denumerable, then there is an onto function $h : A \to B$. By Exercise 25 of Section 6.2, $\mathcal{N} = \cup\{A_m : m \in \mathcal{N}\}$, and the sets A_m are denumerable and pairwise disjoint. Now, for each $m \in \mathcal{N}$, there are onto functions $g_m : A_m \to B_m$. Let H_m be the set of all such onto functions. Use the axiom of choice to obtain a choice function for $\{H_m : m \in N\}$, thus obtaining a collection of functions $\Omega = \{f_m : m \in \mathcal{N}$ and $f_m : A_m \to B_m$ is onto$\}$. Then $h = \cup\Omega$ is an onto function with domain $\mathcal{N}$ and codomain $\cup\Upsilon$. Use Theorem 9 of Section 6.2 to conclude that $\cup\Upsilon$ is countable.

Chapter 6 Review

1. Prove that the function suggested in the hint is bijective.

3. There is a denumerable subset $\{a_0, a_1, a_2, \ldots\}$ of A. Define a function $f : A \cup \{x\} \to A$ by $f(x) = a_0$, $f(a_j) = a_{j+1}$ for $j = 0, 1, 2, \ldots$, and $f(a) = a$ for all other elements a of A. Prove that f is bijective.

5. There is a bijective function $f : A \cup \{x\} \to A$. Use f to construct a bijective function $g : A \to A - \{f(x)\}$, thus showing that A is Dedekind infinite.

7. The partition Π of A is a collection of nonempty sets. Use the axiom of choice to obtain a choice function g for Π. Let $B = g[\Pi]$. Show that $B \cap X = \{g(X)\}$ for each $X \in \Pi$.

Section 7.1

1. **(a)** No, since $2*3 = 6$.
   (b) Yes.

3. **(a)** $(x + y')x' = x'y'$
   (b) $(x \cdot 1 + 0)x' = 0$
   (c) If $x + y = 0$, then $x = 0$ and $y = 0$.

5. **(a)** No, since $3' = 8$.
   (b) No. The zero element of B is 1, but $2 \cdot 2' = \gcd(2, 12) = 2 \neq 1$.

7. $xx = xx + 0 = xx + xx' = x(x + x') = x \cdot 1 = x$

9. **(a)** $0 + 1 = 1$ and $0 \cdot 1 = 0$ implies $1 = 0'$, by the unique complement law.
   (b) $x' + x = 1$ and $x' \cdot x = 0$ implies $x = (x')'$.

11. **(a)** Yes
    (b) Yes
    (c) No

13. **(a)** No for D_{40}; yes for D_{42}.
    (b) $(D_n, +, \cdot, ', 1, n)$ is a Boolean algebra if and only if n is not divisible by the square of any prime. For the proof, notice that p^2 divides n implies $n = kp^2$. It follows that p and $p' = n/p = kp$ are elements of D_n. However, $\gcd(p, p') = p \neq 1$. In other words, $p \cdot p'$ is not the zero element of the Boolean algebra. For the converse, note that $\gcd(k, n/k) = 1$ for all $k \in D_n$ if n is not divisible by the square of any prime.

15. Proof of Theorem 7b: $(x + y) \cdot (x + y') = x + yy' = x + 0 = x$.

17. $x' + y' = [(x' + y')']' = (x'' \cdot y'')' = (x \cdot y)'$

19. **(a)** $(x + 0)(y + 1) = x(y + 1) = x \cdot 1 = x$
    (b) $[(xy)z + (xy)z'] + xy' = xy(z + z') + xy' = xy \cdot 1 + xy' = xy + xy' = x(y + y') = x \cdot 1 = x$

21. **(a)** $w + xyz = w + x(yz) = (w + x)(w + yz) = (w + x)(w + y)(w + z)$
    (b) $w(x + y + z) = wx + wy + wz$

23. $xy + (xy)'z + z' = [xy + (xy)'][xy + z] + z' = 1 \cdot (xy + z) + z' = (xy + z) + z' = xy + (z + z') = xy + 1 = 1$

Section 7.2

1. **(a)** $0 \leq x$ since $0 \cdot x = 0$.
   (b) $x \leq 1$ since $x \cdot 1 = x$.

3. **(a)** $xy \leq y$ since $(xy)y = x(yy) = xy$.
   (b) The proof of part b is the dual of the proof of part a.

5. Assume $1 \leq y$. Then, $1 \cdot y = 1$, and hence, $y = 1$.

7. $x \leq y$ if and only if $xy' = 0$, by Theorem 2. Finish by showing that $xy' = 0$ if and only if $x' + y = 1$.

9. Assume $x \leq y$ and $y \leq x$. Then $xy = x$ and $yx = y$. Hence, $x = y$.

11. Assume $x \leq y$ and $w \leq z$.
    (a) By Theorem 2, $x + y = y$ and $w + z = z$. Hence, $(x + w) + (y + z) = (x + y) + (w + z) = y + z$. Therefore, $x + w \leq y + z$.

(b) By definition, $xy = x$ and $wz = w$. Hence, $(xw)(yz) = (xy)(wz) = xw$. Therefore, $xw \leq yz$.

13. Since a_2 is a minimal element in the set of nonzero elements and $a_1 \neq 0$, $a_1 \leq a_2$ implies $a_1 = a_2$.

15. **(a)** Atom$[0] = \emptyset$, since $x \leq 0$ implies $x = 0$.
(b) Atom$[1] = W$, since $x \leq 1$ for all x.
(c) Assume a is an atom. Then Atom$[a] = \{a\}$ by Theorem 8.

17. **(a)** ($\Leftarrow$) follows from Theorem 11a. ($\Rightarrow$) follows from the definition of an atomistic Boolean algebra.
(b) ($\Leftarrow$) follows from the fact that $x = y$ implies $a \leq x$ if and only if $a \leq y$.

19. **(a)** No, since Card(D_{32}) is not a power of 2.
(b) Yes. (See Exercise 13b of Section 7.1.)

21. $P(W) = \{\emptyset, \{3\}, \{5\}, \{7\}, \{3,5\}, \{5,7\}, \{3,7\}, \{3,5,7\}\}$; $f(1) = \emptyset, f(3) = \{3\}, f(5) = \{5\}, f(7) = \{7\}, f(15) = \{3,5\}, f(35) = \{5,7\}, f(21) = \{3,7\}, f(105) = \{3,5,7\}$.

23. See the solution to Exercise 13b of Section 7.1. Show that m^2 divides n implies that $m \cdot m'$ is not the zero element of the Boolean algebra.

25. Assume that $Y = \{a_1, a_2, \ldots, a_k\}$ and $y = a_1 + a_2 + \ldots + a_k$. Show that $Y = \{a_1\} \cup \{a_2\} \cup \ldots \cup \{a_k\} = $ Atom$[a_1] \cup$ Atom$[a_2] \cup \ldots \cup$ Atom$[a_k] = $ Atom$[a_1 + a_2 + \ldots + a_k]$.

27. **(a)** $W = \{\{x\} : x \in \mathcal{N}\}$, and hence, $P(W) = \{\{\{x\} : x \in X\} : X \subseteq \mathcal{N}\}$. Define f by $f(X) = \{\{x\} : x \in X\}$ for all $X \in S$.
(b) The function f is not onto. Note that $\{\{0\}, \{2\}, \{4\}, \ldots\}$ has no preimage, since neither $E = \{0, 2, 4, \ldots\}$ nor $E' = \{1, 3, 5, \ldots\}$ is in S.

Chapter 7 Review

1. The verification of Axiom 3a takes eight cases. To verify Axiom 4b, we note that $0 \cdot 1 = 0$ and $1 \cdot 1 = 1$.

3. No, by the corollary to the proposition at the end of Section 7.2.

5. The conjecture is false. For example, consider $(D_{30}, +, \cdot, 1, 30)$ with $D_{30} = \{1, 2, 3, 5, 6, 10, 15, 30\}$, $x + y = \text{lcm}(x, y)$, $x \cdot y = \gcd(x, y)$, and $x' = 30/x$. Let $x = 15$, $y = 3$, and $z = 5$. Then $x + y = 15 = x + z$.

7. $x(x' + y) + z + y = xx' + xy + z + y = 0 + xy + y + z = (x + 1)y + z = 1 \cdot y + z = y + z$

9. $x'w + x'y' + yz' + x'z = x'w + x'y' + yz' + x'z + yz' = x'w + x'y' + (x + x')yz' + x'z + yz' = (x'w + x'y' + x'yz' + x'z) + (xyz' + yz') = x'[w + y' + (yz' + z)] + yz' = x'(w + y' + y + z) + yz' = x' \cdot 1 + yz' = x' + yz'$

11. Any Boolean algebra of the form $(P(U), \cup, \cap, ', \emptyset, U, \subseteq)$, where U is infinite, will do.

13. Let Atom$[x] = \{b_1, b_2, \ldots, b_j\}$. Then Atom$[x] = \{b_1\} \cup \{b_2\} \cup \ldots \cup \{b_j\}$. Prove that Atom$[x] = $ Atom$[b_1 + b_2 + \ldots + b_j]$ and, hence, $x = b_1 + b_2 + \ldots + b_j$ by Theorem 14b of Section 7.2.

Section 8.1

1. **(a)** $u = u + 0 = 0 + u = 0$
(c) $w = w + 0 = w + (x + u) = (w + x) + u = (x + w) + u = 0 + u = u$

3. $-(-x) = -(-x) + 0 = -(-x) + (x + -x) = -(-x) + (-x + x) = [-(-x) + -x] + x = [-x + -(-x)] + x = 0 + x = x$

5. Assume $ax = ay$ and $a \neq 0$. Then $a(x - y) = ax + a(-y) = ay + a(-y) = a(y + -y) = a(0) = 0$. By Axiom 6, $a = 0$ or $x - y = 0$. Since $a \neq 0$, $x - y = 0$, and we conclude that $x = y$.

7. We show that $(-x)y = -xy$. The proof that $x(-y) = -xy$ is similar. $(-x)y = (-x)y + 0 = (-x)y + (xy - xy) = [(-x)y + xy] - xy = [y(-x) + yx] - xy = y(-x + x) - xy = y(0) - xy = 0 - xy = -xy$.

9. Let x and y be arbitrary elements of D. By Axiom 9, either $x - y = 0$ or $x - y \in D^+$ or $-(x - y) \in D^+$. Since additive inverses are unique and $(x - y) + (y - x) = 0$, $-(x - y) = y - x$. Hence, either $x = y$ or $y \leq x$ or $x \leq y$.

11. (a) Note that $y - x \in D^+$ if and only if $(y + w) - (x + w) \in D^+$.
   (b) If $x = y$, then $xw = yw$. Since $w > 0$, $w \in D^+$. If $y - x \in D^+$, then $(y - x)w \in D^+$ since D^+ is closed under multiplication. By Exercise 6 and commutativity, $yw - xw \in D^+$. Therefore, $xw \leq yw$.

13. For the basis step $(n = 1)$, we use the definition to conclude that $1 \cdot 1 = (0 + 1) \cdot 1 = 0 \cdot 1 + 1 = 1$. But 1 is positive by the corollary to Theorem 5. Assume $k \cdot 1$ is positive. Then $(k + 1) \cdot 1 = k \cdot 1 + 1$. Since D^+ is closed under addition, $(k + 1) \cdot 1$ is positive.

15. First, use mathematical induction to prove that $m \cdot (a + b) = m \cdot a + m \cdot b$ for all $m \in \mathcal{N}$. For the basis step, $0 \cdot (a + b) = 0 = 0 + 0 = 0 \cdot a + 0 \cdot b$. Assume $k \cdot (a + b) = k \cdot a + k \cdot b$. Then $(k + 1) \cdot (a + b) = k \cdot (a + b) + (a + b) = (k \cdot a + k \cdot b) + (a + b) = (k \cdot a + a) + (k \cdot b + b) = (k + 1) \cdot a + (k + 1) \cdot b$.

   For m a negative integer, $m \cdot (a + b) = (-m) \cdot [-(a + b)] = (-m) \cdot [(-a) + (-b)] = (-m) \cdot (-a) + (-m) \cdot (-b) = m \cdot a + m \cdot b$.

17. First, use mathematical induction to prove that $m \cdot (ab) = (m \cdot a)b = a(m \cdot b)$ for all $m \in \mathcal{N}$. The basis step is easy. Assume that $k \cdot (ab) = (k \cdot a)b = a(k \cdot b)$. Then $(k + 1) \cdot (ab) = k \cdot (ab) + ab = (k \cdot a)b + ab = (k \cdot a + a)b = [(k + 1) \cdot a]b$. Similarly, $(k + 1) \cdot (ab) = a[(k + 1) \cdot b]$.

   For m a negative integer, $m \cdot (ab) = (-m) \cdot (-ab) = (-m) \cdot [(-a)b] = [(-m) \cdot (-a)]b = (m \cdot a)b$. Similarly, $m \cdot (ab) = a(m \cdot b)$.

19. We only need to prove that $D \subseteq \{n \cdot 1 : n \text{ is an integer}\}$. Let $x \in D$. By Axiom 9, either $x = 0$ or $x \in D^+$ or $-x \in D^+$. If $x = 0$, then $0 = 0 \cdot 1$ and $x \in \{n \cdot 1 : n \text{ is an integer}\}$. If $x \in D^+$, then $x \in \{n \cdot 1 : n \text{ is an integer}\}$ by part a. If $-x \in D^+$, then $-x = m \cdot 1$ for some positive integer m. Hence, $x = -(m \cdot 1) = -[1(m \cdot 1)] = (-1)(m \cdot 1) = [(-1)m] \cdot 1 = (-m) \cdot 1 \in \{n \cdot 1 : n \text{ is an integer}\}$.

Section 8.2

1. Assume $xy = 1$. Then $|x||y| = |xy| = 1$. It follows that $|x| \neq 0$ and $|y| \neq 0$. By Theorem 7 of Section 8.1, $|x| \geq 1$ and $|y| \geq 1$. If $|x| > 1$, then $|xy| > 1$. Hence, $|x| = 1$. Similarly, $|y| = 1$. Therefore, $x = \pm 1$ and $y = \pm 1$. If either $x = -1$ and $x = 1$ or $x = 1$ and $y = -1$, then $xy = -1$. Since this is a contradiction, $x = 1 = y$ or $x = -1 = y$.

3. (b) $a = 1 \cdot a$ (c) $d = 1 \cdot d$

5. Assume $e = md$ and $b = ne$. Then $b = n(md) = (nm)d$.

7. (a) $q = 0$, $r = 17$ (b) $q = -7$, $r = 2$

9. For the induction step, assume $k^2 + k = 2m$. Then $(k + 1)^2 + (k + 1) = k^2 + 2k + 1 + k + 1 = (k^2 + k) + 2k + 2 = 2(m + k + 1)$.

11. For the induction step, assume $k^4 + 3k^2 = 4m$. Then $(k + 1)^4 + 3(k + 1)^2 = k^4 + 4k^3 + 6k^2 + 4k + 1 + 3k^2 + 6k + 3 = (k^4 + 3k^2) + (4k^3 + 4k^2 + 8k + 4) + (2k^2 + 2k)$. The first two summands are divisible by 4. The third summand, $2k^2 + 2k = 2(k^2 + k)$, is also divisible by 4 since $k^2 + k$ is divisible by 2 (Exercise 9).

13. We prove part a, $x^n x^m = x^{n+m}$. Parts b and c are similar. Fix m and use mathematical induction on n. The basis step is easy. For the induction step, assume $x^k x^m = x^{k+m}$. Then $x^{k+1} x^m = (x^k x)x^m = (x^k x^m)x = (x^{k+m})x = x^{k+m+1} = x^{k+1+m}$.

15. Assume $0 \le a < d$ and $0 \le b < d$. Then, $-d < -b \le 0$. Verify that adding the inequalities $0 \le a < d$ and $-d < -b \le 0$ yields $-d < a - b < d$. Hence, $|a - b| < d$ (See Exercise 11 of Section 2.1).

17. We have integers r and q such that $a = qd + r$ and $0 \le r < |d|$. Assume there exist integers r' and q' such that $a = q' \cdot d + r'$ and $0 \le r' < |d|$. Since $qd + r = q'd + r'$, $|r' - r| = |q - q'||d|$ and $|d|$ divides $|r' - r|$. By Exercise 15, $|r' - r| < |d|$. By Exercise 16, $|r' - r| = 0$. Hence, $r' = r$, and it follows that $q' = q$.

19. Assume $a = qd + r_1$ with $0 \le r_1 < |d|$, and $q = ks + r_2$ with $0 \le r_2 < |k|$. Then $a = (ks + r_2)d + r_1 = (kd)s + (dr_2 + r_1)$. We need only show that $dr_2 + r_1 < |kd|$. Since $r_1 \le |d| - 1$ and $r_2 \le |k| - 1$, $dr_2 + r_1 \le (|k| - 1)|d| + |d| - 1 = |kd| - 1$.

Section 8.3

1. By Property a, $g|a$ and $g|b$. By Property d, $g|d$. Similarly, $d|g$. By Theorem 4b of Section 8.2, $d = g$.

3. Assume $a|b$. Then a is a common divisor of a and b. Also, if $m|a$ and $m|b$, then $m|a$. Hence, $a = \gcd(a, b)$.

5. $\gcd(155, 20) = 5$, $m = -1$, $n = 8$.

7. $\gcd(32, 76) = 4$, $m = -7$, $n = 3$.

9. $\gcd(2, 765, 145, 299) = 7$, $m = 8,513$, $n = -162$.

11. (a) No, by the corollary to Theorem 2 since $\gcd(42, 66) = 6$ and 6 does not divide 3.
    (b) Yes, by the corollary to Theorem 2 since $\gcd(11, 47) = 1$.

13. No, by the corollary to Theorem 2.

15. Use the corollary to Theorem 2.

17. Assume $\gcd(a, b) = d$, $a = cd$, and $b = fd$. Use the corollary to Theorem 2. There exist integers m and n such that $d = ma + nb = mcd + nfd$. Hence, $1 = mc + nf$. Therefore, $\gcd(c, f) = 1$.

19. Assume $\gcd(a, b) = d$ and $b|ag$. By Exercise 17, $a = cd$, $b = fd$, and $\gcd(c, f) = 1$. Since $b|ag$, $fd|cdg$, and hence, $f|cg$. By Exercise 18, $f|g$. Therefore, $fd|dg$, and since $b = fd$, $b|dg$.

21. Assume $\gcd(a, b) = 1$. Let $u = \gcd(a + b, a - b)$. Then u divides $2a = (a + b) + (a - b)$ and u divides $2b = (a + b) - (a - b)$. Since $\gcd(a, b) = 1$, $1 = am + bn$ for some integers m and n. Hence, $2 = 2am + 2bn$ and $u|2$. Therefore, $\gcd(a + b, a - b) = 1$ or $\gcd(a + b, a - b) = 2$.

Section 8.4

1. Assume that every integer m such that $2 \le m < k$ is the product of primes. By Theorem 1, k has a prime factor q. So $k = qn$. If $n = 1$, then k is a prime, and we are done. Otherwise, $2 \le n < k$. By the induction hypothesis, n is the product of primes. Therefore, $k = qn$ is the product of primes.

3. $m = 2, n = -3$

5. (a) $2 \cdot 3 \cdot 5 \cdot 7$

    (b) 2^{10}

7. By Theorem 1, $m = 1 + p_1 p_2 \cdots p_k$ has a prime factor q. If $q = p_i$ for some $i = 1, 2, \ldots, k$, then q is a factor of $1 = m - p_1 p_2 \cdots p_k$. But this is a contradiction, and hence, $q \neq p_i$ for any $i = 1, 2, \ldots, k$.

9. This proof is similar to the proof in Example 2.

11. Suppose $\sqrt[3]{p}$ is rational. Then there are positive integers m and n such that $pn^3 = m^3$. Now, the number of p's on the right side of this equation is divisible by 3, but the number of p's on the left is not, a contradiction.

13. Suppose $\log_2 p$ is rational. Then there are positive integers m and n such that $2^{m/n} = p$. Hence, $2^m = p^n$, contradicting the prime factorization theorem.

15.

$$\operatorname{lcm}(40, 68) = \frac{40 \cdot 68}{\gcd(40, 68)} = \frac{40 \cdot 68}{4} = 680$$

17.

$$\operatorname{lcm}(196, 302) = \frac{196 \cdot 302}{2} = 29,596$$

19.

$$\operatorname{lcm}(2, 765, 145, 299) = \frac{2,765 \cdot 145,299}{7} = 57,393,105$$

21. (a) $45 = 3^2 \cdot 5, 225 = 3^2 \cdot 5^2, \operatorname{lcm}(45, 225) = 3^2 \cdot 5^2$

    (b) $\gcd(45, 225) = 45$

23. Use Exercise 3 of Section 8.3 and Theorem 8.

25. Let $a = p_1{}^{k_1} \cdot p_2{}^{k_2} \cdots p_m{}^{k_m}$ and $b = p_1{}^{j_1} \cdot p_2{}^{j_2} \cdots p_m{}^{j_m}$, where the p_i are distinct primes. Let $v = p_1{}^{u_1} \cdot p_2{}^{u_2} \cdots p_m{}^{u_m}$, where $u_i = \max(k_i, j_i)$. Then $a | v$ and $b | v$. If $a | c$ and $b | c$, then $p_i{}^{u_i} | c$ for $i = 1, 2, \ldots, m$, and hence, $v | c$. Therefore, v is a least common multiple of a and b.

27. Let $a = p_1{}^{k_1} \cdot p_2{}^{k_2} \cdots p_m{}^{k_m}$ and $b = p_1{}^{j_1} \cdot p_2{}^{j_2} \cdots p_m{}^{j_m}$, where the p_i are distinct primes. By Theorems 6 and 7, $\operatorname{lcm}(a, b) = p_1{}^{u_1} \cdot p_2{}^{u_2} \cdots p_m{}^{u_m}$, where $u_i = \max(k_i, j_i)$. By Theorem 5, $\gcd(a, b) = p_1{}^{v_1} \cdot p_2{}^{v_2} \cdots p_m{}^{v_m}$, where $v_i = \min(k_i, j_i)$. It follows that $ab = \gcd(a, b) \cdot \operatorname{lcm}(a, b)$.

Section 8.5

1. (a) $r = 0$

    (b) $r = 7$

3. (a) 5

    (b) 4

5. (a) 3

    (b) 3

7. Assume $x \equiv_d y$. Then $x - y = md$ for some m. Hence, $wx - wy = (wm)d$. Therefore, $w \cdot x \equiv_d w \cdot y$.

9. The conjecture is false. Let $a = 6$ and $b = 2$. Then $6 \cdot 2 \equiv_4 0$, but it is not the case that $6 \equiv_4 0$ or $2 \equiv_4 0$.

11. By the remarks preceding Example 4, we need only prove that Axiom 6 holds if and only if d is a prime. ($\Rightarrow$) Suppose d is not a prime. Then $d = mn$, where $1 < m < d$ and $1 < n < d$. Hence, $[m] *_d [n] = 0$, but $[m] \neq 0$ and $[n] \neq 0$. ($\Leftarrow$) Assume d is a prime and $[x] *_d [y] = 0$. Then $d|xy$, and by Theorem 3 of Section 8.4, $d|x$ or $d|y$. Hence, $[x] = 0$ or $[y] = 0$.

13. Assume $ac \equiv_n bc$ and c and n have no common divisors other than ± 1. Then $n|(a - b)c$ and $\gcd(c, n) = 1$. By Exercise 18 of Section 8.3, $n|(a - b)$. Hence, $a \equiv_n b$.

15. Define a function $h : \mathcal{Z} \to Z_5$ by $h(x) = [x]$. Verify that h is a homomorphism from $(\mathcal{Z}, +, \cdot, 0, 1)$ to $(Z_5, +_5, *_5, 0, 1)$.

Chapter 8 Review

1. For the induction step, note that $8^{k+1} - 3^{k+1} = 3(8^k - 3^k) + 5 \cdot 8^k$.

3. Assume $\gcd(a, b) = 1$, $a|c$, and $b|c$. Then $1 = ma + nb$, and it follows that $mac + nbc = c$. Since $a|c$, $ab|nbc$, and since $b|c$, $ab|mac$. Therefore, $ab|c$.

5. $n^5 - n = (n^3 - n)(n^2 + 1)$. By Exercise 4, $6|(n^3 - n)$. By Exercise 3, we only need show that $5|(n^5 - n)$. Consider the cases $n \equiv_5 i$, for $i = 0, 1, 2, 3, 4$.

7. The conjecture is false. Let $a = 10$, $b = 12$, and $c = 5$. Then $d = \gcd(a, b) = 2$ and $ad|cb$ since $20|60$. But a is not a factor of c.

9. Let $v = \gcd(a, b)$ and $u = \gcd(ac, bc)$. Then vc is a common divisor of ac and bc. Hence, $vc|u$. Also $vc = acm + bcn$ since $v = am + bn$. Hence, $u|vc$ since $u|ac$ and $u|bc$. Therefore, $u = vc$.

11. Assume $f(m, n) = f(k, j)$. Then $2^m 3^n = 2^k 3^j$. By the prime factorization theorem, $m = k$ and $n = j$. Therefore, $(m, n) = (k, j)$.

13. Since the definitions of gcd and lcm are symmetric, it follows that $*$ and $\ominus$ are commutative.

To prove $*$ is associative, we prove $\gcd(\gcd(a, b), c) = \gcd(a, \gcd(b, c))$. We may assume that $a = p_1^{k_1} \cdot p_2^{k_2} \cdots p_m^{k_m}$, $b = p_1^{j_1} \cdot p_2^{j_2} \cdots p_m^{j_m}$, and $c = p_1^{l_1} \cdot p_2^{l_2} \cdots p_m^{l_m}$, where the p_i are distinct primes. Then, by Theorem 5 of Section 8.4, $\gcd(a, b) = p_1^{v_1} \cdot p_2^{v_2} \cdots p_m^{v_m}$, where $v_i = \min(k_i, j_i)$, and $\gcd(b, c) = p_1^{u_1} \cdot p_2^{u_2} \cdots p_m^{u_m}$, where $u_i = \min(j_i, l_i)$. Similarly, $\gcd(\gcd(a, b), c) = p_1^{x_1} \cdot p_2^{x_2} \cdots p_m^{x_m}$, where $x_i = \min(v_i, l_i)$ and $\gcd(a, \gcd(b, c)) = p_1^{y_1} \cdot p_2^{y_2} \cdots p_m^{y_m}$, where $y_i = \min(k_i, u_i)$. The desired result follows from the fact that $x_i = y_i$ for all i.

15. Verify that $\operatorname{lcm}(a, \gcd(b, c)) = \gcd(\operatorname{lcm}(a, b), \operatorname{lcm}(a, c))$. Use Theorems 5 and 6 of Section 8.4.

Section 9.1

1. Let $\epsilon > 0$. We need to find an n such that

$$\left| \frac{5x - 1}{x + 2} - 5 \right| < \epsilon \text{ if } x > n$$

So we find an n such that

$$\frac{11}{x + 2} < \epsilon \text{ if } x > n$$

Choose

$$n > \frac{11}{\epsilon} - 2$$

3. Assume

$$\epsilon \le \frac{3 + 2\pi}{2}$$

and choose

$$n > \sqrt{\frac{3 + 2\pi}{\epsilon}} - 2$$

5. The proof in Example 1 applies to this function.

7. (a) $(\forall \epsilon > 0)(\exists n \in \mathcal{N})(\forall x \in \text{dom}(f))(x > n \Rightarrow |f(x) - L| < \epsilon)$
  (b) $(\exists \epsilon > 0)(\forall n \in \mathcal{N})(\exists x \in \text{dom}(f))(x > n \wedge |f(x) - L| \ge \epsilon)$

9. *Conjecture:* $\lim\limits_{x \to \infty} (3 + 1/x) = 3$. The proof in Example 1 applies here.

11. *Conjecture:* $\lim\limits_{x \to \infty} 1/x^2 = 0$. For arbitrary $\epsilon > 0$, choose $n > \sqrt{1/\epsilon}$.

13. *Conjecture:* $\lim\limits_{x \to \infty} \dfrac{\pi x - 3}{7x + 2} = \pi/7$. For arbitrary $\epsilon > 0$, choose

$$n > \frac{21 + 2\pi}{49\epsilon} - \frac{2}{7}$$

Section 9.2

1. The definitions of $\lim\limits_{x \to \infty} f(x) = L$ and $\lim\limits_{x \to \infty} (f(x) - L) = 0$ are identical.

3. Assume $\lim\limits_{x \to \infty} \sin x = L$ for some L. Choose $\epsilon = \frac{1}{2}$. Then $|\sin x - L| < \frac{1}{2}$ whenever $x > n$ for some n. But there exists an m such that $2\pi m > n$. Hence, $|\sin 2\pi m - L| < \frac{1}{2}$ and $|\sin(2\pi m + \pi/2) - L| < \frac{1}{2}$. It follows that $|L| < \frac{1}{2}$ and $|1 - L| < \frac{1}{2}$. Since $1 - |L| \le |1 - L|, 1 - |L| < \frac{1}{2}$. Hence, $|L| < \frac{1}{2}$ and $\frac{1}{2} < |L|$, a contradiction.

5. Assume $\lim\limits_{x \to \infty} f(x) = L$ for some L. Choose $\epsilon = \frac{1}{2}$. Then $|f(x) - L| < \frac{1}{2}$ whenever $x > n$ for some n. If $x > n$ and $x \in \mathcal{N}$, then $|1/x - L| < 1/2$. Also, $|2 - L| < \frac{1}{2}$ if $x > n$. Therefore, $|L| - |1/x| < \frac{1}{2}$ and $2 - |L| < \frac{1}{2}$. Now, note that $1/x \le 1$. Hence, $|L| < \frac{1}{2} + 1/x < \frac{1}{2} + 1 = \frac{3}{2}$ and $\frac{3}{2} < |L|$, a contradiction.

7. Use Theorem 5.

9. Use the product theorem and the reciprocal theorem.

11. First prove that $\lim\limits_{x \to \infty} g(x) = k$, where $g(x) = k$ for all x. Then use the product theorem.

13. Assume $\epsilon > 0$. Since $\lim\limits_{x \to \infty} f(x) = L$, there exists an m such that $z > m$ implies $|f(z) - L| < \epsilon$. We may choose $m > k$. Now, if $x > m - k$, then $x + k > m$, and it follows that $|f(x + k) - L| < \epsilon$. Therefore, $\lim\limits_{x \to \infty} f(x + k) = L$.

15. Let $\epsilon = |L|/2$. Since $\lim\limits_{x \to \infty} f(x) = L$, there exists an n such that $x > n$ implies $|f(x) - L| < |L|/2$. Hence, $|L|/2 < |f(x)|$ whenever $x > n$. Therefore, $|1/f(x)| < 2/|L|$ whenever $x > n$, and $1/f(x)$ is bounded on $[n + 1, \infty)$.

17. Use mathematical induction. For the basis step, use Example 4 in Section 9.1. For the induction step, use the product theorem.

19. (a) Because the limits of the numerator and denominator do not exist.
  (b) Use Example 1 of Section 9.1, Exercise 11, the sum theorem, and the quotient theorem.

21. Follows from Exercise 17 since $(1 + 1/x)^{px} = [(1 + 1/x)^x]^p$.

23. Use Exercise 21 and the quotient theorem.

25. Note that $(1 + p/x)^x = \{[1 - 1/(-x/p)]^{(-x/p)}\}^{-p}$, and use Exercises 22, 12, and 17.

Section 9.3

1. $(\forall \epsilon > 0)(\exists \delta > 0)(\forall x \in \mathrm{dom}(f))(0 < |x - a| < \delta \Rightarrow |f(x) - L| < \epsilon)$

3. *Case 1.* $a = 0$. The proof is routine.

   Case 2. $a \neq 0$. Let $\epsilon > 0$. First, assume $\delta < |a|$. Then $|x - a| < \delta$ implies $-|a| < x - a < |a|$, which in turn implies $-3|a| = -2|a| - |a| \leq 2a - |a| < x + a < 2a + |a| \leq 2|a| + |a| = 3|a|$. Hence, $|x - a| < \delta$ implies $|x^2 - a^2| = |x + a||x - a| < 3|a||x - a|$. Choose $\delta = \min(|a|, \epsilon/3|a|)$.

5. The proof is similar to the corresponding proof in Section 9.2.

7. The proof is similar to the corresponding proof in Section 9.2.

9. Choose δ so that $0 < |x - a| < \delta$ implies $|g(x) - K| < |K|/2$. Then $|K|/2 > |K| - |g(x)|$. Hence, $|g(x)| > |K|/2$ whenever $0 < |x - a| < \delta$.

11. Use the product theorem and the reciprocal theorem.

13. The proof is similar to the proof in Example 3.

15. Use mathematical induction. For the basis step, use Example 2. For the induction step, use the product theorem.

17. $\lim_{x \to 0} (1 + x)^{2/x} = e^2$. For the proof, use Exercise 15.

19. Note that $(1 + 2x)^{1/x} = [(1 + 2x)^{1/2x}]^2$, and use Exercise 16.

21. Suppose $\lim_{x \to 0} 1/x$ exists. Then by Theorem 4, $1/x$ is bounded at 0. Hence, there exist q and δ such that $|1/x| < q$ for all x such that $0 < |x| < \delta$. Let $w = \min(1/(q + 1), \delta/2)$. Then $1/w < q$ and $1/w \geq q + 1$, a contradiction.

23. Use Theorem 5.

25. Use the squeezing theorem and the fact that $\lim_{x \to 0} \cos x = 1$.

27. $\lim_{x \to 1} g(x) = 0$. For the proof, consider Example 4.

29. $\lim_{x \to 0} \sin \pi x / x = \pi$. For the proof, note that

$$\frac{\sin \pi x}{x} = \pi \cdot \frac{\sin \pi x}{\pi x}$$

and use Exercise 16.

Section 9.4

1. (a) Let $\epsilon = b/a$ and $m = q$ in the ϵ-property.
    (b) Let $b = \epsilon$, $a = 1$, and $q = m$ in the Archimedean property.

3. Let x be an irrational number and r be a rational number. Suppose $y = x + r$ is rational. It is easy to verify that $-r$ is rational. Then, by Exercise 2, $y - r = x$ is rational, a contradiction.

5. Suppose the interval (a, b) contains finitely many rationals; say $r_1, r_2, \ldots, r_n$. We may assume $r_1 < r_2 < \ldots < r_n$. By Theorem 2, there is a rational number r in (r_1, r_2). So we have a contradiction, since r is in (a, b).

7. Assume $\lim_{x \to a} f(x) = L$ for some a and some L. Choose $\epsilon = \frac{1}{2}$. Then there exists $\delta > 0$ such that $|f(x) - L| < \frac{1}{2}$ whenever $0 < |x - a| < \delta$. Since there is a rational number x

such that $0 < |x - a| < \delta$, $|1 - L| < \frac{1}{2}$, and since there is an irrational number x such that $0 < |x - a| < \delta$, $|-1 - L| < \frac{1}{2}$. But $|1 - L| < \frac{1}{2}$ implies $L > \frac{1}{2}$, and $|-1 - L| < \frac{1}{2}$ implies $L < -\frac{1}{2}$, a contradiction.

9. (a) Since $x^2 < 2$ implies $x < 3$; 3 is an upper bound of B.
   (b) $\sqrt{2}$ is the least upper bound of B.

11. If $p/q = 1/m$ for some positive integer m, we are done. (Why?) So assume p/q is not the reciprocal of any positive integer. We may assume that both p and q are positive. By the Archimedean property, there exists a positive integer m such that $mp > q$. Let $n + 1$ be the least such positive integer. Then $(n + 1)p > q$, and so $p/q > 1/(n + 1)$. It also follows that $1/n > p/q$.

13. Assume A is a nonempty set that is bounded below. Verify that $B = \{x : -x \in A\}$ is a nonempty set that is bounded above. By the least upper bound property, B has a least upper bound b. Verify that $-b$ is the greatest lower bound of A.

15. Let $B = \{r \in Q : r^2 < 3\}$.

   (1) B is bounded above by, for example, 2. Hence, B has a least upper bound u.

   (2) Since $(1.1)^2 < 3$, $1.1 \in B$, and hence, $u > 1$.

   (3) Suppose $u^2 < 3$. Then $3 - u^2 > 0$, and there exists a positive integer m such that $1/m < (3 - u^2)/3u$. Since $u > 1$ and $m^2 > m$, $1/m^2 < u/m$. It follows that $(u + 1/m)^2 = u^2 + 2u/m + 1/m^2 < u^2 + 3u/m < u^2 + (3 - u^2) = 3$. This contradicts the fact that u is the least upper bound of B. Therefore, $u^2 \geq 3$.

   (4) Suppose $u^2 > 3$. Then there exists a positive integer m such that $1/m < (u^2 - 3)/2u$. Hence, $(u - 1/m)^2 = u^2 - 2u/m + 1/m^2 > u^2 - (u^2 - 3) + 1/m^2 > 3$, a contradiction. Therefore, $u^2 \leq 3$.

Chapter 9 Review

1. Divide the numerator and denominator of $(3x^3 - 2x + 1)/(5x^3 + 2)$ by x^3 and then use appropriate limit theorems.

3. Assume $\lim_{x \to \infty} f(x) = L$ for some L. Choose $\epsilon = \frac{1}{2}$. Then there exists an n such that $|f(x) - L| < \frac{1}{2}$ whenever $x > n$. The rest of the proof is similar to the proof in Exercise 7 of Section 9.4.

5. Assume $\lim_{x \to a} f(x) = L$, where $L > 0$, and $\lim_{x \to a} f(x)^{1/q} = K$. Let $\epsilon = L/2$. Then there exists a $\delta > 0$ such that $-L/2 < f(x) - L < L/2$ whenever $0 < |x - a| < \delta$. Hence, $f(x) > L/2 > 0$ whenever $0 < |x - a| < \delta$. Let $g(x) = f(x)^{1/q}$. Then $\lim_{x \to a} g(x) = K$, and so $\lim_{x \to a} g(x)^q = K^q$, by Exercise 15 in Section 9.3. Since $g(x)^q = f(x)$, $K^q = L$, and hence, $K = L^{1/q}$.

7. No, the sum of two irrationals is not necessarily irrational. For example, $(1 - \sqrt{2}) + \sqrt{2}$ is rational, but $\sqrt{2}$ is irrational, and by Exercise 3 of Section 9.4, $1 - \sqrt{2}$ is irrational. Also, the product of two irrationals is not necessarily irrational. Why?

9. Assume $\lim_{x \to 0} f(x) = L$, and let $\epsilon > 0$. Hence, there exists a $\delta > 0$ such that $|f(z) - L| < \epsilon$ whenever $0 < |z| < \delta$. In particular, $|f(z) - L| < \epsilon$ whenever $0 < z < \delta$. Assume $x > 0$ and let $z = 1/x$. Then $|f(1/x) - L| < \epsilon$ whenever $1/x < \delta$. Choose $n > 1/\delta$. Then $x > n$ implies $1/x < \delta$, which in turn implies $|f(1/x) - L| < \epsilon$. Therefore, $\lim_{x \to \infty} f(1/x) = L$.

11. (a) $\text{lub}(A) = 1$

(b) Let $u = \text{lub}(A)$. Since $1 - 1/(n + 1) < 1$ for all $n \epsilon \mathcal{N}$, 1 is an upper bound of A, and hence, $u \leq 1$. Suppose $u < 1$. Then by the ϵ- property, there exists an m such that $1/(m + 1) < 1 - u$. Hence, $u < 1 - 1/(m + 1)$, contradicting the fact that $u > 1 - 1/(n + 1)$ for all $n \epsilon \mathcal{N}$.

Index

320